Tumors of the Central Nervous System

Tumors of the Central Nervous System
Volume 13

For further volumes:
http://www.springer.com/series/8812

Tumors of the Central Nervous System
Volume 13

Tumors of the Central Nervous System

Types of Tumors, Diagnosis, Ultrasonography, Surgery, Brain Metastasis, and General CNS Diseases

Edited by

M.A. Hayat
Distinguished Professor
Department of Biological Sciences
Kean University, Union, NJ, USA

Springer

Editor
M.A. Hayat
Department of Biological Sciences
Kean University
Room 213, Library building
Morris Avenue 1000
Union, NJ 07083, USA

ISSN 2215-096X ISSN 2215-0978 (electronic)
ISBN 978-94-007-7601-2 ISBN 978-94-007-7602-9 (eBook)
DOI 10.1007/978-94-007-7602-9
Springer Dordrecht Heidelberg New York London

Library of Congress Control Number: 2013956518

Printed on acid-free paper

Springer is part of Springer Science+Business Media (www.springer.com)

Although touched by technology, surgical pathology always has been, and remains, an art. Surgical pathologists, like all artists, depict in their artwork (surgical pathology reports) their interactions with nature: emotions, observations, and knowledge are all integrated. The resulting artwork is a poor record of complex phenomena.

Richard J. Reed, MD

Preface

It is recognized that scientific journals and books not only provide current information but also facilitate exchange of information, resulting in rapid progress in the medical field. In this endeavor, the main role of scientific books is to present current information in more details after careful additional evaluation of the investigational results, especially those of new or relatively new therapeutic methods and their potential toxic side-effects.

Although subjects of diagnosis, drug development, therapy and its assessment, and prognosis of tumors of the central nervous system, cancer recurrence, and resistance to chemotherapy are scattered in a vast number of journals and books, there is need of combining these subjects in single volumes. An attempt will be made to accomplish this goal in the projected fourteen-volume series of handbooks.

In the era of cost-effectiveness, my opinion may be minority perspective, but it needs to be recognized that the potential for false-positive or false-negative interpretation on the basis of a single laboratory test in clinical pathology does exist. Interobservor or intraobservor variability in the interpretation of results in pathology is not uncommon. Interpretative differences often are related to the relative importance of the criteria being used.

Generally, no test always performs perfectly. Although there is no perfect remedy to this problem, standardized classifications with written definitions and guidelines will help. Standardization of methods to achieve objectivity is imperative in this effort. The validity of a test should be based on the careful, objective interpretation of the tomographic images, photo-micrographs, and other tests. The interpretation of the results should be explicit rather than implicit. To achieve accurate diagnosis and correct prognosis, the use of molecular criteria and targeted medicine is important. Equally important are the translation of molecular genetics into clinical practice and evidence-based therapy. Translation of medicine from the laboratory to clinical application needs to be carefully expedited. Indeed, molecular medicine has arrived.

The contents are divided into six parts: *types of tumors, diagnosis, ultrasonography, surgery, brain metastasis,* and *general CNS Diseases*, for the convenience of the readers. Molecular characterization and other aspects of a large number of tumor types, including embryonal tumors, oligodendrogliomas, hemangiopericytoma, schwannomas, gliosarcoma, mesenchymal chondrosarcoma, retinoblastoma, and supratentorial primitive neuroectodermal tumors, are discussed. Advantages and limitations of using computer systems for cell counting in histopathologic slides of tumors of the CNS are explained. Also is explained the advantage of using intraoperative power Doppler ultrasonography for intracranial tumors. Factors responsible for local recurrence of brain metastasis are discussed. The details of the application of intraoperative confocal microscopy technology in conjunction with surgical resection for metastatic brain tumors are presented.

Treatments for brainstem cavernomas and differentiation choroid plexus from metastatic carcinomas are discussed. Immunotherapies for brain cancer, including human trials, are explained. Alexander disease, which is a fatal CNS degenerative condition of infants, and lipoma, which is a benign neoplasm of angiogenesis in brain cancer development, are clarified. Whether or not the use of mobile phones presents risk of brain cancer is objectively discussed.

This is the thirteenth volume in the series, *Tumors of the Central Nervous System*. As in the case of the 12 previously published volumes, this volume mainly contains information on the diagnosis, therapy, and prognosis of brain and spinal cord tumors. Various aspects of a large number of tumor types, including neuroblastoma, medulloblastoma, meningioma, and chordoma, are discussed. The contents are divided into four parts: molecular mechanisms, children's cancer, treatments, and radiosurgery, for the convenience of the readers. Molecular profiling of brain tumors to select appropriate therapy in clinical trials of brain tumors is discussed in detail. The classification/diagnosis of brain tumors based on function analysis is presented. CDK6 as the molecular regulator of neuronal differentiation in the adult brain and the role of aquaporins in human brain tumor growth are explained. Children's tumors, including neuroblastoma and medulloblastoma, are discussed. Molecular genetic alterations in medulloblastoma are explained. Survival differences between children and adults with medulloblastoma are pointed out. The use of various types of imaging methods to diagnose brain tumors is explained. Important, effective treatments for patients with brain and spinal tumors are included. Treatments, such as stereotactic radiosurgery, endoscopic neurosurgery, electrochemotherapy, transsphenoidal surgery, focal ablation, whole brain radiation therapy, and recraniotomy, are detailed. The remaining volumes in this series will provide additional recent information on these and other aspects of CNS malignancies.

By bringing together a large number of experts (oncologists, neurosurgeons, physicians, research scientists, and pathologists) in various aspects of this medical field, it is my hope that substantial progress will be made against this terrible disease. It would be difficult for a single author to discuss effectively the complexity of diagnosis, therapy, and prognosis of any type of tumor in one volume. Another advantage of involving more than one author is to present different points of view on a specific controversial aspect of the CNS cancer. I hope these goals will be fulfilled in this and other volumes of this series. This volume was written by 78 contributors representing 14 countries. I am grateful to them for their promptness in accepting my suggestions. Their practical experience highlights their writings, which should build and further the endeavors of the reader in this important area of disease. I respect and appreciate the hard work and exceptional insight into the nature of cancer provided by these contributors. The contents of the volume are divided into three parts: pineal tumors, pituitary tumors, and spinal tumors for the convenience of the reader.

It is my hope that the current volume will join the preceding volumes of the series for assisting in the more complete understanding of globally relevant cancer syndromes. There exists a tremendous, urgent demand by the public and the scientific community to address to cancer diagnosis, treatment, cure, and hopefully prevention. In the light of existing cancer calamity, financial funding by governments must give priority to eradicating this deadly malignancy over military superiority.

I am thankful to Dr. Dawood Farahi and Philip Connelly for recognizing the importance of medical research and publishing through an institution of higher education. I am also thankful to my students for their contribution to the preparation of this volume.

Union, NJ, USA M.A. Hayat

Contents

Part II Diagnosis

Part VI General CNS Diseases

Contents of Volume 1

Contents of Volume 2

Contents of Volume 3

Contents of Volume 4

Contents of Volume 5

Contents of Volume 6

Contents of Volume 7

Contents of Volume 8

Contents of Volume 9

Contents of Volume 10

Contents of Volume 11

Contents of Volume 12

Contributors

Adib A. Abla Department of Neurosurgical Oncology, Barrow Brain Tumor Research Center, Barrow Neurological Institute, St. Joseph's Hospital and Medical Center, Phoenix, AZ, USA

Takenori Akiyama Department of Neurosurgery, Keio University School of Medicine, Shinjuku-ku, Tokyo, Japan

İlkser Akpolat Department of Pathology, School of Medicine, Ondokuz Mayıs University, Samsun, Turkey

Abbas Amirjamshidi Department of Neurosurgery, Sina Hospital, Tehran University of Medical Sciences, Tehran, Iran

Gregory D. Arnone Department of Neurosurgery, University of Illinois at Chicago, Chicago, IL, USA

Hikmat Assi Department of Neurosurgery, Department of Cell and Developmental Biology, School of Medicine, University of Michigan, Ann Arbor, MI, USA

Mariela Moreno Ayala Instituto de Investigaciones Biomédicas, National Council of Science and Technology – University of Buenos Aires, Buenos Aires, Argentina

Gianna Baroni Department of Biomedicine, Careggi Hospital, Florence, Italy

Rudi Beschorner Institute of Pathology and Neuropathology, Department of Neuropathology, University Hospital Tübingen, Tübingen, Germany

Peter C. Brooks Center for Molecular Medicine, Maine Medical Center Research Institute, Scarborough, ME, USA

Anna Maria Buccoliero Pathology Unit, Meyer Hospital, Florence, Italy

Bilge Can Meydan Department of Pathology, School of Medicine, Ondokuz Mayıs University, Samsun, Turkey

Marianela Candolfi Instituto de Investigaciones Biomédicas, INBIOMED, Facultad de Medicina, Universidad de Buenos Aires, Buenos Aires, Argentina

Francesca Castiglione Department of Biomedicine, Careggi Hospital, Florence, Italy

Maria G. Castro Department of Neurosurgery, Department of Cell and Developmental Biology, School of Medicine, University of Michigan, Ann Arbor, MI, USA

Duccio Rossi Degl'Innocenti Department of Biomedicine, Careggi Hospital, Florence, Italy

Rajiv D. Desai Center for Molecular Medicine, Maine Medical Center Research Institute, Scarborough, ME, USA

Charlotte J. Dommering Department of Clinical Genetics, VU University Medical Center, Amsterdam, The Netherlands

Nancy Doolittle Division of Pediatric Neurosurgery, Department of Neurosurgery, Oregon Health and Science University, Portland, OR, USA

María Elena Erro Aguirre Department of Neurology, Hospital Complex of Navarra, Pamplona, Spain

Maria Feychting Institute of Environmental Medicine, Karolinska Institutet, Stockholm, Sweden

Robert Friesel Center for Molecular Medicine, Maine Medical Center Research Institute, Scarborough, ME, USA

Lorenzo Genitori Neurosurgery Unit, Meyer Hospital, Florence, Italy

Daniel Guillaume Division of Pediatric Neurosurgery, Department of Neurosurgery, Oregon Health and Science University, Portland, OR, USA

Adrian Häne Department of Neurology and Brain Tumor Center, Cantonal Hospital Aarau, Aarau, Switzerland

Maryam Hamidi Department of Human Oncology and Medical Physics, School of Medical and Public Health, University of Wisconsin-Madison, Madison, WI, USA

Lidewij Henneman Department of Clinical Genetics, VU University Medical Center, Amsterdam, The Netherlands

Elena Hernández M. de Lapiscina Department of Neurology, Hospital Complex of Navarra, Pamplona, Spain

Shawn L. Hervey-Jumper Department of Neurosurgery, School of Medicine, University of Michigan, Ann Arbor, MI, USA

Joseph Hobbs Department of Neurosurgery, Center for Cranial Nerve Disorders, Allegheny General Hospital, Pittsburgh, PA, USA

Bujung Hong Department of Neurosurgery, Hannover Medical School, Hannover, Germany

Takashi Horiguchi Department of Neurosurgery, Keio University School of Medicine, Shinjuku-ku, Tokyo, Japan

Steven N. Kalkanis Department of Neurosurgery, Henry Ford Center for Cancer Surgery, Detroit, MI, USA

Hiroshi Kanno Department of Neurosurgery, Yokohama City University School of Medicine, Yoohama, Japan

Kourosh Karimi Yarandi Department of Neurosurgery, Sina Hospital, Tehran University of Medical Sciences, Tehran, Iran

Takeshi Kawase Department of Neurosurgery, Keio University School of Medicine, Shinjuku-ku, Tokyo, Japan

Yushan Kim Department of Bioengineering and California, Institute for Quantitative Biosciences, University of California, Berkeley, CA, USA

Kevin Kozak Department of Human Oncology and Medical Physics, School of Medical and Public Health, University of Wisconsin-Madison, Madison, WI, USA

Wojciech Kozlowski Military Institute of Medicine, Warsaw, Poland

Joachim K. Krauss Department of Neurosurgery, Hannover Medical School, Hannover, Germany

Sanjay Kumar Department of Bioengineering and California, Institute for Quantitative Biosciences, University of California, Berkeley, CA, USA

Lisa Lin Department of Tissue Pathology, Level 3, ICPMR, Westmead Hospital, Westmead, NSW, Australia

Pedro R. Lowenstein Department of Neurosurgery, Department of Cell and Developmental Biology, School of Medicine, University of Michigan, Ann Arbor, MI, USA

Gina Lungu Veterinary Pathobiology, Texas A&M University, College Station, TX, USA

Vasiliy N. Manskikh Faculty of Bioengineering and Bioinformatics, Lomonosov Moscow State University, Moscow, Russia

Tomasz Markiewicz Warsaw University of Technology and Military Institute of Medicine, Warsaw, Poland

Rosalia Mendez-Otero Instituto de Biofísica Carlos Chagas Filho, Universidade Federal do Rio de Janeiro, Rio de Janeiro, Brazil

Dorothea Miller Department of Neurosurgery, University Hospital Essen, Essen, Germany

Yohei Mineharu Department of Neurosurgery, Department of Cell and Developmental Biology, School of Medicine, University of Michigan, Ann Arbor, MI, USA

Aliasgar Moiyadi Neurosurgery Services, Department of Surgical Oncology, Tata Memorial Hospital, Mumbai, India

Daniela Moncini Department of Biomedicine, Careggi Hospital, Florence, Italy

John Moody Department of Human Oncology and Medical Physics, School of Medical and Public Health, University of Wisconsin-Madison, Madison, WI, USA

Masanori Nakagawa Department of Neurology, Graduate School of Medical Science, Kyoto Prefectural University of Medicine, Kyoto, Japan

Makoto Nakamura Department of Neurosurgery, Hannover Medical School, Hannover, Germany

Edward Neuwelt Division of Pediatric Neurosurgery, Department of Neurosurgery, Oregon Health and Science University, Portland, OR, USA

Thomas Ng Department of Tissue Pathology, Level 3, ICPMR, Westmead Hospital, Westmead, NSW, Australia

Stanislaw Osowski Warsaw University of Technology and Military University of Technology, Warsaw, Poland

Aqueel Pabaney Department of Neurosurgery, Henry Ford Center for Cancer Surgery, Detroit, MI, USA

Milena Paglierani Department of Biomedicine, Careggi Hospital, Florence, Italy

Akash J. Patel Department of Neurosurgery-442, The University of Texas, M. D. Anderson Cancer Center, Houston, TX, USA

Vladimir M. Perelmuter Pathological Anatomy and Cytology Department, Cancer Research Institute, Siberian Branch of the Russian Academy of Medical Sciences, Tomsk, Russia

Pedro M. Pimentel-Coelho Instituto de Biofísica Carlos Chagas Filho, Universidade Federal do Rio de Janeiro, Rio de Janeiro, Brazil

Ulrich Roelcke Department of Neurology and Brain Tumor Center, Cantonal Hospital Aarau, Aarau, Switzerland

Martin Röösli Department of Epidemiology and Public Health, Swiss Tropical and Public Health Institute, Basel, Switzerland

Jennifer Roth Center for Molecular Medicine, Maine Medical Center Research Institute, Scarborough, ME, USA

Nader Sanai Department of Neurosurgical Oncology, Barrow Brain Tumor Research Center, Barrow Neurological Institute, St. Joseph's Hospital and Medical Center, Phoenix, AZ, USA

Raymond Sawaya Department of Neurosurgery-442, The University of Texas, M. D. Anderson Cancer Center, Houston, TX, USA

Joachim Schüz Section of Environment and Radiation, International Agency for Research on Cancer (IARC), Lyon, France

Raymond F. Sekula Jr. Department of Neurosurgery, Center for Cranial Nerve Disorders, Allegheny General Hospital, Pittsburgh, PA, USA

Antonella Simoni Department of Biomedicine, Careggi Hospital, Florence, Italy

Robert F. Spetzler Department of Neurosurgical Oncology, Barrow Brain Tumor Research Center, Barrow Neurological Institute, St. Joseph's Hospital and Medical Center, Phoenix, AZ, USA

George Stoica Veterinary Pathobiology, Texas A&M University, College Station, TX, USA

Gian Luigi Taddei Department of Biomedicine, Careggi Hospital, Florence, Italy

Winny Varikatt Department of Tissue Pathology, Level 3, ICPMR, Westmead Hospital, Westmead, NSW, Australia

Thomas J. Wilson Department of Neurosurgery, School of Medicine, University of Michigan, Ann Arbor, MI, USA

Kazunari Yoshida Department of Neurosurgery, Keio University School of Medicine, Shinjuku-ku, Tokyo, Japan

Tomokatsu Yoshida Department of Neurology, Graduate School of Medical Science, Kyoto Prefectural University of Medicine, Kyoto, Japan

Part I

Types of Tumors

Embryonal Tumor: Molecular Characterization

1

Anna Maria Buccoliero, Francesca Castiglione, Duccio Rossi Degl'Innocenti, Daniela Moncini, Milena Paglierani, Gianna Baroni, Antonella Simoni, Lorenzo Genitori, and Gian Luigi Taddei

Contents

A.M. Buccoliero (✉)
Pathology Unit, Meyer Hospital, Florence, Italy
e-mail: ambuccoliero@unifi.it

F. Castiglione • D.R. Degl'Innocenti • D. Moncini • M. Paglierani • G. Baroni • A. Simoni • G.L. Taddei
Department of Biomedicine, Careggi Hospital, Florence, Italy

L. Genitori
Neurosurgery Unit, Meyer Hospital, Florence, Italy

Abstract

Central Nervous System (CNS) embryonal tumors are malignant neoplasms that share the predilection for the pediatric age, the tendency to early disseminate throughout the CNS via the cerebro spinal fluid, the high mortality and the significant long-term morbidity for survivors.

The pathological diagnose of these tumors on the basis of the morphological analysis may be sometimes difficult, and also not all of these tumors, with an equal stage and treatment, have the same behavior. The application of basic research findings to clinical oncology has led to new approaches to the diagnosis, prognosis, and treatment of these tumors. In this chapter we discuss the molecular biology of the CNS embryonal tumor with particular emphasis on that aspects having clinical impact.

Introduction

Primary central nervous system (CNS) tumors have traditionally been classified on the basis of their morphological features. The advent of immunohistochemistry has subsequently improved our ability to distinguish, and thus classify this tumors. In recent years, the morphological (including ultrastructural) and immunohistochemical analyses have been gradually joined by molecular studies (Louis et al. 2007a, b). Molecular characterization of tumors is now universally considered as a useful tool not only for the diagnosis but also for the treatment planning and prognostic predicting.

M. Hayat (ed.), *Tumors of the Central Nervous System, Volume 13*,
DOI 10.1007/978-94-007-7602-9_1, © Springer Science+Business Media Dordrecht 2014

Histological diagnosis may sometimes be potentially subjective. In fact, depending on the weight given to each morphological and even immunohistochemical features of a single lesion, inter-observer variability is possible. The knowledge of the molecular traits of the tumors offers ancillary and more objective diagnostic criteria. An example of a tumor for which the molecular biology plays a diagnostic key role is the atypical teratoid rhabdoid tumor. It is now universally accepted that, as discussed below, the demonstration of alterations in *INI-1*gene is essential in the diagnosis of this malignant tumor.

Current therapeutic strategies and prognosis were in the past mainly determined on the basis of some clinical (i.e. patient age, extent of the surgery, spread of the tumor) and histological variable (i.e. histotype and grade). In recent years the molecular alterations have assumed a great importance for both the therapy and the prognosis. In this regard it should be just remember the prognostic and therapeutic impact of the evaluation of *MYC* genes amplification in medulloblastomas.

Techniques

It's been several decades since molecular biologists have begun to learned to characterize, isolate, and manipulate the molecular components, DNA, RNA, and proteins, of the cells.

Many techniques are available to study the molecular characteristics of the cells. Some of these are especially used in basic research. The techniques most commonly used in clinical oncology are those with good versatility, easy to use and applicable to routinely formalin fixed and paraffin embedded tissue samples. Among these there are Fluorescence In Situ Hybridation (FISH) and the related Chromogenic In Situ Hybridation (CISH), Polymerase Chain Reaction (PCR) and Reverse Transcriptase PCR (RT-PCR), Real Time PCR, DNA sequencing, and immunohistochemistry.

FISH is based on the use of fluorescence-labeled oligonucleotide probes that specifically bind to their complementary DNA sequence target and label that region with fluorescence color. The labeled region is visualized under a fluorescence microscope. CISH, utilizes oligonucleotide probes conjugated to enzymes which catalyze a color producing reaction thus, analogously to immunocyto-histochemistry (see below), the reactions are visualized under a standard bright-field microscope.

PCR is a technique to amplify a single piece of DNA. The method relies on thermal cycling, consisting of cycles of repeated heating and cooling of the reaction for DNA melting and enzymatic DNA replication. Short DNA fragments (primers) containing sequences complementary to the target region and the enzyme TAQ DNA polymerase play the key role to obtain selective and exponential amplification of the target. The result of the reaction can be visualized through gel electrophoresis (amplified DNA copies are identical in electrical charge and molecular weight thus migrate simultaneously forming a single band on the gel) or through capillary electrophoresis (the primers are labeled with fluorescent dye therefore the PCR product can then analyzed in a capillary electrophoresis instrument which tracks the fluorescence of the identical PCR sequences as the migrate). RT-PCR allows to amplify an RNA target sequence that is first converted to a double-stranded nucleic acid sequence (cDNA) by using a reverse transcriptase enzyme obtained from a retrovirus.

Real-time PCR is quantitative PCR method for the computerized determination of copy number of PCR templates such as DNA or cDNA in a PCR reaction. It enables both detection and quantification (as absolute number of copies or relative amount when normalized to DNA input or additional normalizing genes) of one or more specific sequences in a DNA sample. A fluorescent reporter molecule is used to monitor in real time the PCR as it progresses assuming that the amount of fluorescence is proportional to the number of copies of the amplification target.

Gene sequencing consists in the determination of the order of the nucleotide bases in a genomic tract amplified by PCR. The technique involves the use of automated instruments which read the nucleotidic sequence of the gene of interest allowing to the evaluation of the presence of mutations. After a first PCR round, the sequencing reaction inserts nucleodites each one labeled with different fluorescent dye. Through a capillary electrophoresis the instrument registers and allocates the different fluorescence to each nucleotide.

Immunocyto-histo-chemistry is a method of detecting cellular antigen of interest (e.g. proteins) on cytological or histological samples by means of specific antibodies. The visualization of the antibody-antigen interaction is commonly accomplished in two ways. The most widely used method uses antibodies conjugated to an enzyme, such as peroxidase, that catalyze a color producing reaction. Alternatively, the antibody may be conjugated to a fluorophore. In the first case the reaction is visualized under a standard bright-field microscope, in the second case the reaction is visualized under a fluorescence microscope. Immunocyto-histo-chemistry has the advantage of being able to show exactly where a given protein is located within the tissue examined. The demonstration of the presence, absence and localization of a specific protein in a cell (i.e. nucleus, cytoplasm, cell membrane) or tissue (i.e. focal, diffuse) has diagnostic, prognostic and therapeutic implications. In addition, as we shall see, it may indirectly give us information on the genetic profile of that cell or tissue.

Classification and Histopathology

CNS embryonal tumors include a broad spectrum of malignancies that share the predilection for the pediatric age, the proclivity to early disseminate throughout the CNS via the cerebro spinal fluid, the high mortality and the significant long-term morbidity for survivors. The current World Health Organization (WHO) classification of the CNS system tumors distinguishes three main entities which are medulloblastoma, CNS Primitive Neuroectodermal Tumors (CNS PNETs) and Atypical Teratoid Rhabdoid Tumor (AT/RT) (Louis et al. 2007a, b).

Furthermore, in this family of tumors we could also consider pineoblastoma. Indeed, although this tumor has not been included by the WHO in the group of embryonal tumors (it is part of the tumors of the pineal region together with pineocytoma, pineal tumor of intermediate differentiation and papillary tumor of the pineal region), it shares with the embryonal CNS tumors the same embryonal features. However, pineoblastomas are distinct from the others embryonal tumors for their peculiar photosensory differentiation.

To all embryonal tumors it has been assigned the highest histological grade (WHO IV).

Medulloblastoma

Medulloblastoma represents the largest group of CNS embryonal tumor. Several familial cancer syndromes are associated with an increased risk to develop medulloblastoma including Turcot (*APC* gene mutation) and Gorlin (*PTCH* gene mutation) syndromes. Medulloblastoma exclusively arises in the cerebellum (usually in the vermis in younger children and in the cerebellar hemispheres in the older children and in young adults) where it is believed to originate from transformed granule cell precursors. The WHO identifies, besides the classical form ("classic" medulloblastoma), four different histological variants designated as desmoplastic/nodular medulloblastoma, medulloblastoma with extensive nodularity, large cell medulloblastoma and anaplastic medulloblastoma and two morphological patterns designated as medullomyoblastoma and melanotic medulloblastoma. Classic medulloblastoma is composed of densely packed cells with round–to-oval hyperchromatic nuclei surrounded by scanty cytoplasm. Rosettes, particularly neuroblastic (Homer Wright) rosettes which consist of tumor cell nuclei arranged in a circular fashion around tangled cytoplasmic processes may be observed. Less frequently tumoral cells are arranged in parallel rows (spongioblastic architecture).

The desmoplastic/nodular medulloblastoma variant has a lobular architecture with nodular reticulin-free areas containing small neurocytic cells in a fibrillary background. These areas are surrounded by reticulin rich inter-nodular compartment composed of densely packed and highly proliferative small undifferentiated cells. Medulloblastoma with extensive nodularity differs from desmoplastic/nodular medulloblastoma for the predominance of the reticulin free nodules.

Large cell medulloblastoma consists of large spherical cells displaying discohesive growth, large nuclei, vesicular chromatin, prominent nuclei and abundant apoptotic and mitotic (often atypical) figures. Anaplastic medulloblastoma when compared with large cell medulloblastoma is composed of more pleomorphic and less discohesive cells. Since both, large cell and anaplastic medulloblastomas, often coexist in the same tumor, they are merged into a combined large cell/anaplastic category (Eberhart 2011). In addition, in the fourth edition of International Classification of Diseases for Oncology (ICD-O) anaplastic medulloblastoma and large cell medulloblastoma share the same code 9474/3 (Louis et al. 2007a, b).

Medullomyoblastoma and melanotic medulloblastoma shows selectively muscular (striated muscle fibers, myogenic and rhabdomyoblastic cells) and melanotic (pigmented cells) differentiation.

A dismal prognosis has been associated with anaplastic/large cell medulloblastoma so that aggressive therapeutical regimes have been employed in these cases. Conversely, desmoplastic medulloblastoma and medulloblastoma with extensive nodularity are associated with a better prognosis (Ellison et al. 2011).

Central Nervous System Primitive Neuroectodermal Tumors (CNS PNETs)

CNS PNETs constitute a heterogeneous group of commonly non cerebellar undifferentiated or poorly differentiated tumors which may show divergent differentiation along neuronal, astrocytic and ependymal lines. In this group of tumors have been included CNS/supratentorial PNET and four histological variants which are CNS neuroblastoma, CNS ganglioneuroblastoma, medulloepithelioma, and ependimoblastoma. However, this family of tumors still remains in a state of flux. For example, a rare pediatric embryonal tumor called neuroblastic tumor containing abundant neuropil and true rosettes has been some years ago described (Eberhart et al. 2000). It is still not clear whether this lesion is a variant of other embryonal tumors or is a distinct tumoral entity of its own (Judkins and Ellison 2010).

CNS PNET is a tumor most commonly found in the cerebrum. It is typically composed of very poorly small cells showing round regular nuclei and high nucleus/cytoplasm rations.

Tumors better differentiated along a neuronal/ganglionic lineage are termed CNS neuroblastoma and CNS ganglioneuroblastoma.

Medulloepithelioma may develop both in the supra- and infra-tentorial compartments. It may also occur outside the CNS i.e. orbital cavity, nerve trunks, pelvic cavity and may also occur associated with others CNS embryonal tumors (Buccoliero et al. 2010). Medulloepithelimoma morphologically mimics the embryonic neuronal tube (pseudostratified tall epithelium delimitated by a basement membrane and disposed in glands-, tubules- or canals-like structures).

Ependymoblastoma is characterized by the presence of multilayered ependymoblastic rosettes consisting of undifferentiated cells arranged around a central lumen.

Neuroblastic tumor containing abundant neuropil and true rosettes is morphologically characterized by undifferentiated neuroepithelial cells arranged in clusters, cords and several types of rosettes (i.e. ependymoblastic, Homer-Wright, Flexner-Wintersteiner and perivascular) in a neuropil-rich background (Fig. 1.1). Exceptionally. Neuroblastic tumor containing abundant neuropil and true rosettes may also display a composite morphology including medulloepithelioma, mesenchymal and epithelial areas (Buccoliero et al. 2010).

Atypical Teratoid Rhabdoid Tumor (AT/ RT)

AT/RT is a rare and highly malignant polyphenotypic neoplasm of uncertain origin that often occurs in the posterior fossa of early childhood. It can be sporadic or can occur in the setting of

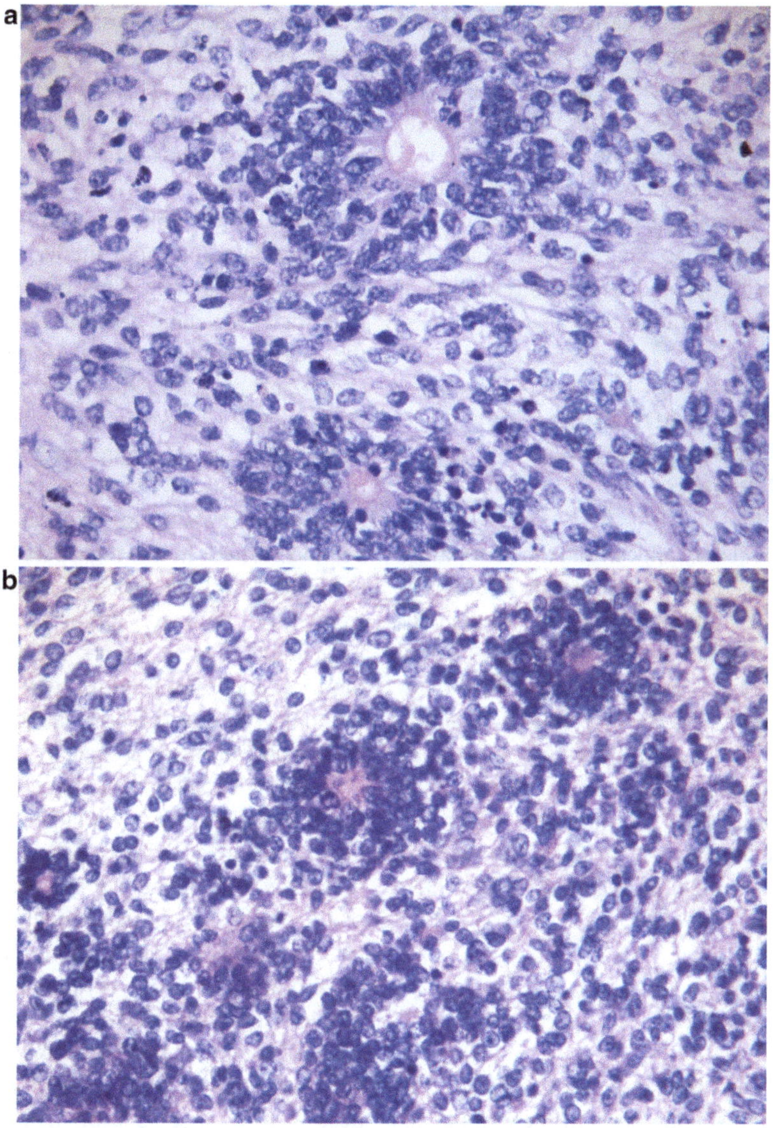

Fig. 1.1 Neuroblastic tumor containing abundant neuropil and true rosettes. These tumors characteristically show numerous rosettes, i.e. ependymoblastic (**a**) and Homer-Wright (**b**). HE; original magnification (**a, b**) ×200

the rhabdoid tumor predisposition syndrome. This syndrome is characterized by an increased risk to develop malignant rhabdoid tumor in CNS and outside CNS (i.e. kidney, head and neck, soft tissue, heart, mediastinum and liver). AT/RTs are morphologically heterogeneous lesions characterized microscopically by a variable presence of small/embryonal cell, mesenchymal and epithelial features. The most typical, even if inconstant, morphological aspect is the presence of rhabdoid cells with vesicular chromatin, eccentric nuclei, prominent nucleoli, abundant cytoplasm containing inclusion-like mass of filaments.

Pineoblastoma

Pineoblastoma is a rare malignant brain tumor which arises from the pineal gland. It can be sporadic or occur in association with retinoblastoma (ectopic intracranial retinoblastoma) in the setting of the trilateral retinoblastoma syndrome (Blach et al. 1994). Pineoblastoma is a highly cellular neoplasm composed of undifferentiated cells. Pineoblastomas may exhibit histopathological features indicative of photosensory differentiation ranging from immunohistochemical expression of retinal S-antigen to characteristic Flexner–Wintersteiner rosettes and fleurettes reminiscent of retinoblastic differentiation.

Molecular Characterization of Central Nervous System Embryonal Tumors

From the above it may appear that the histopathological diagnosis of the different type of embryonal tumors is easy and that all these malignancies are prognostically and therapeutically equivalent tumors. Actually, it is sometimes difficult to diagnose these tumors with the morphological analysis only, and also not all of these tumors, with an equal stage and treatment, have the same behavior. For these reasons, a molecular approach either for the diagnosis and for the prognostic and therapeutical stratification of these entities, have been advocated.

Medulloblastoma

Many studies have been conducted with the aim of evaluating the genetic basis of medulloblastomas and so obtaining data useful for the diagnosis, the prognosis and the personalization of the therapeutic strategy.

The most common chromosomal aberrations observed in medulloblastomas involve the chromosome 17. Changes in chromosome 17 have been identified in several types of human cancer. In particular, a peculiar chromosomal abnormality called isochromosome 17q (i17q) frequently occurs in some cancers and among these medulloblastomas where has been found in 30–40% of the cases (Pfister et al. 2010). This abnormal version of chromosome 17 combines loss of 17p and gain of 17q. As a result, the chromosome has an extra copy of some genes and is missing copies of other genes. i17p correlates with histological variants and predicts the survival. i17q is more frequently observed in classic and anaplastic/large cell medulloblastoma and has been associated with a poor clinical outcome when compared with that of desmoplastic/nodular medulloblastoma and medulloblastoma with extensive nodularity. This observation suggests that this cytogenetic alteration may contribute to the development of aggressive histotypes (Lamont et al. 2004; Gilbertson and Ellison 2008; Gulino et al. 2008). It has been postulated the existence of tumor suppressor gene(s) on 17p and of oncogene(s) on 17q. In reality, specific genes recurrently mutated in i17q cases have not yet been identified. In this regard it is interestingly to note that *TP53* gene, one of the best known and frequently mutated tumor suppressor gene, located at 17p13.1 is not mutated more commonly in medulloblastomas with i17q than those without i17q. On the other hand, the possibility of a inactivation of a specific gene at the 17p breakpoint has been considered unlikely because the breakpoint has a variable localization and occurs in a gene poor chromosomal region (Pfister et al. 2010). Notably, loss of 17p and gain of 17q can also occur independently. Stratification by chromosome 17 data alone demonstrated an unfavorable outcome in patients with isolated 17p loss and suggested a best survival in patients with isolated 17q gain (McCabe et al. 2011).

Alterations in *MYC* genes have also been described as about 5% of medulloblastomas in mixed cohorts (Lamont et al. 2004; Aldosari et al. 2002). The *MYC* oncogene family consists of at least three genes: *c*- (on 8q24), *N*- (on 2p24), and *L*- *MYC*(on 1p32). These proto-oncogenes are among the most powerful activators of tumorigenesis and are frequently overexpressed in several tumors. However, it is still unclear how *MYCs* promote tumor formation and confer aggressive phenotypes and metastatic potential. It is believed that they act by targeting diverse cellular processes, including cell cycle, proliferation, differentiation, apoptosis, telomere maintenance, cell adhesion and angiogenesis. What is certain is that *MYC* genes code for proteins that bind to the

DNA of other genes and are therefore a transcription factors. They act as activator or, in some cases, repressor, of gene expression (Zitterbart et al. 2011; Takei et al. 2009). The MYC oncoproteins play a central role as regulators of tumorigenesis in numerous different human cancers. Deregulated expression of *MYCs* is often associated with aggressive, poorly differentiated tumors. High-level amplification (>10-fold) of *MYC* oncogenes is particularly associated with the large cell/anaplastic phenotype. A moderate *MYC* oncogenes amplification (>5-fold but <10-fold), particularly *c-MYC* and *N-MYC*, is sometimes detected in other histological subtypes and may depend on the presence a large cell/anaplastic component in this medulloblastomas (Eberhart et al. 2004; Stearns et al. 2006; Takei et al. 2009). M*YC* amplification may also be acquired in recurrent medulloblastomas. Survival analysis showed, independently of histological subtype, a decreasing survival in *MYC* non-amplified (<5-fold), *MYC* moderately amplified (<10-fold) and *MYC* highly amplified (>10-fold) medulloblastomas. Actually, no widely accepted cut-off has been established for defining *MYC* amplification when, as usual, it is evaluated by FISH method. In a recent report on a large series it has been suggested to considered *MYC* amplification as clinical relevant when more than 10% of tumor cells exhibit either more than ten signals or innumerable thigh clusters of signals of the respective probe (Pfister et al. 2009). However, lower cut off (*MYC* copy number >4–5 fold) has been also suggested (Takei et al. 2009) (Fig. 1.2). In any cases it is remarkable that it has been proposed to use the evaluation of the *MYC* amplification as a molecular stratification factor in the *Société Internationale d'Oncologie Pédiatrique en Europe* (SIOP Europe) PNET 5 and 6 MB trials. Lastly, some preclinical investigations analyzed the value of *MYC* expression (particularly c-*MYC*) in the response of tumors to anticancer therapy. In particular, c-*MYC* over-expression seems to be associated with enhanced susceptibility of medulloblastoma to radiotherapy- and to etoposide-, doxorubicin-, and cisplatin-induced apoptosis. Consequently, more aggressive treatment approach might be considered in the future to improve the outcome in those patients (von Bueren et al. 2011).

The occurrence of medulloblastoma as a possible manifestation of a number of familial syndromes persuaded many authors to investigate whether alterations in the same molecular pathway are implicated in the development of sporadic medulloblastomas.

Particular interest was given to the genetic abnormalities responsible for the type 2 Turcot's syndrome, in which patients have a predisposition to colon cancer due to mutations in the *APC* tumor suppressor gene lies on chromosome 5q21. APC protein is a negative regulator of the Wnt/Wingless (Wnt/Wg) pathway. It forms a complex with glycogen synthase kinase 3b and axin and together they control the activity of beta-catenin. Beta-catenin is encoded by the *CTNNB1* gene that maps to chromosome 3p22. Beta-catenin binds the actin of the cytoskeleton and is part of a complex of proteins that constitute adherens junctions. Through these skills it is implicated in the transmission the contact inhibition signal that causes cells to stop dividing once the epithelial sheet is complete and in the creation and maintenance of epithelial cell layers by regulating cell growth and adhesion between cells. Sporadic medulloblastomas may show mutation in the Wnt/Wg pathway at different levels: β-catenin, axin and *APC* genes. Wnt/Wg pathway activation destabilizes the protein complex, upregulating levels of beta-catenin and enhancing its translocation to the nucleus. Here, it cooperate in the regulation of cell cycle progression, apoptosis, and differentiation. Thus, nuclear accumulation of beta-catenin is a marker for physiologic or abnormal Wnt/Wg pathway activation. Mutations of the Wnt/Wg pathway occur in approximately 15% of sporadic medulloblastomas (mainly as a consequence of a mutation in the *CTNNB1* gene) and are predicted to cause aberrant pathway activation (Rubin and Rowitch 2002). Children with medulloblastomas that showed a nucleopositive beta-catenin immunophenotype had significantly better overall and event-free survivals than children with tumors that showed either membranous/cytoplasmic beta-catenin immunoreactivity or no immunoreactivity (Rubin and Rowitch

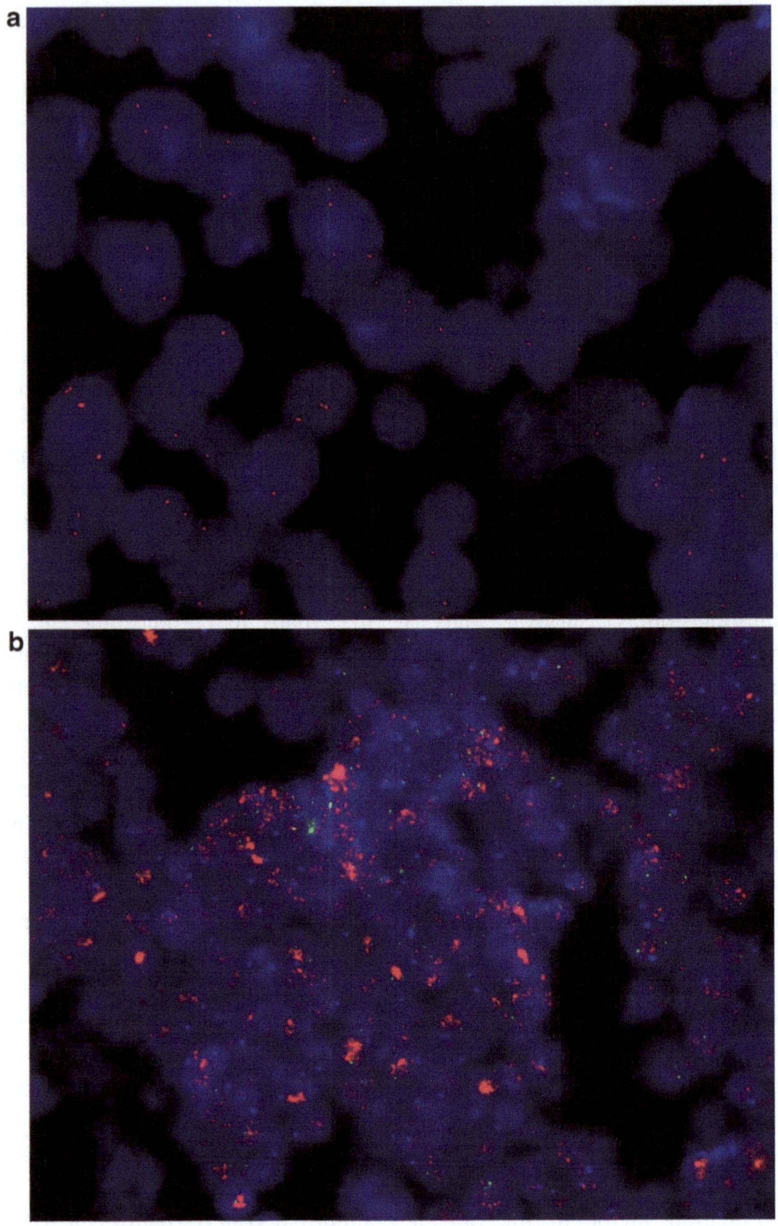

Fig. 1.2 FISH analyses showing not amplified c-MYC gene (less than five *red signals* for each cells) (**a**) and amplified c-MYC gene (more than ten signals or innumerable thigh clusters of *red signals*; *green signals* correspond to the gene centromere) (**b**)

2002; Ellison et al. 2005; Fatet et al. 2009). This finding might be unexpected due to studies of colonic, lung, and hepatocellular carcinomas indicate that nuclear beta-catenin immunoreactivity is associated with both enhanced cell proliferation and an aggressive behavior (Lugli et al. 2007; Martensson et al. 2007). On this regard, we can suppose that multifactorial molecular interactions are involved in the Wnt/Wg pathway in different tumors.

Inactivating germline mutations on tumor suppressor *PTCH* gene, mapping on chromosome 9q and coding for a transmembrane protein (Ptch), are responsible for Gorlin syndrome (also called nevoid basal cell carcinoma syndrome) in which patients have a predisposition to developmental disorders and benign and malignant tumors. Affected individuals develop multiple basal cell carcinomas, odontogenic keratocysts of the jaws, palmar and plantar dyskeratoses, skeletal anomalies, intracranial calcifications, macrocephaly, dysgenesis of the corpus callosus, congenital hydrocephalus and medulloblastomas particularly of the desmoplastic variant. Somatic mutations on this gene have been also identified in about 8% of sporadic medulloblastomas. These mutations, both germline and sporadic, determine the truncation of the Ptch protein. Ptch protein is a receptor for secreted hedgehog protein family of signaling molecules. It acts as inhibitor of the hedgehog pathway and through complex mechanisms controls the proliferation and differentiation of progenitor cells. Ptch has an essential role in embryonic patterning. This may explain the congenital anomalies associated with Gorlin syndrome. As known medulloblastomas, or at least a subset of them, arise from the external granular cell layer of the developing cerebellum. Interestingly, hedgehog is secreted by Purkinje cells and received by external granule cells for which it acts as mitogen. The lack of inhibition of hedgehog by Ptch may contribute to the onset of medulloblastomas (Pfister et al. 2010; Raffel et al. 1997; Eberhart 2011).

Central Nervous System Primitive Neuroectodermal Tumors (CNS PNETs)

Genetic traits of PNETs are less known than those of medulloblastomas. However, genetic analyses are greatly contributing to our understanding these tumors. Indeed, based on molecular studies, it has been recently demonstrated that neuroblastic tumor containing abundant neuropil and true rosettes and ependymoblastomas comprise a single biological entity. Indeed, both showed the same specific cytogenetic aberration consisting in a highly specific focal amplification at chromosome band 19q13.42 containing a cluster of mi-RNA-coding genes (Korshunov et al. 2010).

Atypical Teratoid Rhabdoid Tumor (AT/ RT)

AT/RT is characterized by deletions and/or mutations of *INI-1* tumor-suppressor gene also known as hSNF, SMARCB1 or BAF47 gene on chromosome band 22q11.2 (Pfister et al. 2010). Genomic changes in *INI-1* tumor-suppressor gene have been described in families with an inherited predisposition to rhabdoid tumors (rhabdoid tumor predisposition syndrome). The encoded protein, INI-1 protein, is part of a multiprotein complex involved in chromatin remodeling, an essential process in the cell nucleus for regulation of gene expression. This protein is normally expressed in all tissues. Genetic alterations of *INI-1* gene cause loss of INI-1 protein expression which can easily demonstrated by immunohistochemistry (Fig. 1.3). However, loss of INI-1 nuclear staining is typical but not specific of tumor displaying rhabdoid morphology. Indeed, immunohistochemical negative nuclear staining for INI-1 protein may also be observed in non rhabdoid tumor such as epithelioid sarcoma, myxoid chondrosarcoma and in medullary carcinoma (Hornick et al. 2009) while positive nuclear staining is observed in rhabdoid meningiomas (Buccoliero et al. 2011).

Pineoblastoma

Molecular profile of pineoblastoma are still largely unknown because of the extreme rarity of this tumor. In the few cases studied were found monosomy of the chromosome 22 and missense *INI-1* mutations (Biegel et al. 2000). It have been also referred recurrent rearrangements in short arm of chromosome 1 and unbalanced gain of chromosome 17q (Brown et al. 2006).

Germinal mutation in *RB1* gene on 13q typically occur in heritable pineoblastomas. The *RB1* product, the retinoblastoma protein (pRB), is a nuclear protein that plays a critical role in cell cycle regulation, cellular differentiation, senescence, apoptosis, and embryonic development. In quiescent cells pRB binds and inactivates the E2F protein. The E2F protein normally activates

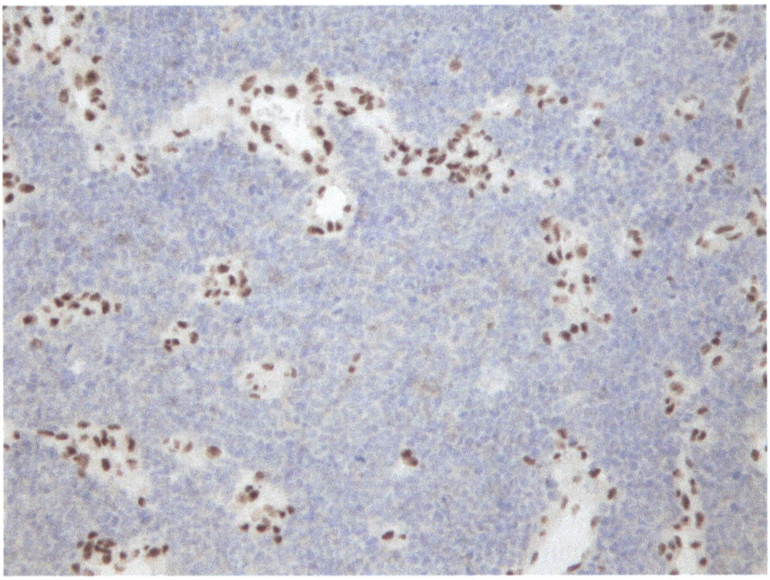

Fig. 1.3 AT/RT: Loss of immunohistochemical expression of INI-1 in nuclei of neoplastic cells with retained expression in intratumoral blood vessels. Original magnification ×200

gene transcription by binding to consensus positive regulatory DNA elements in gene promoters. *RB1* mutations make the pRB unable to bind E2F, causing continued cell growth. Although children with germ line mutations of *RB1* gene are at high risk to develop pineoblastoma, losses on 13q are not commonly seen in sporadic pineoblastoma.

Conclusion

In conclusion, the recent decades there were radical changes in our way of approaching the pathological diagnosis of tumors. Sophisticated tools and methodologies have progressively supported the old optical microscope thus increasing the resolving power of our diagnostic capabilities. This allowed us to classify tumors, not only on morphological basis but also on molecular basis with major implications both prognostic and therapeutic.

References

Aldosari N, Bigne SH, Burger P, Becker L, Kepner JL, Friedman HS, McLendon RE (2002) MYCC and MYCN oncogene amplification in medulloblastoma. A fluorescence in situ hybridization study on paraffin sections from the Children's Oncology Group. Arch Pathol Lab Med 126:540–544

Biegel JA, Fogelgren B, Zhou JY, James CD, Janss AJ, Allen JC, Zagzag D, Raffel C, Rorke LB (2000) Mutations of the INI1 rhabdoid tumor suppressor gene in medulloblastoma and primitive neuroectodermal tumors of the central nervous system. Clin Cancer Res 6:279–2763

Blach LE, McCormick B, Abramson DH, Ellsworth RM (1994) Trilateral retinoblastoma-incidence and outcome: a decade of experience. Int J Radiat Oncol Biol Phys 429:729–733

Brown AE, Leibundgut K, Niggli FK, Betts DR (2006) Cytogenetics of pineoblastoma: four new cases and a literature review. Cancer Genet Cytogenet 170:175–179

Buccoliero AM, Castiglione F, Degl'Innocenti DR, Franchi A, Paglierani M, Sanzo M, Cetica V, Giunti L, Sardi I, Genitori L, Taddei GL (2010) Embryonal tumor with abundant neuropil and true rosettes: morphological, immunohistochemical, ultrastructural and molecular study of a case showing features of medulloepithelioma and areas of mesenchymal and epithelial differentiation. Neuropathology 1:84–91

Buccoliero AM, Castiglione F, Degl'Innocenti DR, Franchi A, Sanzo M, Cetica V, Giunti L, Sardi I, Mussa F, Giordano F, Genitori L, Taddei GL (2011) Pediatric rhabdoid meningioma: a morphological, immunohistochemical, ultrastructural and molecular case study. Neuropathology 31:59–65

Eberhart CG (2011) Molecular diagnostics in embryonal brain tumor. Brain Pathol 21:6–104

Eberhart CG, Brat DJ, Cohen KJ, Burger PC (2000) Pediatric neuroblastic brain tumors containing abundant neuropil and true rosettes. Pediatr Dev Pathol 3:346–352

Eberhart CG, Kratz J, Wang Y, Summers K, Stearns D, Cohen K, Dang CV, Burger PC (2004) Histopathological and molecular prognostic markers in medulloblastoma: c-myc, N-myc, TrkC, and anaplasia. J Neuropathol Exp Neurol 63:441–449

Ellison DW, Onilude OE, Lindsey JC, Lusher ME, Weston CL, Taylor RE, Pearson AD, Clifford SC, United Kingdom Children's Cancer Study Group Brain Tumour Committee (2005) Beta-catenin status predicts a favorable outcome in childhood medulloblastoma: the United Kingdom Children's Cancer Study Group Brain Tumour Committee. J Clin Oncol 23:7951–7957

Ellison DW, Kocak M, Dalton J, Megahed H, Lusher ME, Ryan SL, Zhao W, Nicholson SL, Taylor RE, Bailey S, Clifford SC (2011) Definition of disease-risk stratification groups in childhood medulloblastoma using combined clinical, pathologic, and molecular variables. J Clin Oncol 28:1400–1407; Epub ahead of print

Fattet S, Haberler C, Legoix P, Varlet P, Lellouch-Tubiana A, Lair S, Manie E, Raquin MA, Bours D, Carpentier S, Barillot E, Grill J, Doz F, Puget S, Janoueix-Lerosey I, Delattre O (2009) Beta-catenin status in paediatric medulloblastomas: correlation of immunohistochemical expression with mutational status, genetic profiles, and clinical characteristics. J Pathol 218:86–94

Gilbertson RJ, Ellison DW (2008) The origins of medulloblastoma subtypes. Annu Rev Pathol 3:241–365

Gulino A, Arcella A, Giangaspero F (2008) Pathological and molecular heterogeneity of medulloblastoma. Curr Opin Ocol 20:668–675

Hornick JL, Dal Cin P, Fletcher CD (2009) Loss of INI1 expression is characteristic of both conventional and proximal-type epithelioid sarcoma. Am J Surg Pathol 33:542–550

Judkins AR, Ellison DW (2010) Ependymoblastoma: dear, damned, distracting diagnosis, farewell! Brain Pathol 20:133–139

Korshunov A, Remke M, Gessi M, Ryzhova M, Hielscher T, Witt H, Tobias V, Buccoliero AM, Sardi I, Gardiman MP, Bonnin J, Scheithauer B, Kulozik AE, Witt O, Mork S, von Deimling A, Wiestler OD, Giangaspero F, Rosenblum M, Pietsch T, Lichter P, Pfister SM (2010) Focal genomic amplification at 19q13.42 comprises a powerful diagnostic marker for embryonal tumors with ependymoblastic rosettes. Acta Neuropathol 120:253–260

Lamont J, McManamy CS, Pearson AD, Clifford SC, Ellison DW (2004) Combined histopathological and molecular cytogenetic stratification of medulloblastoma patients. Clin Cancer Res 10:5482–5493

Louis DN, Ohgak IH, Wiestler OD, Cavenee WK (2007a) WHO classification of tumours of the central nervous system. International Agency for Research on Cancer, Lyon

Louis DN, Ohgaki H, Wiestler OD, Cavenee WK, Burger PC, Jouvet A, Scheithauer BW, Kleihues P (2007b) The 2007 WHO classification of tumours of the central nervous system. Acta Neuropathol 114:97–109

Lugli A, Zlobec I, Minoo P, Baker K, Tornillo L, Terracciano L, Jass JR (2007) Prognostic significance of the wnt signaling pathway molecules APC, beta-catenin and E-cadherin in colorectal cancer: a tissue microarray-based analysis. Histopathology 50:453–464

Martensson A, Oberg A, Jung A, Cederquist K, Stenling R, Palmqvist R (2007) Beta-catenin expression in relation to genetic instability and prognosis in colorectal cancer. Oncol Rep 17:447–452

McCabe MG, Backlund LM, Leong HS, Ichimura K, Collins VP (2011) Chromosome 17 alterations identify good-risk and poor-risk tumors independently of clinical factors in medulloblastoma. Neuro Oncol 13:376–383

Pfister S, Remke M, Benner A, Mendrzyk F, Toedt G, Felsberg J, Wittmann A, Devens F, Gerber NU, Joos S, Kulozik A, Reifenberger G, Rutkowski S, Wiestler OD, Radlwimmer B, Scheurlen W, Lichter P, Korshunov A (2009) Outcome prediction in pediatric medulloblastoma based on DNA copy-number aberrations of chromosomes 6q and 17q and the MYC and MYCN loci. J Clin Oncol 27:1627–1636

Pfister S, Korshunov A, Kool M, Hasselblatt M, Eberhart C, Taylor MD (2010) Molecular diagnostics of CNS embryonal tumors. Acta Neuropathol 120:553–566

Raffel C, Jenkins RB, Frederik L, Hebrink D, Alderete B, Fults DW, James CD (1997) Sporadic medulloblastomas contain *PTCH* mutations. Cancer Res 57:842–845

Rubin JB, Rowitch DH (2002) Medulloblastoma: a problem of developmental biology. Cancer Cell 2:7–8

Stearns D, Chaudhry A, Abel TW, Burger PC, Dang CV, Eberhart CG (2006) c-myc overexpression causes anaplasia in medulloblastoma. Cancer Res 66:673–681

Takei H, Nguyen Y, Mehta V, Chintagumpala M, Dause RRC, Adesina AM (2009) Low-level copy gain versus amplification of myc oncogenes in medulloblastoma: utility in predicting prognosis and survival. Laboratory investigation. Neurosurg Pediatr 3:61–65

von Bueren AO, Oehler C, Shalaby T, von Hoff K, Pruschy M, Seifert B, Gerber NU, Warmuth-Metz M, Stearns D, Eberhart CG, Kortmann RD, Rutkowski S, Grotzer MA (2011) c-MYC expression sensitizes medulloblastoma cells to radio- and chemotherapy and has no impact on response in medulloblastoma patients. B M C Cancer 11:74

Zitterbart K, Filkova H, Tomasikova L, Necesalova E, Zambo I, Kantorova D, Slamova I, Vranova V, Zezulkova D, Pesakova M, Pavelka Z, Veselska R, Kuglik P, Sterba J (2011) Low-level copy number changes of MYC genes have a prognostic impact in medulloblastoma. J Neurooncol 102:25–33

Oligodendroglial Tumors: Intra-arterial Chemotherapy

2

Daniel Guillaume, Nancy Doolittle, and Edward Neuwelt

Contents

Abstract

Brain tumors with an oligodendroglial component represent up to 20% of all primary brain tumors. Many of these tumors are sensitive to procarbazine, lomustine and vincristine (PCV), or temozolomide (TMZ) chemotherapy. In patients with oligodendroglial tumors that are resistant to PCV or TMZ chemotherapy, few treatment options are available and improved therapy is needed. One strategy, administration of chemotherapy via the intra-arterial (IA) route, can result in a higher concentration of drug delivery to the brain and brain tumor cells, with a lower systemic exposure, resulting in increased tumor-specific cytotoxicity, and avoidance of systemic toxicities. Osmotic blood-brain barrier disruption (BBBD), achieved by IA infusion of a hyperosmotic agent such as mannitol, can further intensify drug delivery to the tumor and surrounding brain, particularly in smaller, less permeable tumors and when using higher-molecular weight chemotherapy agents. General experience with IA chemotherapy, and IA chemotherapy combined with BBBD in patients with many tumor types is extensive. In the case of aggressive oligodendroglial tumors, hopeful results have been reported using combination delivery of IA chemotherapy (IA melphalan, IA carboplatin and IV etoposide phosphate) in conjunction with BBBD, with acceptably low toxicity and encouraging response data in many patients. Delivery of chemotherapy via the IA route, in conjunction with BBBD may be a good option in those with aggressive

D. Guillaume (✉) • N. Doolittle • E. Neuwelt
Division of Pediatric Neurosurgery, Department of Neurosurgery, Oregon Health and Science University, Portland, OR 97221, USA
e-mail: guillam@ohsu.edu

M. Hayat (ed.), *Tumors of the Central Nervous System, Volume 13*,
DOI 10.1007/978-94-007-7602-9_2,

oligodendroglial tumors. Further work is required to uncover the true potential of IA therapy in the setting of aggressive oligodendroglial tumors.

Introduction

Oligodendroglial brain tumors are now felt to represent up to 20% of primary brain tumors using expanded criteria (Bromberg and van den Bent 2009). Diagnosis of oligodendrogliomas as a separate entity from astrocytoma is important for a few important reasons. First, oligodendroglial tumors are often highly sensitive to procarbazine, lomustine and vincristine (PCV), or temozolomide (TMZ) chemotherapy (Cairncross et al. 1994; van den Bent et al. 1998). Second, the combined loss of the short arm of chromosome 1 (1p) and the long arm of chromosome 19 (19q) as genetic lesions occurring in 60–90% of oligodendrogliomas is associated with an improved response to chemotherapy (Cairncross et al. 1998; Ino et al. 2001; van den Bent et al. 2003), a more indolent clinical course, and an enhanced response to radiotherapy (Bromberg and van den Bent 2009) compared to tumors which have intact 1p/19q.

Classic oligodendroglioma is a well-differentiated WHO Grade II glioma. Although nuclear atypia and occasional mitosis can be noted in classic WHO grade II oligodendrogliomas, marked mitotic activity, microvascular proliferation and or necrosis are hallmarks of anaplastic oligodendroglioma (AO), a WHO Grade III glioma or glioblastoma, a WHO grade IV glioma. Oligoastrocytoma (OA), on the other hand, is a mixed glioma, containing both neoplastic astrocytes and neoplastic oligodendroglia. Again, the identification of oligodendroglial differentiation within malignant gliomas is essential, as these tumors may respond better to chemotherapy compared to those without oligodendroglial differentiation (Perry et al. 1999).

Treatment Options for Aggressive Oligodendroglial Tumors

Aggressive and recurrent oligodendroglial tumors are challenging to treat. Although extensive surgical resection has been shown, in two prospective studies, to be associated with increased survival (Cairncross et al. 2006; van den Bent et al. 2006), further adjuvant treatment (radiation or chemotherapy) is required in patients with these higher-grade tumors to prevent recurrence (van den Bent 2007).

Since oligodendroglial tumors were found to be sensitive to PCV chemotherapy in the mid 1990s (Cairncross et al. 1994; van den Bent et al. 1998), there has been increased interest in optimizing adjuvant chemotherapy regimens for these tumors. Most prospective uncontrolled single arm studies evaluating chemotherapy for recurrent oligodendroglial tumors have involved PCV or TMZ chemotherapy. While grouping of oligodendroglial tumors with other gliomas in clinical trials makes it difficult to draw conclusions regarding oligodendroglial tumors unequivocally, two trials specifically evaluated therapies for aggressive oligodendroglial tumors. A Radiation Therapy Oncology Group (RTOG) Phase III trial for grade III oligodendroglioma randomized 289 patients to receive either radiation alone or neoadjuvant dose intense PCV chemotherapy followed by radiation. With 3-year follow-up the median survival times were similar (4.9 years in the group receiving PCV + radiation therapy versus 4.7 years in the group receiving radiation therapy alone). However, PFS time was increased in the group receiving PCV (2.6 years) compared to radiation therapy alone (1.7 years, $P=0.004$). Unfortunately, 65% of patients receiving chemotherapy experienced grade 3 or 4 toxicity and one patient died (Cairncross et al. 2006). In another similar multicenter randomized controlled trial, 368 patients with anaplastic oligodendroglioma were randomized to receive either radiation therapy alone, or radiation followed by standard PCV chemotherapy. Results in this trial were similar, showing no significant difference in median survival, but increased PFS in the group receiving PCV (van den Bent et al. 2006). These studies suggest that, although chemotherapy in addition to radiation can prolong PFS in patients with aggressive oligodendroglial tumors, it is associated with significant toxicity that limits dose escalation.

Because the PCV regimen is associated with significant hematologic and GI toxicity, and most patients do not tolerate the six cycles intended, TMZ has developed a role as therapy for aggressive oligodendroglial tumors. Compared to PCV chemotherapy, TMZ is a better-tolerated oral agent with modest myelosuppression and easily controlled nausea and vomiting. The reported response rate of recurrent oligodendroglioma to TMZ after failure of radiation therapy is up to 50% with a median PFS of 10–12 months (van den Bent et al. 2003). Although no formal comparison between PCV and TMZ in recurrent oligodendroglial tumors is yet available, TMZ is often employed in practice because it is better tolerated by patients and more easily administered (van den Bent 2007).

For cases of aggressive oligodendroglial tumors that are resistant to PCV or TMZ chemotherapy, few treatment options are available. These patients are often poor candidates for additional surgery due to tumor size and location, often involving eloquent brain tissue. And, as mentioned earlier, simple dose escalations are limited by systemic toxicities. Improved innovative therapy with low toxicity is required for treatment of these refractory tumors.

Many innovative approaches have been investigated with the goal of increasing direct tumor cytotoxic effects while avoiding nonspecific systemic toxicities. Myeloablative chemotherapy with autologous peripheral blood progenitor cell rescue is one approach that was developed in order to achieve higher dose intensity and to avoid the toxicities associated with radiation therapy. For treatment of aggressive oligodendroglial tumors, this strategy was investigated several years ago (Mohile et al. 2008). In this study, 20 patients with aggressive oligodendroglial tumors were treated with four cycles of PCV chemotherapy every 6 weeks. Of the 20 treated patients, 15 demonstrated a response. The 15 responders were treated with myeloablative chemotherapy followed by autologous peripheral blood progenitor cell rescue. Fourteen patients underwent transplantation, with median disease-free and overall survival of at least 36 weeks (not yet reached at time of report). Although this approach allowed deferral of radiation for at least 3 years, a major limitation was the acute toxicity associated with both the induction and consolidation regimens, making this a poor option in most cases.

Intra-arterial Chemotherapy

Administration of a chemotherapeutic agent directly into an artery was first investigated as a method to treat brain tumors more than 50 years ago (Newton 2006). The aim of this approach is to achieve higher dose intensity of chemotherapeutic agent directly to tumor cells while achieving a low systemic exposure, avoiding many of the systemic toxicities that occur with dose escalations. This strategy increases the intra-tumoral cellular concentration of chemotherapeutic drug and subsequently can improve tumor cell death. Because primary and metastatic brain tumors are localized within brain tissue and receive an arterial blood supply that can be accessed using standard interventional techniques, the regional strategy of IA therapy is ideally suited.

The benefits of IA over IV administration of a chemotherapeutic agent, from a pharmacology viewpoint, occurs with the "first pass" of drug through the brain and tumor circulation, resulting in a higher concentration of drug delivered to brain and tumor tissue (Eckman et al. 1974). After the first pass, the drug continues to circulate through the blood until it is cleared. Certain drugs with a high extraction fraction, or high lipid solubility, maximize the first pass effect when delivered using this approach. Chemotherapeutic drugs with this favorable pharmacologic profile also possess rapid systemic metabolism and/or excretion. Evidence from several studies suggests that nitrosourea drugs, such as carmustine, are ideal for IA administration. In one early study, up to fivefold delivery of 14C-labeled carmustine was demonstrated with IA administration compared to IV in monkeys (Levin et al. 1978). In another study, super selective IA infusion of 11C-labeled carmustine into the middle cerebral artery of patients with recurrent gliomas resulted in up to a 50-fold increase in drug delivered compared to IV administration (Tyler et al. 1986).

Chemotherapeutic agents possessing a rapid total body clearance are more appropriate for IA delivery than those with slower clearance (Eckman et al. 1974). In this regard, the drugs most appropriate for IA chemotherapy include BCNU, cisplatin, carboplatin, and etoposide. Unfortunately, the two drugs most effective for IA delivery, BCNU and cisplatin, also have the most significant neuro-toxicity. Although there are pharmacological differences making one agent more appropriate for IA delivery than another, in general, administration of chemotherapy via the IA route for the treatment of brain tumors in both animal studies and clinical trials has demonstrated a higher concentration of drug to the tumor and brain surrounding tumor (Dropcho et al. 1992; Cloughesy et al. 1997) with less associated systemic toxicity compared with IV administration. Several investigators have assessed the toxicity and/or efficacy of specific antineoplastic agents administered IA to treat many types of brain tumors (Dropcho et al. 1992; Mortimer et al. 1992; Shapiro et al. 1992; Stewart et al. 1992). Experience with IA chemotherapy in patients with high-grade gliomas is extensive (Newton 2006). In studies treating patients with newly diagnosed gliomas using IA chemotherapy, administration has typically occurred just prior to or in combination with radiation. Single agents that have been used in this context include carmustine, nimustine, HeCNU, cisplatin and 5-fluorouracil. In a review of 15 studies evaluating a total of 395 patients, the median time to progression (TTP) of newly diagnosed patients receiving single-agent IA chemotherapy ranged from 12 to 32 weeks, with a median survival of approximately 1 year (range of 32–73 weeks) (Newton 2006). Combination IA chemotherapy in patients with newly diagnosed gliomas, in which one or more drugs were administered via the IA route has yielded similar results. In a review of eight studies, evaluating a total of 542 patients, the median TTP of combination IA regimens ranged from 33 to greater than 50 weeks and the median survival for combination regimens is approximately 1 year, with a range of 40–228 weeks (Newton 2006).

Naturally, in addition to treatments for newly diagnosed gliomas, single agents administered by the IA route have been studied as therapy for previously treated gliomas that were refractory to standard therapies. The majority of published studies evaluated either a nitrosourea derivative or a platinum analog. In a review of data from 18 studies, involving a total of 348 patients, median TTP ranged from 13 to 40 weeks in responders, and median survival ranged from 8 to 50 weeks in responders. Percent response ranged from no response to 80%. IA chemotherapy approaches utilizing multiple IA agents or the combination of IA agent with oral or IV drug has also been extensively evaluated for recurrent gliomas. Most regimens focused on the use of IA carmustine in combination with other drugs. Data from 11 studies, with a total of 291 patients showed percent response ranging from NR to 68%, with median TTP of 14 to 24 weeks, and median survival of 14–56 weeks. Taken together, these extensive studies in patients with high-grade gliomas provide hope for the use of IA delivery in general for treatment of glial tumors. None of these studies evaluated IA therapy specifically for oligodendroglial tumors.

IA Chemotherapy in Conjunction with BBBD

One strategy that can be used to promote increased drug delivery to brain tumor tissue, further intensifying tumor cytotoxicity while avoiding systemic toxicity, is osmotic disruption of the blood-brain barrier (BBB). The normal intact BBB prevents the passage of ionized water-soluble compounds with molecular weights greater than about 180 Dalton (D). As most chemotherapeutic agents have molecular weights between 200 and 1,200 D, the BBB is a major obstacle, preventing chemotherapy access to the CNS and tumor tissue residing here (Kroll et al. 1998b). Transient osmotic disruption of the BBB and blood-tumor barriers can be achieved by IA infusion of the hyperosmotic agent, mannitol (25%).

In numerous basic and clinical studies, osmotic BBBD has been shown to further intensify delivery of chemotherapeutic agent to brain tumors and surrounding brain tissue (Kroll and Neuwelt 1998; Kroll et al. 1998; Doolittle et al. 2000; Angelov et al. 2009) particularly in smaller, less permeable tumors and when using higher-molecular weight agents. In the case of methotrexate, vascular permeability reaches a maximum by 15 min after IA infusion of mannitol, and returns to normal within 2 h (Neuwelt et al. 1998). Compared to IV administration of chemotherapy, IA delivery following BBBD results in a 10–100-fold increase in delivery of chemotherapeutic drugs to brain tumors and tumor-infiltrated brain tissue (Muldoon et al. 2007). In humans BBB permeability continues to be elevated for up to 2 h after disruption', returning to baseline levels by 4 h (Siegal et al. 2000). This strategy has been so successful that centers across the globe continue to obtain the technical expertise to offer this option for treatment of specific tumors. The approach of combining IA administration of chemotherapy with osmotic BBBD is currently used to enhance chemotherapy delivery in brain tumor patients by trained teams at nine centers across the United States, Canada and Israel. With over 6,000 procedures performed to date, this procedure has demonstrated low morbidity and mortality (Doolittle et al. 2000). Numerous ongoing clinical studies are underway to better identify the specific tumors and chemotherapeutic agent combinations to lead to improved outcomes over other strategies.

To date, encouraging results have been obtained using IA chemotherapy in conjunction with BBBD in patients with many tumor types including aggressive oligodendroglioma (Williams et al. 1995; Muldoon et al. 2007; Neuwelt et al. 2008). Justification for use of this treatment gains further support from a recent report, summarizing the multi-institutional experience of this approach in treating newly diagnosed patients with primary central nervous system lymphoma (PCNSL). Although PCNSL is often chemoresponsive, as are aggressive oligodendroglial tumors, chemotherapeutic agents are often ineffective due to limited ability to cross the BBB. Use of radiation therapy for PCNSL is often associated with significant neurotoxicity. Administration of chemotherapy using the IA route after BBBD in 149 consecutive PCNSL patients who had not previously been treated with whole-brain radiotherapy demonstrated exciting results, with 25% of patients alive at 8 years and, in some, survival extending beyond 20 years (Angelov et al. 2009). Importantly, in these patients, survival occurred without the cognitive loss associated with up-front radiation therapy. Together, these encouraging results, as well as preceding studies that span more than 20 years (Williams et al. 1995; Muldoon et al. 2007; Jahnke et al. 2008; Neuwelt et al. 2008), support further investigation of this method of improved delivery to other chemoresponsive tumors, such as aggressive oligodendroglial tumors.

Toxicity of IA Chemotherapy

In general, IA chemotherapy with or without BBBD can be associated with various specific and nonspecific toxicities. Specific toxicities reported include ototoxicity, neurotoxicity, visual toxicity and procedural toxicities. Ipsilateral periorbital erythalgia and visual loss have been described in patients receiving IA chemotherapy. In some cases this was dose limiting (Gebarski et al. 1984; Stewart 1991; Basso et al. 2002). Ototoxicity has been described in patients receiving IA cisplatin and carboplatin (Stewart 1991; Neuwelt et al. 2006) with an incidence of symptomatic hearing loss of 5–15% but much higher rates of hearing loss have been reported, up to 45–62%, when serial audiometric testing was utilized (Dropcho et al. 1992). Neurotoxicity can also occur secondary to IA delivery of chemotherapy. Acute neurological toxicities occur, typically within 48 h of infusion, in 8–12% of patients receiving IA chemotherapy. Acute neurotoxicity is more common with carmustine and other nitrosoureas and cisplatin, and less frequent with carboplatin and methotrexate. Symptoms include seizure, confusion, headaches, and focal deficits. This is associated ipsilateral hemispheric white matter changes, is seen most frequently after IA treatment with carmustine, and clinically presents as mild cognitive deficits (Stewart 1991).

When BBBD is combined with IA chemotherapy, specific toxicities can occur which have been thoroughly studied and well described (Doolittle et al. 2000). Adverse effects included catheter-related complications and disruption-associated complications. Complications related to IA catheter placement include groin hematoma, subintimal tear and vessel thrombosis, although the frequency of these complications is low. Reported toxicities related to BBB disruption include seizures and brain edema. Importantly, in contrast to radiation therapy, the BBBD procedure does not appear to be associated with any decline in formal neurocognitive assessment. Overall, this approach has proven to be safe in multiple clinical studies (Doolittle et al. 2000).

IA Chemotherapy with BBBD for Aggressive Oligodendroglial Tumors

Because aggressive oligodendroglial tumors have demonstrated responsiveness to PCV and TMZ chemotherapy regimens, rationale exists to further evaluate the role of IA chemotherapy in conjunction with BBBD as a means to improve delivery of known cytotoxic agents directly to tumor cells. The responsiveness of aggressive oligodendroglial tumors to this therapy is currently being investigated in an ongoing prospective clinical trial. This trial is aimed to evaluate combination chemotherapy (IA melphalan, IA carboplatin and IV etoposide phosphate) in conjunction with BBBD in patients with aggressive oligodendroglial tumors. The safety and preliminary efficacy of this study was recently published (Guillaume et al. 2010). Thirteen subjects with aggressive oligodendroglial tumors refractory to TMZ were included. As expected, this drug combination showed very low toxicity as well as encouraging response data. Five of thirteen patients demonstrated a response, five remained stable and only three developing progressive disease. In the three patients who demonstrated progression, this occurred within 1 month from first IA/BBBD treatment, suggesting severe advanced disease. In the ten subjects demonstrating a response or stable disease, the median PFS using Kaplan-Meier estimation was 19 months. Overall PFS in all 13 subjects was 11 months. Survival from the date of first treatment ranged from 1 to greater than 27 months, suggesting there is a subgroup of patients who respond well to this therapy. Consistent with other studies utilizing different treatment paradigms, the five patients with complete or partial response demonstrated deletion of both 1p and 19q. Conversely, the two patients in whom both 1p and 19q were intact developed progressive disease within 1 month and died at 1 and 8 months following first treatment. Interestingly, the two patients achieving a complete response did not receive radiation therapy either prior to or after BBBD treatment. These preliminary results investigating IA chemotherapy in conjunction with BBBD for aggressive oligodendroglial tumors are encouraging, and results of the Phase II portion of the ongoing study are eagerly anticipated.

Conclusions

Administration of many chemotherapeutic agents via the IA route can lead to improved delivery of drug to tumor cells and brain tissue, resulting in increased tumor specific cytotoxicity with avoidance of non-specific systemic toxicities. In many circumstances, the addition of BBBD can augment this delivery. Efforts are underway to identify candidates for these therapies. As clinical studies continue to provide evidence of efficacy, applications continue to grow. This has led to increasing numbers of clinical teams across the globe obtaining the technical expertise necessary to offer this approach. In the case of aggressive oligodendroglial tumors, there is evidence that this approach is safe, can lead to a good clinical response in patients who are refractory to other therapies, and that a clinical response can occur without radiation therapy. However, further studies, which are ongoing, are required to fully evaluate this.

Acknowledgements The authors thank Shirley McCartney, Ph.D., for professional editing and manuscript preparation.

References

Angelov L, Doolittle ND, Kraemer DF, Siegal T, Barnett GH, Peereboom DM, Stevens G, McGregor J, Jahnke K, Lacy CA, Hedrick NA, Shalom E, Ference S, Bell S, Sorenson L, Tyson RM, Haluska M, Neuwelt EA (2009) Blood-brain barrier disruption and intra-arterial methotrexate-based therapy for newly diagnosed primary CNS lymphoma: a multi-institutional experience. J Clin Oncol 27:3503–3509

Basso U, Lonardi S, Brandes AA (2002) Is intra-arterial chemotherapy useful in high-grade gliomas. Expert Rev Anticancer Ther 2:507–519

Bromberg JE, van den Bent MJ (2009) Oligodendrogliomas: molecular biology and treatment. Oncologist 14:155–163

Cairncross G, Macdonald D, Ludwin S, Lee D, Cascino T, Buckner J, Fulton D, Dropcho E, Stewart D, Schold C Jr, Wainman N, Eisenhauer E (1994) Chemotherapy for anaplastic oligodendroglioma. National Cancer Institute of Canada Clinical Trials Group. J Clin Oncol 12:2013–2021

Cairncross JG, Ueki K, Zlatescu MC, Lisle DK, Finkelstein DM, Hammond RR, Silver JS, Stark PC, Macdonald DR, Ino Y, Ramsay DA, Louis DN (1998) Specific genetic predictors of chemotherapeutic response and survival in patients with anaplastic oligodendrogliomas. J Natl Cancer Inst 90:1473–1479

Cairncross G, Berkey B, Shaw E, Jenkins R, Scheithauer B, Brachman D, Buckner J, Fink K, Souhami L, Laperierre N, Mehta M, Curran W (2006) Phase III trial of chemotherapy plus radiotherapy compared with radiotherapy alone for pure and mixed anaplastic oligodendroglioma: Intergroup Radiation Therapy Oncology Group Trial. J Clin Oncol 24:2707–2714

Cloughesy TF, Gobin YP, Black KL, Vinuela F, Taft F, Kadkhoda B, Kabbinavar F (1997) Intra-arterial carboplatin chemotherapy for brain tumors: a dose escalation study based on cerebral blood flow. J Neurooncol 35:121–131

Doolittle ND, Miner ME, Hall WA, Siegal T, Jerome E, Osztie E, McAllister LD, Bubalo JS, Kraemer DF, Fortin D, Nixon R, Muldoon LL, Neuwelt EA (2000) Safety and efficacy of a multicenter study using intraarterial chemotherapy in conjunction with osmotic opening of the blood-brain barrier for the treatment of patients with malignant brain tumors. Cancer 88:637–647

Dropcho EJ, Rosenfeld SS, Morawetz RB, Vitek J, Brothers M, Gorum T, Bell S, Gillespie GY, Glantz M, Mahaley MS Jr et al (1992) Preradiation intracarotid cisplatin treatment of newly diagnosed anaplastic gliomas. The CNS Cancer Consortium. J Clin Oncol 10:452–458

Eckman WW, Patlak CS, Fenstermacher JD (1974) A critical evaluation of the principles governing the advantages of intra-arterial infusions. J Pharmacokinet Biopharm 2:257–285

Gebarski SS, Greenberg HS, Gabrielsen TO, Vine AK (1984) Orbital angiographic changes after intracarotid BCNU chemotherapy. AJNR Am J Neuroradiol 5:55–58

Guillaume DJ, Doolittle ND, Gahramanov S, Hedrick NA, Delashaw JB, Neuwelt EA (2010) Intra-arterial chemotherapy with osmotic blood-brain barrier disruption for aggressive oligodendroglial tumors: results of a phase I study. Neurosurgery 66:48–58

Ino Y, Betensky RA, Zlatescu MC, Sasaki H, Macdonald DR, Stemmer-Rachamimov AO, Ramsay DA, Cairncross JG, Louis DN (2001) Molecular subtypes of anaplastic oligodendroglioma: implications for patient management at diagnosis. Clin Cancer Res 7:839–845

Jahnke K, Kraemer DF, Knight KR, Fortin D, Bell S, Doolittle ND, Muldoon LL, Neuwelt EA (2008) Intraarterial chemotherapy and osmotic blood-brain barrier disruption for patients with embryonal and germ cell tumors of the central nervous system. Cancer 112:581–588

Kroll RA, Neuwelt EA (1998) Outwitting the blood-brain barrier for therapeutic purposes: osmotic opening and other means. Neurosurgery 42:1083–1099

Kroll RA, Pagel MA, Muldoon LL, Roman-Goldstein S, Fiamengo SA, Neuwelt EA (1998) Improving drug delivery to intracerebral tumor and surrounding brain in a rodent model: a comparison of osmotic versus bradykinin modification of the blood-brain and/or blood-tumor barriers. Neurosurgery 43:879–886

Levin VA, Kabra PM, Freeman-Dove MA (1978) Pharmacokinetics of intracarotid artery 14C-BCNU in the squirrel monkey. J Neurosurg 48:587–593

Mohile NA, Forsyth P, Stewart D, Raizer JJ, Paleologos N, Kewalramani T, Louis DN, Cairncross JG, Abrey LE (2008) A phase II study of intensified chemotherapy alone as initial treatment for newly diagnosed anaplastic oligodendroglioma: an interim analysis. J Neurooncol 89:187–193

Mortimer JE, Crowley J, Eyre H, Weiden P, Eltringham J, Stuckey WJ (1992) A phase II randomized study comparing sequential and combined intraarterial cisplatin and radiation therapy in primary brain tumors. A Southwest Oncology Group study. Cancer 69:1220–1223

Muldoon LL, Soussain C, Jahnke K, Johanson C, Siegal T, Smith QR, Hall WA, Hynynen K, Senter PD, Peereboom DM, Neuwelt EA (2007) Chemotherapy delivery issues in central nervous system malignancy: a reality check. J Clin Oncol 25:2295–2305

Neuwelt EA, Barnett PA, McCormick CI, Remsen LG, Kroll RA, Sexton G (1998) Differential permeability of a human brain tumor xenograft in the nude rat: impact of tumor size and method of administration on optimizing delivery of biologically diverse agents. Clin Cancer Res 4:1549–1555

Neuwelt EA, Gilmer-Knight K, Lacy C, Nicholson HS, Kraemer DF, Doolittle ND, Hornig GW, Muldoon LL (2006) Toxicity profile of delayed high dose sodium thiosulfate in children treated with carboplatin in conjunction with blood-brain-barrier disruption. Pediatr Blood Cancer 47:174–182

Neuwelt E, Abbott NJ, Abrey L, Banks WA, Blakley B, Davis T, Engelhardt B, Grammas P, Nedergaard M, Nutt J, Pardridge W, Rosenberg GA, Smith Q, Drewes LR (2008) Strategies to advance translational research into brain barriers. Lancet Neurol 7:84–96

Newton H (2006) Intra-arterial chemotherapy. In: Newton H (ed) Handbook of brain tumor chemotherapy. Elsevier, London

Perry JR, Louis DN, Cairncross JG (1999) Current treatment of oligodendrogliomas. Arch Neurol 56:434–436

Shapiro WR, Green SB, Burger PC, Selker RG, VanGilder JC, Robertson JT, Mealey J Jr, Ransohff J, Mahaley MS Jr (1992) A randomized comparison of intra-arterial versus intravenous BCNU, with or without intravenous 5-fluorouracil, for newly diagnosed patients with malignant glioma. J Neurosurg 76:772–781

Siegal T, Rubinstein R, Bokstein F, Schwartz A, Lossos A, Shalom E, Chisin R, Gomori JM (2000) In vivo assessment of the window of barrier opening after osmotic blood-brain barrier disruption in humans. J Neurosurg 92:599–605

Stewart D (1991) Intra-arterial chemotherapy of primary and metastatic brain tumors. In: Rottenberg D (ed) Neurological complications of cancer treatment. Butterworth-Heinimann, Boston, pp 143–170

Stewart DJ, Belanger JM, Grahovac Z, Curuvija S, Gionet LR, Aitken SE, Hugenholtz H, Benoit BG, DaSilva VF (1992) Phase I study of intracarotid administration of carboplatin. Neurosurgery 30:512–516

Tyler JL, Yamamoto YL, Diksic M, Theron J, Villemure JG, Worthington C, Evans AC, Feindel W (1986) Pharmacokinetics of superselective intra-arterial and intravenous [11C]BCNU evaluated by PET. J Nucl Med 27:775–780

van den Bent MJ (2007) Anaplastic oligodendroglioma and oligoastrocytoma. Neurol Clin 25:1089–1109, ix–x

van den Bent MJ, Kros JM, Heimans JJ, Pronk LC, van Groeningen CJ, Krouwer HG, Taphoorn MJ, Zonnenberg BA, Tijssen CC, Twijnstra A, Punt CJ, Boogerd W (1998) Response rate and prognostic factors of recurrent oligodendroglioma treated with procarbazine, CCNU, and vincristine chemotherapy. Dutch Neuro-oncology Group. Neurology 51:1140–1145

van den Bent MJ, Taphoorn MJ, Brandes AA, Menten J, Stupp R, Frenay M, Chinot O, Kros JM, van der Rijt CC, Vecht Ch J, Allgeier A, Gorlia T (2003) Phase II study of first-line chemotherapy with temozolomide in recurrent oligodendroglial tumors: the European Organization for Research and Treatment of Cancer Brain Tumor Group Study 26971. J Clin Oncol 21:2525–2528

van den Bent MJ, Carpentier AF, Brandes AA, Sanson M, Taphoorn MJ, Bernsen HJ, Frenay M, Tijssen CC, Grisold W, Sipos L, Haaxma-Reiche H, Kros JM, van Kouwenhoven MC, Vecht CJ, Allgeier A, Lacombe D, Gorlia T (2006) Adjuvant procarbazine, lomustine, and vincristine improves progression-free survival but not overall survival in newly diagnosed anaplastic oligodendrogliomas and oligoastrocytomas: a randomized European Organisation for Research and Treatment of Cancer phase III trial. J Clin Oncol 24:2715–2722

Williams PC, Henner WD, Roman-Goldstein S, Dahlborg SA, Brummett RE, Tableman M, Dana BW, Neuwelt EA (1995) Toxicity and efficacy of carboplatin and etoposide in conjunction with disruption of the blood-brain tumor barrier in the treatment of intracranial neoplasms. Neurosurgery 37:17–27

Metastatic Oligodendroglioma: Diagnosis with Fine-Needle Aspiration Cytology

3

Bilge Can Meydan and İlkser Akpolat

Contents

B. Can Meydan (✉) • İ. Akpolat
Department of Pathology, School of Medicine,
Ondokuz Mayıs University, 55139 Samsun, Turkey
e-mail: canbilge@yahoo.com

Abstract

Extracranial/systemic metastasis of a primary central nervous system glial tumor is very rare, but the incidence is ever-increasing. Oligodendroglial tumors are uncommon; however, after high-grade astrocytomas, they constitute an important group of tumors which are reported to cause extracranial metastases. In the majority of literature-reported cases, the likelihood of a second primary tumor is considered as the first approach; therefore metastases are overlooked. Fine-needle aspiration cytology (FNAC) is currently accepted as a reliable, accurate, cost-effective and rapid diagnostic method for possible metastatic lesions in a patient with known primary tumor. FNAC findings of metastatic oligodendroglioma and small-round-cell tumors (SRCT) are quite similar, therefore differential diagnosis requires combination of the clinical and radiological data. In addition, application of immunochemistry and detection of 1p/19q co-deletion on cytological material confirm the diagnosis.

Introduction

Oligodendroglioma (WHO Grade II) is a slow-growing glial tumor, although often resulting multiple recurrence and/or progression, that is classified as anaplastic oligodendroglioma (WHO grade III), and has longer median-survival-time compared to astrocytoma (Reifenberger et al. 2007; Ohgaki and Kleihues 2005). Especially following high-grade

M. Hayat (ed.), *Tumors of the Central Nervous System, Volume 13*,
DOI 10.1007/978-94-007-7602-9_3, © Springer Science+Business Media Dordrecht 2014

progression, due to leptomeningeal involvement, it may lead to multiple intraneural spread via cerebrospinal fluid. Extracranial metastases have been reported in rare cases (<50 case reports). Given the autopsy cases, however, the true incidence is higher (Jellinger 2009; Krijnen et al. 2010). In the largest series published to date, oligodendroglial tumors followed high-grade astrocytoma, medulloblastoma, and ependymoma with regards to the incidence of extracranial metastases (Liwnicz and Rubinstein 1979).

Although the mechanism of extracranial metastases is not clearly understood, two different patterns can be mentioned: (1) following repeated craniotomy, development of scalp and/or regional lymph node metastases, and subsequent distant metastases after many years, (2) development of direct distant metastases via hematogenous route in the early period (Macdonald et al. 1989; Ordonez et al. 1981). The most common target organs are bone, bone marrow, and lymph nodes (Zustovich et al. 2008). Metastases to liver, lung, and other visceral organs, subcutis, mesothelial cavities have also been rarely reported (Uzuka et al. 2007). Unlike astrocytic tumors, which often metastasize to lymph node and lung, oligodendroglial tumors show significant affinity to bone and bone marrow metastasis (Zustovich et al. 2008; Krijnen et al. 2010). Reported bone metastases are most often osteoblastic (osteosclerotic) type (Jellinger et al. 1969; Wu et al. 2011). Again, according to the post-mortem examination reports, extracranial metastases related with haematogenous spread are remarkably multisystemic (James and Pagel 1951; Uzuka et al. 2007). These cases with multiple occult metastases have a rapid clinical course and poor prognosis (Morrison et al. 2004; Merrell et al. 2006; Zustovich et al. 2008; Noshita et al. 2010; Cordiano et al. 2011). Immediate diagnosis and treatment of metastases is also important for chemosensitive character of these tumors. Therefore, as a routine approach, whole body bone scan and bone marrow biopsy, if hematologic abnormalities unexplained by medication exist (e.g., anemia, pancytopenia, etc.), may should be recommended. Furthermore, considering the fact that these younger patients with occult metastases, when they fail to survive, are potential organ donors, transmission of the tumor to organ recipient may occur (Volavsek et al. 2009). As a controversial point, limited publication is also available indicating that metastatic lesions may show grade II morphology, but regardless of morphology, may result in poor prognosis (Ng et al. 2006).

In this context, the following two questions can be referred in a short paragraph: (1) Why are the extracranial metastases of neuroglial tumor so rare? Some answers may be suggested: absence of lymphatic vessels in the intracranial parenchyma, blood-brain barrier, narrowing of veins due to mechanical impact in parallel to the growing tumor, failure in vascular penetration of tumor cells, inappropriate microenvironment of non-brain tissues and short survival time. (2) Why is the incidence ever-increasing? It is pointed out that increased survival time and/or changes in tumor behavior as a result of treatments may explain this consequence (Perry 2004; Giordana et al. 2004). Furthermore, one of the main factors that facilitate metastases is recurrent craniotomies which may provide access to the parenchyma, dura (subarachnoid and meningeal) or vascular channels within the scalp (Ordonez et al. 1981). Parenthetically, glioblastoma but not oligodendroglial tumor cases have been also reported that led spontaneous metastasis without craniotomy (Hulbanni and Goodman 1976). Chemoradiation may also contribute to the development of metastasis by inhibiting acquisition of immunological defense against tumor. The most known mechanism of metastasis is ventriculoperitoneal shunts.

Practically, any palpable or radiologically localized mass lesion can be evaluated by FNAC. The procedure is minimally invasive and accurate. The technique usually provides enough cellular, even tissue material for microscopic examination and auxiliary diagnostic tests. Therefore, the appropriate specimen triage is definitely required the immediate interpretation of the cytopathologist, who is aware of the clinical history.

FNAC of Metastatic Oligodendroglioma

Generally, some of the FNAC smears are air-dried and stained by Diff-Quik method (DQ) and additional smears are immediately fixed in alcohol (95% ethanol) and stained by Papanicolaou method (PAP). The remaining part of the sample, if present, should be placed in a vial containing cell-preservative medium and kept for further studies, which may be required.

Satisfactory data to determine a uniform criterion for the adequacy of specimens are not yet available. Due to rich vascular network structure, metastatic lesions may be hemorrhagic; still, cytological smears are always hypercellular. However, bone marrow aspirations are suggested to yield dry-tap results (Choon and Roepke 2004). A few case reports highlighting difficulties of diagnosis with bone marrow aspiration smear findings are available (Dawson 1997; Anand et al. 2003).

Cytological preparations show poorly cohesive and singly dispersed small cells that are fairly uniform (Figs. 3.1, 3.2 and 3.3). The cells have generally small, but larger than latent lymphocyte, round, hyperchromatic nucleus and scant to moderate cytoplasm with indistinct borders. Individual cells may have large nuclei with relatively regular contours. Prominent nuclear pleomorphism or multinucleation are not seen. These cells usually lack nucleoli. If the mini-gemistocytic component or/and gliofibrillary oligodendrocytes are not frequent, the cytoplasmic projections are not conspicuous (Lee et al. 2006; Mitsuhashi et al. 2007; Can et al. 2012). When examined the loose-cohesive groups, pseudorosette-like arrangement can be easily recognized, but prominent nuclear angulation or molding is not seen. The presence of fibrillary background varies, generally sparse, and can be observed with DQ stain. Thin-walled branching vasculature is prominent (Can et al. 2012; Lopez-Rios et al. 2000). Endothelial proliferation, which is considered as a high-grade finding, has not been reported so far, and this may be speculated as insufficient paracrine signaling within the target organ. As in our case, prominent apoptosis and necrotic debris on the background may not be seen, while may be detected as important and frequent finding in the other SRCT group. Consequently, high mitotic activity is the most important clue for malignancy. Microcalcifications on the background and the bland-looking cells with eccentrically located nucleus and/or relatively large, distinct cytoplasm that called mini-gemistocytes, has been described in primary brain oligodendroglial tumors, and theoretically may also occur in metastases, but have not been reported so far (Can et al. 2012). Again, intracytoplasmic eosinophilic granules described as a clue feature for primary oligodendroglioma cannot be seen in metastatic lesion (Kojima et al. 2008).

As a result, squash, smear or touch preparation cytology findings described in the primary oligodendroglial tumors are slightly different from FNAC findings of their metastatic lesions (Watson and Hajdu 1977; Can et al. 2012). For example, significant fibrillary ground, mini-gemistocytes and microcalcifications may not exist in metastases. The reason for this can be speculated as inappropriate microenvironment conditions of the target organ and/or insufficient time to generate their unique tissue pattern.

Cytological features of metastatic oligodendrogliomas, particularly of pure ones, necessitate differential diagnosis with SRCT group and it is quite difficult. In the literature, very few case reports in which definitive diagnosis has been made by cytological findings are available, though immunocytochemistry (ICC) is generally required (Lopez-Rios et al. 2000; Can et al. 2012). ICC can be performed on cytospin slides or smear preparations prepared from FNA material. Immunohistochemistry (IHC) can also be performed on parafin-embedded cell block samples prepared from FNA material. As oligodendroglial tumors are usually observed among adult age group, differential diagnosis of SRCT includes small cell neuroendocrin carcinoma, Ewing's sarcoma/primitive neuroectodermal tumor (EWS/PNET), hematolymphoid malignancies,

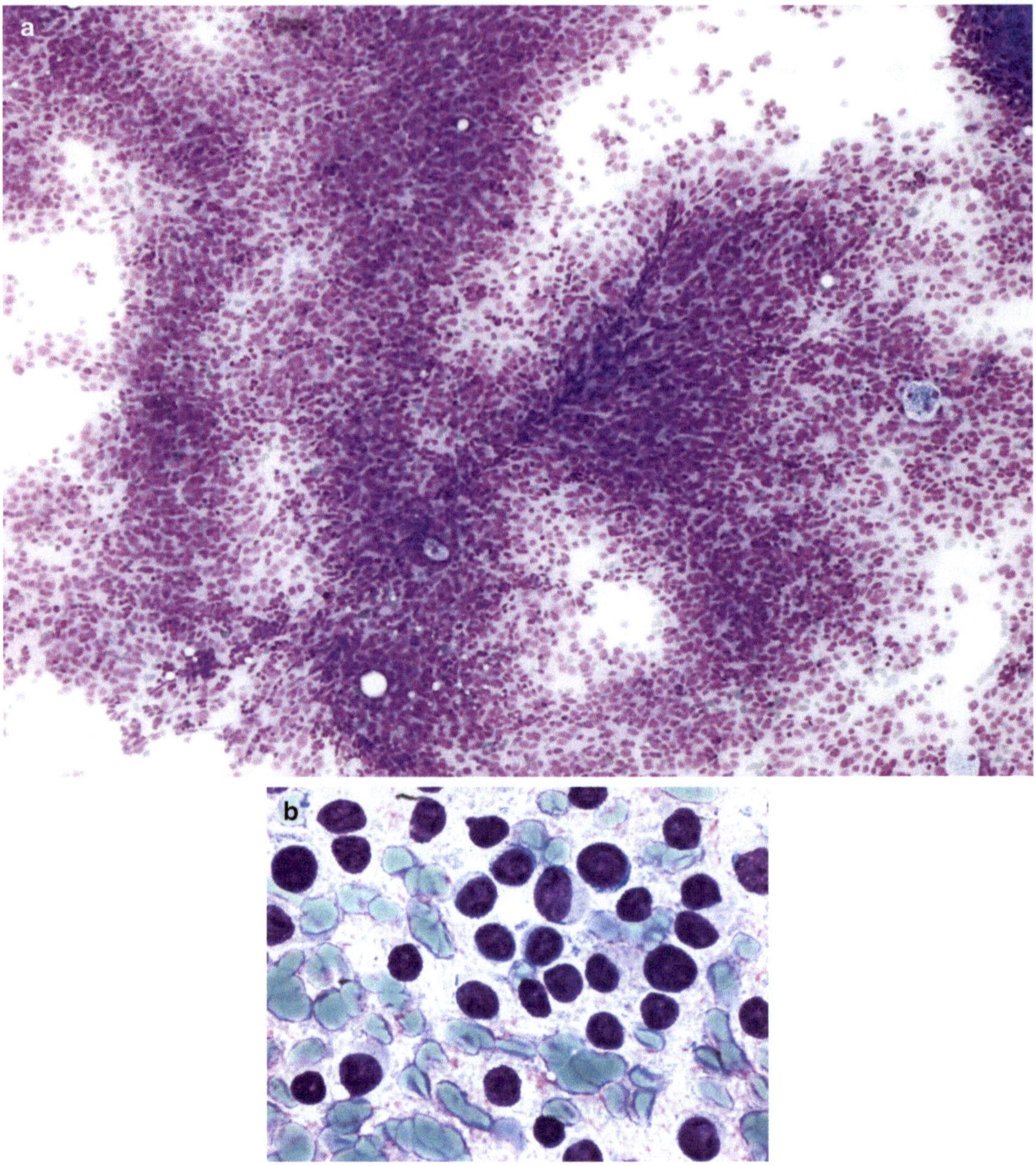

Fig. 3.1 The FNA smear. Poorly cohesive cells condensed around thin vasculature. Fairly uniform cells with moderate anisonucleosis (*inset*) (×100 and ×1,000, DQ)

and rarely small cell variant-melanoma, small-cell osteosarcoma, mesenchymal chondrosarcoma. Any positivity of glial fibrillary acidic protein (GFAP), if present, is very helpful in diagnosis; however, in contrast, focally positive results are noted to be obtained particularly with polyclonal markers in some of the non-glial tumors (Budka 1986). Thus, microtubule-associated protein 2 (MAP-2), SOX10 and glial-lineage markers, such as Olig-1 and Olig-2, may be more helpful in the diagnosis than GFAP staining, although there is insufficient literature data regarding their expression in other SRCTs (Reifenberger et al. 2007). Neoplastic oligodendroglial cells are also positive for neuron specific enolase (NSE), synaptophysin, CD57 (leu-7), and vimentin, but

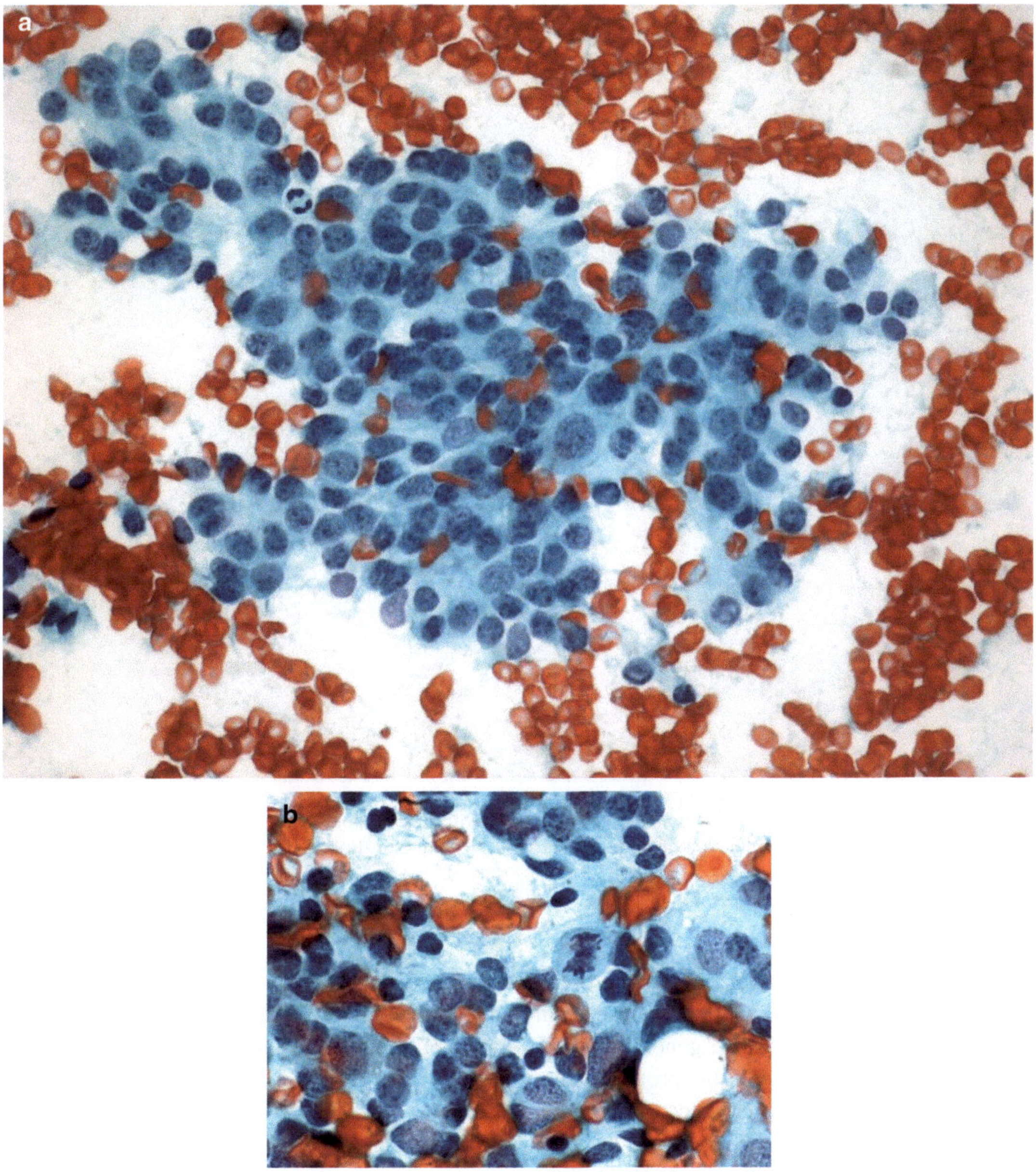

Fig. 3.2 The FNA smear. Round nuclei with evenly distributed chromatin and minimal membrane irregularity. Frequent mitosis (*inset*) (×400 and ×1,000, PAP)

these markers have lower specificity, especially when PNET is considered in the differential diagnosis. Again, oligodendroglial tumor cells are positive for nuclear and cytoplasmic S-100 protein, but this marker cannot be used for the differentiation of PNET, metastatic melanoma or chondrosarcoma. In the very limited literature data, oligodendrogliomas are negative for CD99, as a marker of PNET. Negative results for leukocyte common antigen (LCA), CD20, CD3, CD138 or CD38, myeloperoxidase (MPO), desmin, HMB45, epithelial membrane antigen (EMA), and cytokeratin can also be used in differential diagnosis. However, to know patient's history will substantially prevent unnecessary and confusing immunochemistry applications.

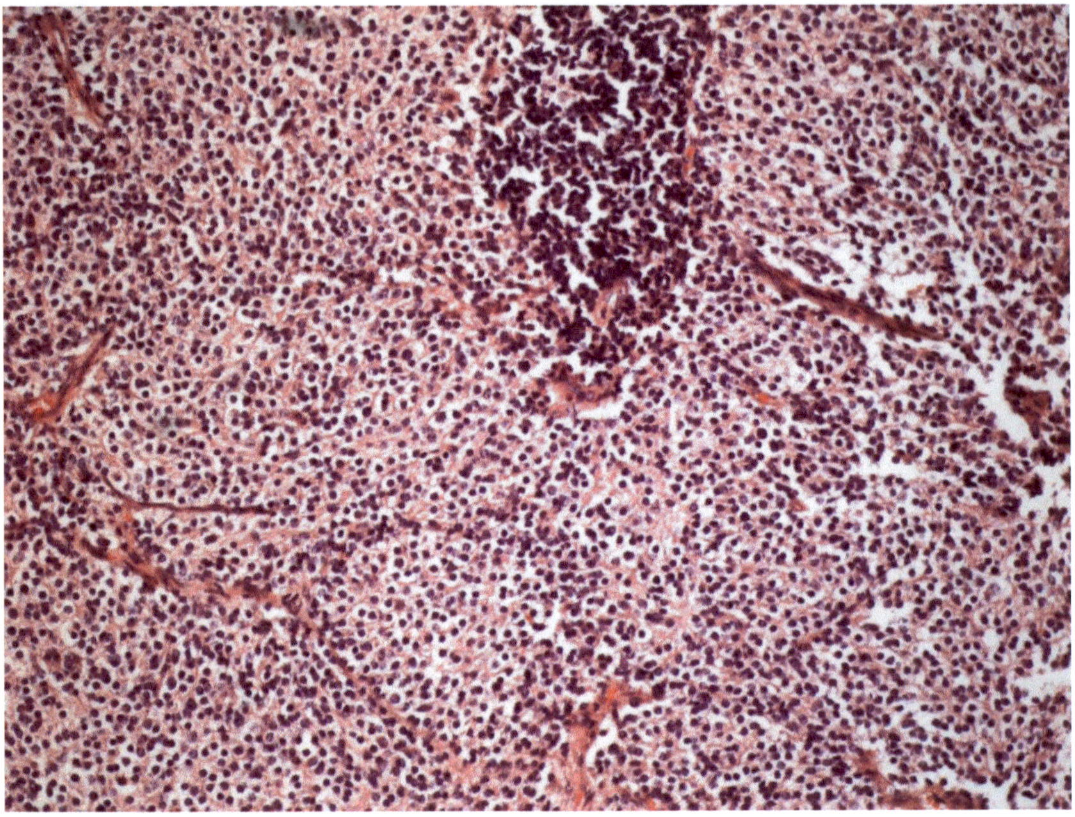

Fig. 3.3 Histologic section. Sheets of monotonous, small cells with clear cytoplasm and round nucleus, consistent with oligodendroglial tumor and residual lymph node tissue (×200, Hematoxylin-Eosin)

Detection of 1p/19q co-deletion by fluorescent in situ hybridization (FISH), polymerase chain reaction (PCR) or comparative genomic hybridization techniques has diagnostic, prognostic and even predictive value for oligodendroglial tumors. 1p/19q co-deletion was also observed in recurrence and metastatic (in few case reports) oligodendroglial tumors (Wang et al. 2004; Bruggers et al. 2007; Campbell et al. 2008; Krijnen et al. 2010; Noshita et al. 2010; Can et al. 2012). For the first time, Wang et al. (2004) demonstrated 1p/19q co-deletion by using capillary electrophoresis and PCR techniques on a cell-bock specimen obtained from FNAC material of an oligodendroglioma case with intraparotid lymph node metastasis. Also, 1p/19q co-deletion by FISH can be performed on cytospin slides or cell blocks. However, it should be kept in mind that the presence of co-deletion is considered highly suggestive for the diagnosis, while its absence will not rule out the diagnosis at all.

Result

Clinical and FNAC findings of metastatic oligodendroglial tumors are briefly reviewed. As a take home message, systemic work-up for metastases should be done in the patient with history of primary central nervous tumor, especially when unexplained clinical symptoms are present and FNAC, is one of the rapid and direct diagnostic methods that should be considered for diagnosis of suspected metastatic mass. Metastatic oligodendrogliomas have some cytomorphologic characteristics that may lead to a definitive diagnosis with the combination of clinic-radiologic data. Moreover, some initial ancillary diagnostic studies such as 1p/19q co-deletion by FISH may provide additional prognostic and predictive information.

References

Anand M, Kumar R, Jain P, Gupta R, Ghosal N, Sharma A, Agarwal A, Sharma MC (2003) Metastatic anaplastic oligodendroglioma simulating acute leukemia. A case report. Acta Cytol 47:467–469

Bruggers C, White K, Zhou H, Chen Z (2007) Extracranial relapse of an anaplastic oligodendroglioma in an adolescent: case report and review of the literature. J Pediatr Hematol Oncol 29:319–322

Budka H (1986) Non-glial specificities of immunocytochemistry for the glial fibrillary acidic protein (GFAP). Triple expression of GFAP, vimentin and cytokeratins in papillary meningioma and metastasizing renal carcinoma. Acta Neuropathol 72:43–54

Campbell BA, Horsman DE, Maguire J, Young S, Curman D, Ma R, Thiessen B (2008) Chromosomal alterations in oligodendroglial tumours over multiple surgeries: is tumour progression associated with change in 1p/19q status? J Neurooncol 89:37–45

Can B, Akpolat I, Meydan D, Üner A, Kandemir B, Söylemezoğlu F (2012) Fine-needle aspiration cytology of metastatic oligodendroglioma: case report and literature review. Acta Cytol 56:97–103

Choon A, Roepke JE (2004) Importance of immunohistochemical staining in metastatic anaplastic oligodendroglioma. Arch Pathol Lab Med 128:489–490

Cordiano V, Miserocchi F, Storti M (2011) Bone marrow metastases from anaplastic oligodendroglioma presenting with pancytopenia and hypogammaglobulinemia: a case report. Tumori 97:808–811

Dawson TP (1997) Pancytopaenia from a disseminated anaplastic oligodendroglioma. Neuropathol Appl Neurobiol 23:516–520

Giordana MT, Ghimenti C, Leonardo E, Balteri I, Iudicello M, Duò D (2004) Molecular genetic study of a metastatic oligodendroglioma. J Neurooncol 66:265–271

Hulbanni S, Goodman PA (1976) Glioblastoma multiforme with extraneural metastases in the absence of previous surgery. Cancer 37:1577–1583

James TG, Pagel W (1951) Oligodendroglioma and extracranial metastases. Br J Surg 39:56–65

Jellinger K (2009) Metastatic oligodendrogliomas: a review of the literature and case report. Acta Neurochir (Wien) 151:987

Jellinger K, Minauf M, Salzer-Kuntschik M (1969) Oligodendroglioma with extraneural metastases. J Neurol Neurosurg Psychiatry 32:249–253

Kojima H, Mori K, Fukudome N, Iseki M, Shimizu S (2008) Cytologic characteristics of intracytoplasmic refractile eosinophilic granular bodies in anaplastic oligodendroglioma: a case report. Acta Cytol 52:467–470

Krijnen JL, Fleischeur RE, van Berkel M, Westenend PJ (2010) Metastatic oligodendroglioma: a case report and incidence in The Netherlands. Clin Neuropathol 29:141–146

Lee CC, Jiang JS, Chen ET, Yokoo H, Pan YH, Tsai MD (2006) Cytologic diagnosis of a metastatic oligodendroglioma in a pleural effusion. A case report. Acta Cytol 50:542–544

Liwnicz BH, Rubinstein LJ (1979) The pathways of extraneural spread in metastasizing gliomas: a report of three cases and critical review of the literature. Hum Pathol 10:453–467

Lopez-Rios F, Alberti N, Ballestin C, Pérez-Barrios A, de Agustín P (2000) Extracranial metastases of a glioma: diagnosis by fine needle aspiration and immunocytochemistry. Diagn Cytopathol 23:43–45

Macdonald DR, O'Brien RA, Gilbert JJ, Cairncross JG (1989) Metastatic anaplastic oligodendroglioma. Neurology 39:1593–1596

Merrell R, Nabors LB, Perry A, Palmer CA (2006) 1p/19q chromosome deletions in metastatic oligodendroglioma. J Neurooncol 80:203–207

Mitsuhashi T, Shimizu Y, Ban S, Ogawa F, Matsutani M, Shimizu M, Hirose T (2007) Anaplastic oligodendroglioma: a case report with characteristic cytologic features, including minigemistocytes. Acta Cytol 51:657–660

Morrison T, Bilbao JM, Yang G, Perry JR (2004) Bony metastases of anaplastic oligodendroglioma respond to temozolomide. Can J Neurol Sci 31:102–108

Ng WH, Lim TC, Tan KK (2006) Disseminated spread of recurrent oligodendroglioma (WHO grade II). J Clin Neurosci 13:602–607

Noshita N, Mashiyama S, Fukawa O, Asano S, Watanabe M, Tominaga T (2010) Extracranial metastasis of anaplastic oligodendroglioma with 1p19q loss of heterozygosity – case report. Neurol Med Chir (Tokyo) 50:161–164

Ohgaki H, Kleihues P (2005) Population-based studies on incidence, survival rates, and genetic alterations in astrocytic and oligodendroglial gliomas. J Neuropathol Exp Neurol 64:479–489

Ordonez NG, Ayala AG, Leavens ME (1981) Extracranial metastases of oligodendroglioma: report of a case and review of the literature. Neurosurgery 8:391–396

Perry A (2004) Metastatic oligodendroglioma: a mini-epidemic? Adv Anat Pathol 11:325

Reifenberger G, Kros JM, Louis DN, Collins VP (2007) Oligodendroglial tumours. In: Louis DN, Ohgaki H, Wiestler OD, Cavenee WK (eds) WHO classification of tumours of the central nervous system. International Agency for Research on Cancer Press, Lyon, pp 53–67

Uzuka T, Kakita A, Inenaga C, Takahashi H, Tanaka R, Takahashi H (2007) Frontal anaplastic oligodendroglioma showing multi-organ metastases after a long clinical course. Case report. Neurol Med Chir (Tokyo) 47:174–177

Volavsek M, Lamovec J, Popović M (2009) Extraneural metastases of anaplastic oligodendroglial tumors. Pathol Res Pract 205:502–507

Wang M, Murphy KM, Kulesza P, Hatanpaa KJ, Olivi A, Tufaro A, Erozan Y, Westra WH, Burger PC, Berg KD (2004) Molecular diagnosis of metastasizing oligodendroglioma: a case report. J Mol Diagn 6:52–57

Watson CW, Hajdu SI (1977) Cytology of primary neoplasms of the central nervous system. Acta Cytol 21:40–47

Wu Y, Liu B, Qu L, Tao H (2011) Extracranial skeletal metastasis in anaplastic oligodendroglioma: case report and review of the literature. J Int Med Res 39:960–967

Zustovich F, Della Puppa A, Scienza R, Anselmi P, Furlan C, Cartei G (2008) Metastatic oligodendrogliomas: a review of the literature and case report. Acta Neurochir (Wien) 150:699–703

Management of Hemangiopericytoma

4

Takenori Akiyama, Kazunari Yoshida, Takashi Horiguchi, and Takeshi Kawase

Contents

Abstract

Macroscopically, hemangiopericytoma (HPC) and meningioma are similar. However, treatment strategies for these tumors sometimes differ because of differences in their clinical behavior and structure, principally their vascularity.

HPC must be totally resected at initial surgery, which sometimes requires preoperative embolization, and close radiographic follow-up is mandatory for detection of local recurrence or metastasis. In the case of residual tumors, adjuvant radiotherapy including conventional radiotherapy, stereotactic radiosurgery, and stereotactic radiotherapy should be deployed. Repeat surgery or salvage radiotherapy is effective. Although chemotherapy has not shown definitive efficacy until date, novel approaches for systemic metastasis that cannot be controlled by present modalities are expected.

Introduction

Intracranial hemangiopericytoma (HPC) is an aggressive tumor with macroscopic and radiographic features that are similar to those of meningioma; however, there are distinct differences in the biology and clinical course of these two tumors. HPC is a rare tumor that accounts for 2.5% of all meningeal tumors and <1% of all intracranial tumors (Schiariti et al. 2011). In contrast, meningioma is one of the most frequent primary brain tumors. There is little difference in the gender incidence of HPC, with only a slight

T. Akiyama • K. Yoshida • T. Horiguchi
T. Kawase (✉)
Department of Neurosurgery, Keio University School of Medicine, Shinjuku-ku, Tokyo, Japan
e-mail: akiyamanor@gmail.com

M. Hayat (ed.), *Tumors of the Central Nervous System, Volume 13*,
DOI 10.1007/978-94-007-7602-9_4, © Springer Science+Business Media Dordrecht 2014

male predominance (Lamar and Lesser 2011). HPC occurs most frequently during the fifth decade of life.

HPC can arise in any part of the human body; however, it is most common in the lower extremities and retroperitoneum. Intracranial or meningeal HPCs usually arise from the meninges of the falx, tentorium, and dural sinuses as meningiomas. Supratentorial lesions are more common than infratentorial or spinal lesions (Sibtain et al. 2007). HPC can also occur in the third or lateral ventricle (Abrahams et al. 1999), the pineal region (Stone et al. 1983), Meckel's cave (Muto et al. 2010), and the sellar region (Juco et al. 2007).

Historically, HPC was first described as a primarily soft tissue tumor arising from pericytes, which are contractile spindle cells surrounding capillaries and postcapillary venules (Stout and Murray 1942). The term "meningioma" was first used to define whole tumors arising from the meninges, including tumors originating from meningothelial (arachnoid) and other mesenchymal cells, such as the angioblastic meningioma first described by (Bailey et al. 1928).

Begg and Garret (1954) reported occurrence of HPC in the meninges and termed such masses "meningeal HPC." However, there has long been controversy as to whether these are meningioma variants of the angioblastic meningioma first defined by (Cushing 1938).

After much discussion, the WHO classification was revised in 1993 (Kleihues and Scheithauser 1993) and tumors arising from the meninges but not originating from arachnoid cells were excluded from meningioma and classified into another category. Since this revision, meningeal HPC has been recognized as an independent category with a different origin, biological behavior, and optimal treatment compared with meningioma.

Pathological Findings

Macroscopically, HPC is a firm and nodular tumor attached to the meninges with a clear margin to the cortical surface. It cannot be distinguished from meningioma, except for a high bleeding tendency. Microscopically, a characteristic staghorn pattern of thin-walled vessels, which is the hallmark of HPC tumors, is observed (Middleton et al. 1998). Tumor cells are uniform and polygonal to spindle shaped, often with vesicular nuclei. Reticulin staining reveals a typical pattern of fine fibers surrounding the individual tumor cells (Middleton et al. 1998). HPCs, like meningiomas, are positive for vimentin and CD34. Epithelial membrane antigen (EMA) is generally negative, but focal reactivity for EMA is sometimes found in HPC (Rajaram et al. 2004). Immunohistochemical expression of the vascular endothelial growth factor receptor (VEGFR) has been reported in HPC tumor cells and in the endothelium, with overexpression of vascular growth factors and receptors such as VEGFR1 and VEGFR2 in the endothelium (Hatva et al. 1996). A study by Dietzmann et al. (1997) showed that HPCs frequently overexpress the platelet-derived growth factor receptor (PDGFR), which is detectable by immunohistochemistry.

Radiographical Features

On computed tomography (CT) imaging, hyperostosis or calcification, which are frequently observed in meningiomas, are seldom observed in HPCs (Chiechi et al. 1996; Barba et al. 2001). Instead, bony erosion is often found (Chiechi et al. 1996).

In magnetic resonance (MR) studies, HPCs are typically isointense to grey matter on T1- and T2-weighted images (Sibtain et al. 2007). The intratumoral flow voids suggest high vascularity, which is sometimes found in meningiomas but is more frequent in HPCs (Chiechi et al. 1996; Sibtain et al. 2007).

On angiography, some angioarchitectural patterns have been shown to be common in HPCs and are useful for distinguishing them from meningiomas [19]. HPC is supplied from the internal carotid artery (ICA) or vertebral artery (VA), and external carotid arteries (ECA), with dominant supply from the ICA branches rather than ECA, which is commonly seen in meningiomas. There are many intratumoral, irregular, corkscrew vessels arising from a main feeder; intense fluffy tumor staining rather than the sunburst pattern observed with meningiomas; no early draining veins; and a prolonged tumor circulation time (Marc et al. 1975; Sibtain et al. 2007).

Clinical Course

HPCs have a peculiar biological behavior and prognosis with local aggressiveness, a high rate of recurrence, and a propensity to metastasize to numerous extracranial locations.

Rutkowski et al. (2010) analyzed 277 patients with HPC and reported an overall median survival of 13 years, with 1-, 5-, 10-, and 20-year survival rates of 95, 82, 60, and 23%, respectively. Schiariti et al. (2011) reported overall survival rates of 93, 67, 45, and 23% at 5, 10, 15, and 20 years, respectively. Ecker et al. (2003) reported that the 5-year Kaplan–Meier survival rate among patients treated since 1990 was 93%. The 5-year disease-free survival rate was 89%.

For local control, which is influenced by treatment strategy, the 5-, 10-, and 15-year recurrence rates after surgery without radiation have been reported to be 54, 92, and 100%, respectively (Schiariti et al. 2011).

Many patients develop distant metastases. Because of the small number of cases, the rate of metastasis varies. Though some authors have reported around 20% metastasis (Rutkowski et al. 2012; Olson et al. 2010; Schiariti et al. 2011), Soyuer et al. (2004) reported 5-, 10-, and 15-year distant metastasis-free survival rates of 80, 46, and 21%, respectively. Adjuvant radiation seems to decrease local recurrence, as is discussed later; however, it has no effect on the incidence of metastasis (Dufour et al. 2001; Rutkowski et al. 2012; Schiariti et al. 2011). Distant metastasis was correlated with shorter survival (Kano et al. 2008). Until date, there has been no report of an effective prophylaxis for preventing distant metastasis. The incidence of recurrence or metastasis increases with duration of follow-up (Olson et al. 2010), emphasizing the necessity of long-term follow-up.

Presurgical Embolization

HPC usually has multiple feeders and plenty of intratumoral vascular networks, and thus, presurgical embolization is required. In general, feeder occlusion and intratumoral embolization must be differentiated. Proximal feeder occlusion by detachable coils or concentrated glue is generally safe and easy, especially in ECA feeders. However, proximal feeder occlusion of ECA feeders, which is sometimes useful in meningioma surgery, has little effect on decreasing intraoperative blood loss in cases where cortical (ICA) or vertebral–basilar artery (VA-BA) feeders dominate supply to the tumor. A combination of embolization of intracranial feeders (ICA, VA-BA) and surgical detachment of ECA feeders might be useful, although embolization of intracranial feeders carries greater risks than that for ECA feeders. Ideally, embolization of the tumor bed by particles or liquid materials is required to decrease intraoperative blood loss. However, in cases of hemangioblastomas, which are also rich in vascularity and sometimes contain arteriovenous (AV) shunt-like intratumoral vessels, tumor bleeding after particle embolization has been reported (Cornelius et al. 2007). As in our case 1 (Fig. 4.1), almost complete embolization can make surgical resection safer and more comfortable and successful. However, tumor bleeding and swelling that worsens the preexisting mass effect of the tumor could sometimes lead to a disastrous situation requiring emergent decompression.

Thus, presurgical embolization can be considered in the case of proximal occlusion of a feeding artery that cannot be accessed in the early stages of surgery or tumor bed embolization can be achieved using liquid formulations such as n-butyl cyanoacrylate (NBCA) or Onyx which rarely induces recanalization after embolization. And care should be taken not to avoid pressure injection of contrast materials or embolic agents. Preparations should also be made for emergency situations requiring surgery for worsening peritumoral edema or tumor bleeding after embolization.

Surgery

Surgery represents the mainstay treatment for HPC, especially in the initial setting. It not only offers immediate alleviation of the mass effect but also allows pathological confirmation, thus facilitating differential diagnosis of HPCs from meningiomas.

Most importantly, the greater the extent of resection, the better the prognosis. In a previous study, local control rates [5-year local control rates for patients treated with gross total removal (GTR) and subtotal removal (STR)] were 84 and 38%, respectively ($P=0.003$) (Soyuer et al. 2004). GTR was also associated with increased overall survival (log-rank, $P<0.05$) (Rutkowski et al. 2012) compared with STR (Rutkowski et al. 2010). Although some reports have not reported any benefit from radiation (Rutkowski et al. 2012), consistent benefits have been reported for resection with GTR (Schiariti et al. 2011).

Surgical Concept

As mentioned previously, obtaining a definitive preoperative diagnosis from radiologic findings alone is sometimes difficult as discriminating meningioma from HPC is not easy. Thus, in practice, the strategy for initial surgery is based on that for meningioma. However, we have to bear in mind that HPC is richer in vascularity and is fed by ICA or VA-BA in addition to ECA; furthermore, it has a higher rate of local recurrence than meningioma. HPC surgery is therefore sometimes more challenging than meningioma surgery. The surgical approach chosen must ensure both a wider operative field and easy accessibility to the majority of feeders. In meningioma surgery, the shortest and most direct approach to the point of tumor attachment should be the priority.

Thereafter, the following seven basic steps in meningioma surgery should be strictly applied: (1) proper craniotomy enabling complete exposure of the dura on the tumor attachment; (2) devascularization of transosseous, skin, or dural feeders; (3) dural incision preserving the bridging vein; (4) meticulous dissection of the arachnoid layer away from the tumor capsule; (5) coagulation and detachment of the tumor attachment; (6) internal decompression of the tumor; and (7) removal of the entire tumor.

In addition to these basic steps, in HPC surgery, tumor devascularization becomes more important and is sometimes difficult to perform in high vascularity tumors fed by cortical arteries and not dural feeders. Presurgical embolization of the tumor and its feeding vessels is sometimes helpful, as discussed previously. However, we sometimes encounter a difficult situation where internal decompression is difficult because of the high bleeding tendency of HPC tumors even after surgical detachment of ECA feeders or partial embolization. Difficulties persist until total devascularization including cortical feeders is achieved. In such cases, because piecemeal tumor resection is impossible, meticulous hemostasis of the tumor can be achieved by coagulating all feeders around the tumor; en-bloc resection may be required.

Another possible reason for the limited extent of resection is the tumor location where radical surgery may worsen the patient's quality of life. Patients with tumors in the posterior fossa have been reported to have a significantly lower median survival of 10.75 years versus 15.6 years for those with nonposterior fossa tumors (Rutkowski et al. 2010), suggesting that GTR is relatively difficult in infratentorial tumors. Although there is no evidence that a skull base tumor has an inferior local control rate, HPCs originating from the skull base or unresectable structures may have a relatively high incidence of local recurrence because of the inability to completely detach the tumor. We experienced four cases of HPCs located around the petroclival area from 14 cases of intracranial HPCs. GTR was achieved in two cases, as is shown for case 1 (Fig. 4.1), STR was achieved in one case, and partial removal in the other. Trochlear and abducens nerve palsy appeared in two cases and hemiparesis occurred in 1. Local recurrence was detected in two cases, one each among patients who achieved GTR and STR, and metastasis occurred in one case among our skull base series.

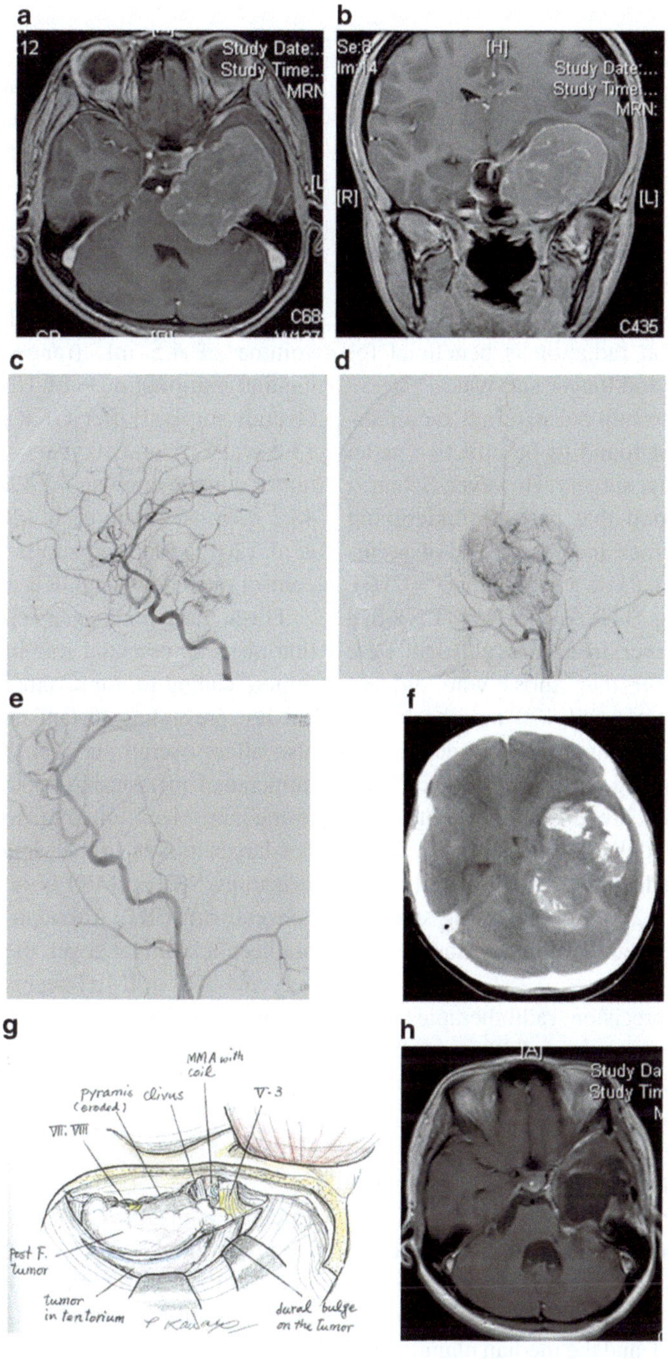

Fig. 4.1 Case 1. A 20-year-old female complained of gait disturbance. Preoperative gadolinium-enhanced MRI of (**a**) axial and (**b**) coronal images showed a dumbbell-type tumor severely compressing the brainstem. Preoperative angiography via (**c**) ICA and (**d**) ECA injection showed marked staining from many feeder branches with intratumoral corkscrew vessels. (**e**) Embolization of the tumor and its feeder vessels with polyvinyl alcohol with tantalum powder and detachable coils led to nearly complete disappearance of the tumor stain on common carotid angiography. (**f**) However, CT revealed a subarachnoid hemorrhage and worsening of peritumoral edema 6 h after embolization. We subsequently performed an emergency tumor removal. (**g**) A transpetrosal approach with osteotomy of the zygomatic arch was selected to obtain direct access and a wider operative field as well as a view of the superior part of the tumor. Drilling of the petrous apex was not necessary because of bone erosion from the tumor. Bleeding from the tumor was not prominent, and softening of the tumor made resection easier than expected. GTR was achieved. The trochlear and abducens nerve could not be preserved because of total encasement by the tumor. (**h**) Postoperative MRI showed no residual tumor. Recurrence has not been detected for 41 months

Radiation

Although some reports have not shown any significant effects of radiation therapy on local control (Soyuer et al. 2004) and have not reported any survival benefit from initial adjuvant external-beam radiation therapy (Ecker et al. 2003), it is widely accepted that radiation is beneficial for better local control and longer survival.

Historically, conventional external radiotherapy (ERT) was first found to be effective as an adjuvant therapy after surgery. However, Schiariti et al. (2011) reported that patients undergoing ERT had a 0.33 times increased risk of recurrence compared with those who did not ($P=0.03$). Patients undergoing GTR followed by ERT had mean local recurrence-free intervals that were 126.3 months longer than those who did not receive ERT ($P=0.04$) and 170 months longer than those undergoing incomplete resection and no ERT after adjusting for resection ($P>0.05$). Radiation responses are dose-dependent; most articles have reported that doses more than 50 Gy are necessary for reducing the risk of local recurrence (Dufour et al. 2001) or providing superior long-term disease-free survival (Bastin and Mehta 1992).

Recently, high-precision radiotherapies such as fractionated stereotactic radiotherapy (SRT) and intensity-modulated radiotherapy (IMRT) have become available. These technologies guarantee the same dose-dependent efficacy as conventional ERT, with less adverse effects as they avoid unnecessary radiation of surrounding vital structures. Combs et al. (2005) reported 37 cases of HPCs treated by SRT or IMRT, with a median follow-up of 34 months. A median total dose of 54 Gy was delivered, and the median planning target volume was 58.2 mL. Overall survival rates at 5 and 10 years were 100 and 64%, respectively. Progression-free survival after radiotherapy was 80 and 61% at 3 and 5 years, respectively.

Stereotactic radiosurgery (SRS) approaches such as gamma knife radiosurgery (GKR) or cyber knife (CK) surgery for relatively small tumors have also been reported. The most common marginal doses range from 15 to 20 Gy, and higher marginal doses are associated with improved progression-free survival (Kano et al. 2008).

Ecker et al. (2003) reported that 93% of 45 tumors treated by GKR were controlled, and 60% of 15 patients were alive at a mean 4.4 years after initial treatment. Kano et al. (2008) reviewed records relating to patients with 29 consecutive tumors treated by SRS, with a median target volume of 4.5 mL (range, 0.07–34.3) and a median marginal dose of 15 Gy (range, 10–20). Overall survival after GKR was 100, 85.9, and 13.8% at 1, 5, and 10 years, respectively, and the tumor control rate was 72.4% after an average of 48.2 months. Olson et al. (2010) and Veeravagu et al. (2011) reported similar survival and tumor control rates between GKR and CK.

Given these findings, it appears clear that radiation plays an essential role in local control of the tumor, leading to improvement in overall survival but not preventing distant metastasis, which can also affect overall survival. SRS is preferred for intracranial metastasis or locally recurrent small-to-medium–sized tumors, as shown in (Fig. 4.2). For larger tumors in the vicinity of radiosensitive structures, SRT or IMRT is theoretically superior to conventional ERT. Although sufficient data have not been accumulated yet, the combination of surgery and SRT or IMRT seems to be reasonable for obtaining the longest survival with the lowest morbidity among patients with STR of large tumors.

Chemotherapy

Chemotherapy provides only marginal benefits for intracranial HPC. As HPC can be categorized as a soft tissue sarcoma (Chamberlain and Glantz 2008), regimens such as CAV (cyclophosphamide, doxorubicin, and vincristine) and ICE (ifosfamide, carboplatin, and etoposide) as well as other chemotherapeutic agents such as methotrexate, cisplatin, and mitomycin have been tested; however, all these have shown limited efficacy. Ecker et al. (2003) reported that only 1 of 11 patients treated with salvage chemotherapy responded to treatment with doxorubicin. In the same patient, however, other metastases developed and the patient subsequently died. All 11 patients

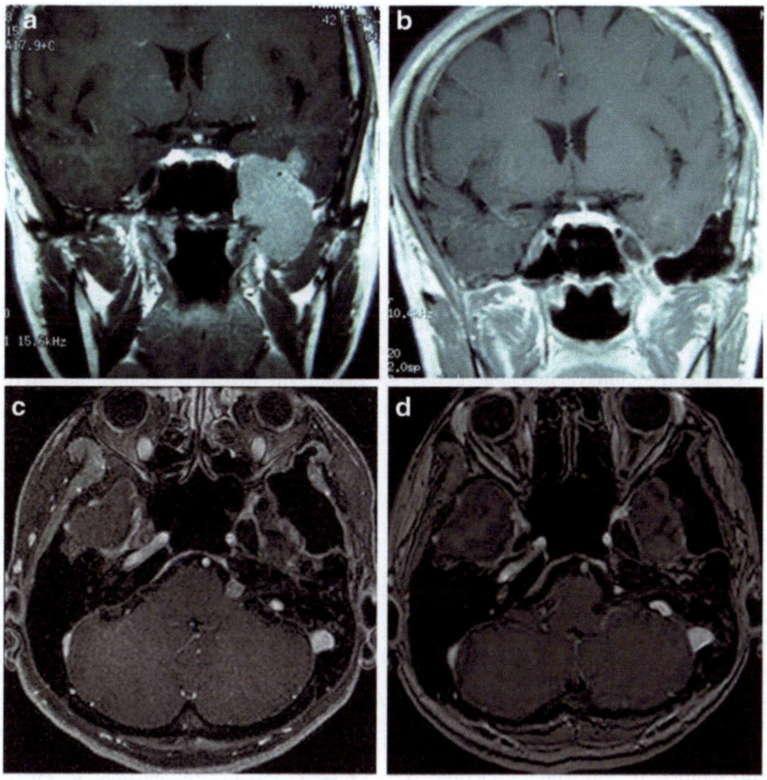

Fig. 4.2 Case 2. A 42-year-old female presented with facial paresthesia. (**a**) Preoperative gadolinium-enhanced MRI (coronal view) showed a tumor in the middle fossa base. The tumor was totally removed via a zygomatic interdural approach. (**b**) Postoperative MRI revealed disappearance of the tumor. Two years after surgery, a local recurrent tumor disappeared after GKR with a marginal dose of 20 Gy. (**c**) Eight years after the surgery, a metastatic lesion appeared in the right cerebellopontine angle (CPA). (**d**) The CPA tumor shrank after GKR with a marginal dose of 20 Gy. The patient has had no neurological deficit for 10 years since the initial diagnosis

died as a result of their tumors. In a retrospective study by Chamberlain and Glantz (2008) 15 patients with recurrent HPC were administered a CAV chemotherapeutic regimen. If this proved ineffective, then interferon (IFN) was administered followed by ICE. The median overall survival among these patients was 14 months. Treatment of recurrent HPC with α-IFN showed the best response (Chamberlain and Glantz 2008). Patients with widely metastatic HPC were treated with α-IFN, and neither experienced disease progression at 18 and 24 months (Kirn and Kramer 1996).

Recent elucidation of the molecular mechanisms behind HPC tumors and angiogenic cells by immunohistochemistry has led to novel chemotherapeutic approaches to HPC. Autocrine and paracrine activation of the VEGF–VEGFR pathway is considered to be involved in the biology of HPC (Park et al. 2010), since VEGFR is expressed in HPC tumor cells and the endothelium (Hatva et al. 1996) and because HPC is also associated with overexpression of PDGFR (Dietzmann et al. 1997). These data suggest that antiangiogenic strategies or targeted therapies may be effective in treating recurrent tumors (Peters et al. 2010; Delgado et al. 2011). Clinical trials of targeted therapies for PDGF and VEGF receptors may change the treatment strategy for patients with recurrent and metastatic HPC refractory to surgery and radiotherapy (Lamar and Lesser 2011), and these trials are continuing even now.

References

Abrahams JM, Forman MS, Lavi E, Goldberg H, Flamm ES (1999) Hemangiopericytoma of the third ventricle. Case report. J Neurosurg 90(2):359–362

Bailey P, Cushing H, Eisenhardt L (1928) Angioblastic meningiomas. Arch Pathol Lab Med 6:953–990

Barba I, Moreno A, Martinez-Perez I, Tate AR, Cabanas ME, Baquero M, Capdevila A, Arus C (2001) Magnetic resonance spectroscopy of brain hemangiopericytomas: high myoinositol concentrations and discrimination from meningiomas. J Neurosurg 94(1):55–60

Bastin KT, Mehta MP (1992) Meningeal hemangiopericytoma: defining the role for radiation therapy. J Neurooncol 14(3):277–287

Begg CF, Garret R (1954) Hemangiopericytoma occurring in the meninges: case report. Cancer 7(3):602–606

Chamberlain MC, Glantz MJ (2008) Sequential salvage chemotherapy for recurrent intracranial hemangiopericytoma. Neurosurgery 63(4):720–726, author reply 726–727

Chiechi MV, Smirniotopoulos JG, Mena H (1996) Intracranial hemangiopericytomas: MR and CT features. AJNR Am J Neuroradiol 17(7):1365–1371

Combs SE, Thilmann C, Debus J, Schulz-Ertner D (2005) Precision radiotherapy for hemangiopericytomas of the central nervous system. Cancer 104(11):2457–2465

Cornelius JF, Saint-Maurice JP, Bresson D, George B, Houdart E (2007) Hemorrhage after particle embolization of hemangioblastomas: comparison of outcomes in spinal and cerebellar lesions. J Neurosurg 106(6):994–998

Cushing H, Eisenhardt L (1938) Meningiomas; their classification, regional behavior, life history, and surgical end results. Charles C. Thomas, Springfield

Delgado M, Perez-Ruiz E, Alcalde J, Perez D, Villatoro R, Rueda A (2011) Anti-angiogenic treatment (sunitinib) for disseminated malignant haemangiopericytoma: a case study and review of the literature. Case Rep Oncol 4(1):55–59

Dietzmann K, von Bossanyi P, Warich-Kirches M, Kirches E, Synowitz HJ, Firsching R (1997) Immunohistochemical detection of vascular growth factors in angiomatous and atypical meningiomas, as well as hemangiopericytomas. Pathol Res Pract 193(7):503–510

Dufour H, Metellus P, Fuentes S, Murracciole X, Regis J, Figarella-Branger D, Grisoli F (2001) Meningeal hemangiopericytoma: a retrospective study of 21 patients with special review of postoperative external radiotherapy. Neurosurgery 48(4):756–762; discussion 762–753

Ecker RD, Marsh WR, Pollock BE, Kurtkaya-Yapicier O, McClelland R, Scheithauer BW, Buckner JC (2003) Hemangiopericytoma in the central nervous system: treatment, pathological features, and long-term follow up in 38 patients. J Neurosurg 98(6):1182–1187

Hatva E, Bohling T, Jaaskelainen J, Persico MG, Haltia M, Alitalo K (1996) Vascular growth factors and receptors in capillary hemangioblastomas and hemangiopericytomas. Am J Pathol 148(3):763–775

Juco J, Horvath E, Smyth H, Rotondo F, Kovacs K (2007) Hemangiopericytoma of the sella mimicking pituitary adenoma: case report and review of the literature. Clin Neuropathol 26(6):288–293

Kano H, Niranjan A, Kondziolka D, Flickinger JC, Lunsford LD (2008) Adjuvant stereotactic radiosurgery after resection of intracranial hemangiopericytomas. Int J Radiat Oncol Biol Phys 72(5):1333–1339

Kirn DH, Kramer A (1996) Long-term freedom from disease progression with interferon alfa therapy in two patients with malignant hemangiopericytoma. J Natl Cancer Inst 88(11):764–765

Kleihues P, Burger PC, Scheithauer BW (1993) The new WHO classification of brain tumours. Brain Pathol 3:255–268

Lamar Z, Lesser GJ (2011) Management of meningeal neoplasms: meningiomas and hemangiopericytomas. Curr Treat Options Oncol 12(3):230–239

Marc JA, Takei Y, Schechter MM, Hoffman JC (1975) Intracranial hemangiopericytomas. Angiography, pathology and differential diagnosis. Am J Roentgenol Radium Ther Nucl Med 125(4):823–832

Middleton LP, Duray PH, Merino MJ (1998) The histological spectrum of hemangiopericytoma: application of immunohistochemical analysis including proliferative markers to facilitate diagnosis and predict prognosis. Hum Pathol 29(6):636–640

Muto J, Kawase T, Yoshida K (2010) Meckel's cave tumors: relation to the meninges and minimally invasive approaches for surgery: anatomic and clinical studies. Neurosurgery 67(3):291–298; discussions 298–299

Olson C, Yen CP, Schlesinger D, Sheehan J (2010) Radiosurgery for intracranial hemangiopericytomas: outcomes after initial and repeat Gamma Knife surgery. J Neurosurg 112(1):133–139

Park MS, Ravi V, Araujo DM (2010) Inhibiting the VEGF-VEGFR pathway in angiosarcoma, epithelioid hemangioendothelioma, and hemangiopericytoma/solitary fibrous tumor. Curr Opin Oncol 22(4):351–355

Peters KB, McLendon R, Morse MA, Vredenburgh JJ (2010) Treatment of recurrent intracranial hemangiopericytoma with SRC-related tyrosine kinase targeted therapy: a case report. Case Rep Oncol 3(1):93–97

Rajaram V, Brat DJ, Perry A (2004) Anaplastic meningioma versus meningeal hemangiopericytoma: immunohistochemical and genetic markers. Hum Pathol 35(11):1413–1418

Rutkowski MJ, Sughrue ME, Kane AJ, Aranda D, Mills SA, Barani IJ, Parsa AT (2010) Predictors of mortality following treatment of intracranial hemangiopericytoma. J Neurosurg 113(2):333–339

Rutkowski MJ, Jian BJ, Bloch O, Chen C, Sughrue ME, Tihan T, Barani IJ, Berger MS, McDermott MW, Parsa AT (2012) Intracranial hemangiopericytoma: clinical experience and treatment considerations in a modern series of 40 adult patients. Cancer 118(6):1628–1636

Schiariti M, Goetz P, El-Maghraby H, Tailor J, Kitchen N (2011) Hemangiopericytoma: long-term outcome revisited. Clinical article. J Neurosurg 114(3): 747–755

Sibtain NA, Butt S, Connor SE (2007) Imaging features of central nervous system haemangiopericytomas. Eur Radiol 17(7):1685–1693

Soyuer S, Chang EL, Selek U, McCutcheon IE, Maor MH (2004) Intracranial meningeal hemangiopericytoma: the role of radiotherapy: report of 29 cases and review of the literature. Cancer 100(7):1491–1497

Stone JL, Cybulski GR, Rhee HL, Bailey OT (1983) Excision of a large pineal region hemangiopericytoma (angioblastic meningioma, hemangiopericytoma type). Surg Neurol 19(2):181–189

Stout AP, Murray MR (1942) Hemangiopericytoma: a vascular tumor featuring Zimmermann's pericytes. Ann Surg 116(1):26–33

Veeravagu A, Jiang B, Patil CG, Lee M, Soltys SG, Gibbs IC, Chang SD (2011) CyberKnife stereotactic radiosurgery for recurrent, metastatic, and residual hemangiopericytomas. J Hematol Oncol 4:26

Role of Cyclooxygenase-2 in the Development and Growth of Schwannomas

5

Bujung Hong, Makoto Nakamura, and Joachim K. Krauss

Contents

B. Hong (✉) • M. Nakamura • J.K. Krauss
Department of Neurosurgery, Hannover Medical School, Carl-Neuberg-Strasse 1, 30625 Hannover, Germany
e-mail: Hong.Bujung@mh-hannover.de

Abstract

Recently, numerous experimental reports as well as clinical trials suggested a possible therapeutic role of selective cyclooxygenase-2 (COX-2) inhibitors in the treatment of various tumors. COX-2 is known to convert procarcinogens to carcinogens and is upregulated in several malignant tumors. Its overexpression seems to correlate with aggressive disease and poor survival. In human schwannomas, COX-2 expression was observed in the cytoplasma and perinuclear regions of tumor cells. The available data suggests involvement of COX-2 in the development and growth of human schwannomas by regulating angiogenic factors and inhibition of tumor cell apoptosis by production of prostaglandins, particularly PGE_2. Furthermore, diverse neurotrophic factors have been also suggested a biological role in development, maintenance, and growth of schwannomas. Selective COX-2 inhibitors have a potential role for targeted therapy against schwannomas.

Introduction

Numerous experimental reports as well as clinical trials have suggested that COX-2 products like prostaglandins lead to stimulation of tumor proangiogenic factors and tumor cell proliferation of various human tumors (Kolev et al. 2007; Perrone et al. 2005; Zhi et al. 2005). Furthermore, prostaglandin products support the release of

M. Hayat (ed.), *Tumors of the Central Nervous System, Volume 13*,
DOI 10.1007/978-94-007-7602-9_5, © Springer Science+Business Media Dordrecht 2014

various neurotrophic factors, which contribute additionally to tumor growth. Overexpression of COX-2 seems to correlate with aggressive disease and poor survival. In tumors of the central nervous system, such as astrocytoma (Perdiki et al. 2007), meningioma (Pistolesi et al. 2007), and glioblastoma (Buccoliero et al. 2006), COX-2 is also expressed and suggests biological aggressiveness. In peripheral nerve disorders, such as inflammatory demyelinating neuropathy (Hu et al. 2003) or in nerve injury (Durrenberger et al. 2006), COX-2 immunoreactivity was also detected. This review will discussed the role of COX-2 in the development and growth of human schwannomas.

Pharmacology of Cyclooxygenase

Cyclooxygenase (COX), or prostaglandin endoperoxide synthase, is an ubiquitous membrane-bound enzyme involved in the synthesis of biologically active lipid compounds, called eicosanoids, from arachidonic acid, a ω-6 polyunsaturated fatty acid (PUFA). COX catalyses the conversion of arachidonic acid to the prostaglandin precursor prostaglandin H_2 (PGH_2). PGH_2 is then converted to other prostaglandins including prostaglandin D_2 (PGD_2), Prostaglandin $F_2\alpha$ ($PGF_2\alpha$), Prostaglandin E_2 (PGE_2), thromboxane A_2 and B_2 (TXA_2 and TXB_2), and prostacyclin (PGI_2). These prostaglandins, particularly PGE_2, play an important role in development and maintenance of pain, fever, as well as inflammatory reaction.

Thus far, two COX isoenzymes have been identified. In 1976, COX-1 was isolated successfully for the first time. This was followed by isolation and sequencing of COX-2 from complimentary desoxyribonucleic acid in 1976. COX-1 is constitutively expressed in normal mammalian tissues with a "housekeeping" function, such as cytoprotection of the gastrointestinal mucosa, regulation of renal blood flow, and control of platelet aggregation. Human COX-1 is localized to the endoplasmic reticulum and contains 599 amino acids. COX-1 is expressed at higher concentrations in tissues and in cells where prostaglandins have specialized signaling functions, such as kidney, stomach, platelets, and vascular endothelium. COX-2 is not regularly present in most mammalian tissues. However, it is inducible in various cell types by pro-inflammatory substances, such as cytokines, endotoxins, interleukins, and phorbolester, by ischemia, and by cell injury. Nevertheless, COX-2 might be expressed in a few specialized tissues in the apparent absence of activation, such as brain, testes, and macula densa of kidney. Human COX-2 is localized to the nuclear membrane and contains 604 amino acids.

Both COX-1 and COX-2 have a homologous biological structure and possess similar kinetic properties. COX-1 and COX-2 showed also similar patterns of expression in inflammatory macrophages and the effect of the expression of either isozyme is also similar, i.e., an enhanced capacity of tissues to produce prostaglandins. While COX-2 inhibitors have been used widely as non-steroidal anti-inflammatory drugs (NSAIDs) for pain treatment (analgesic) and anti-inflammation (antiphlogistic), inhibition of COX-1 seems to be fraught with risk for gastrointestinal bleeding. The regulation of COX-2 expression is a complex process involving multiple signal transduction pathways. In addition to the role of COX-2 in pain and inflammation, many studies have reported the inhibitory effects of COX-2 against diverse human tumors by inhibition of tumor cell proliferation and angiogenesis, blockade of cell cycle progression, and induction of tumor cell apoptosis (Liu et al. 1998; Masferrer et al. 2000; Tsujii et al. 1998).

COX-2 Expression in Human Central Nervous System

In the human central nervous system (CNS), COX-2 is normally expressed in cerebral neurons, glia, cerebrovascular components, and in spinal neurons. In the peripheral nerve, very few scattered COX-2 immunoreactive cells were found in normal nerve tissue of the human brachial plexus (Durrenberger et al. 2006). COX-2 expression was also detected in different peripheral nerve disorders, such as inflammatory demyelinating polyneuropathy (Hu et al. 2003), glaucomatous optic nerve (Neufeld et al. 1997), or during nerve injury (Durrenberger et al. 2006).

Regulation of COX-2 Expression in Human Tumors

There is growing evidence that COX-2 plays a contributory role in the pathogenesis of diverse human tumors, including tumors of the central nervous system, such as astrocytomas (Perdiki et al. 2007), meningiomas (Pistolesi et al. 2007), and glioblastomas (Buccoliero et al. 2006). In these tumors, COX-2 expression suggests biological aggressiveness and worse prognosis. Increase of COX-2 expression in neoplasms has been reported to be associated with loss of gene regulation of transcriptional and post-transcriptional mechanisms due to cellular differentiation. Several neurotrophic factors, such as epidermal growth factor (EGF), transforming growth factor-ß1 (TGF-ß1), and tumor necrosis factor-α (TNF-α), are known to induce COX-2 production (Gately and Li 2004). Overexpression of COX-2 increase releases basic fibroblast growth factor (bFGF) and vascular endothelial growth factor (VEGF). Association of COX-2 overexpression with loss of wild-type of tumor suppressor p53 has been also suggested previously (Subbaramaiah et al. 1999).

COX-2 contributes to the carcinogenesis of human tumors by different mechanisms. Nevertheless, the exact role of COX-2 in tumor development and growth is not clearly known. Increased production of prostaglandins supports cell proliferation of diverse malignant tumors (Bennet 1986) and modulates cell adhesion and motility (Tsujii and DuBois 1995). Furthermore, overexpression of COX-2 inhibits apoptosis and supports the invasiveness of malignant cells (Tsujii and DuBois 1995). COX-2 modulates also inflammation and immune function which supports tumorigenic action. Among those contributions, the stimulation of tumor angiogenesis through products of COX-2 has been suggested to be the most likely mechanism.

The proangiogenic effects of COX-2 are mediated primarily by three products of arachidonic metabolism: thromboxane A_2 (TXA_2), prostaglandin E_2 (PGE_2), and prostaglandin I_2 (PGI_2). TXA_2, PGE_2, and PGI_2 promote specific angiogenic steps and mediators (Fig. 5.1) (Gately and Li 2004). In human vascular endothelial cells, the expression of COX-2 is regulated by

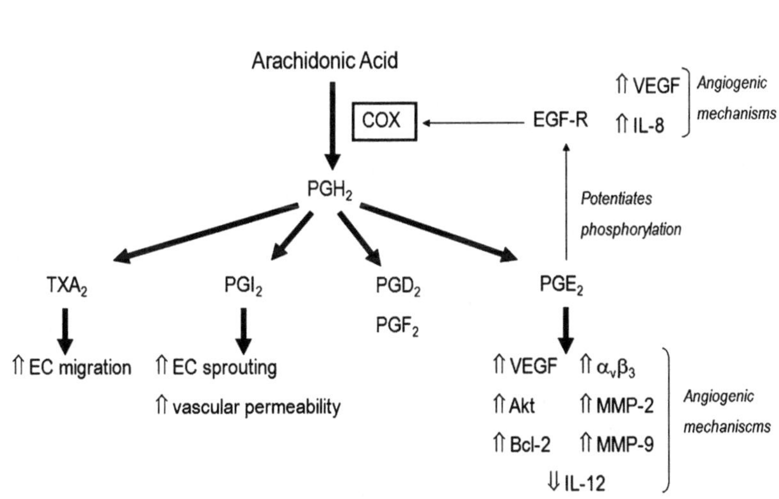

Fig. 5.1 COX-2 induces tumor angiogenesis via multiple pathways (Gately and Li 2004; reprint with permission from Publisher Elsevier)

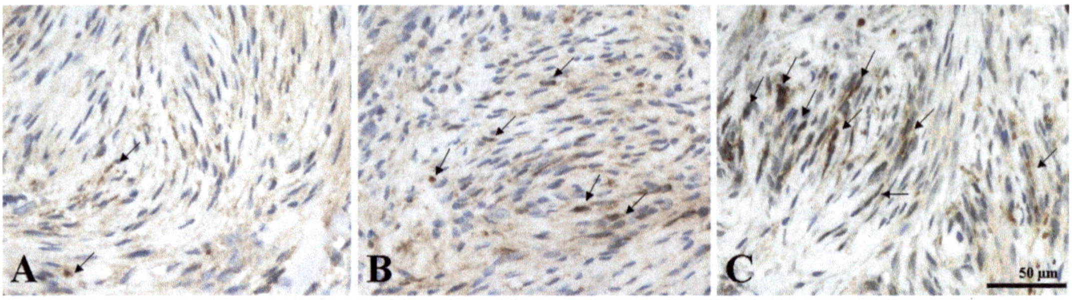

Fig. 5.2 Immunohistochemical staining of cyclooxygenase-2 (COX-2) expression in vestibular schwannomas demonstrated a diffuse or focal brown reactivity in the cytoplasma and perinuclear regions of tumor cells [*arrows*]. (**a**) Weak COX-2 expression; (**b**) moderate COX-2 expression; (**c**) strong COX-2 expression (original magnification, ×400; calibration bar 50 µm) (Hong et al. 2011; reprint with permission from Wolters Kluwer Health)

hypoxia (Schmedtje et al. 1997). Increase of the levels of prostaglandins stimulate directly endothelial cell sprouting, migration and tube formation, und subsequently support tumor progression. COX-2 has been also associated with endothelial cell motility and ability to form capillary-like structures, which can be quantified by measurement of microvascular density (MVD) (Kolev et al. 2007).

COX-2 and Human Schwannomas

Schwannoma is a common tumor of the peripheral nerves and accounts for an estimated 8% of intracranial tumors. Schwannomas are considered benign tumors originating from Schwann cells with a growth rate of about 1–2 mm/year. The average proliferative activity of schwannomas has been reported to be as low as 1–3%. Schwannomas appear sporadically in the cranium, in the spinal axis, or in peripheral nerves. The occurrence of multiple schwannomas as seen in neurofibromatosis type 2 (NF2) is usually associated with an aberration of a tumor suppressor gene on chromosome 22q12. Despite its benign histopathology, progressive growth of schwannoma may lead to a variety of neurological deficits and severe disability. With progressive growth, intracranial schwannomas may be life-threatening due to compression of the brainstem or secondary hydrocephalus.

COX-2 expression has been detected in vestibular schwannoma in different intensities (Fig. 5.2). The COX-2 expression has been localized in the cytoplasma and perinuclear regions of tumor cells (Hong et al. 2011). A significant correlation between COX-2 expression and proliferation index was demonstrated indicating the involvement of COX-2 pathways in the development and growth of schwannomas. In an early study, Kökoğlu et al. (1998) reported elevated concentration of PGE_2 in schwannomas as compared to control brain tissue. Increased levels of prostaglandins TBX_2, PGE_2, and PGF were also detected in retroperitoneal schwannomas (Komiya et al. 1991). Prostacyclin and thromboxane are known to regulate endothelial sprouting as well as VEGF-induced vascular permeability. Nevertheless, there is no report about direct correlations between TXA_2 and schwannomas until now.

The exact role of COX-2 in development and growth of schwannomas is not known. Several pathways have been suggested which link COX-2 and growth of schwannomas. The stimulation of tumor angiogenesis through products of COX-2 resulting in increased release of numerous proangiogenic factors (e.g. vascular endothelial growth factor, VEGF; basic fibroblast growth factor, bFGF) has been suggested to be the most reasonable hypothesis. PGE_2 is known to be associated with angiogenesis in tumor development by increasing various proangiogenic factors, such as VEGF, Akt, Bcl-2, $\alpha_v\beta_3$, MMP-2, MMP-9, and suppressing the production of IL-2 (Gately and Li 2004). Furthermore, numerous neurotrophic factors including nerve growth factor (NGF) (Charabi et al. 1996), TGF-ß1 (Cardillo et al.

1999), bFGF (Murphy et al. 1989), neuregulin (NRG) (Hansen and Linthicum 2004), erythropoietin (EPO) (Dillard et al. 2001), and EGF (Sturgis et al. 1996) have been suggested to have a biological role in development, maintenance, and growth of schwannomas.

It is well established, that angiogenesis is a prerequisite for proliferation and growth of any neoplasms. Even in slow-growing tumors like schwannomas, neovascularization still remains important for tumor growth. Together with MMPs and bFGF, VEGF is a potent mediator for tumor angiogenesis and vessel permeability in human schwannomas (Koutsimpelas et al. 2007). Expression of COX-2 and the resultant eicosanoid products, promote the release of VEGF. VEGF is a diffusible glycoprotein, which is expressed in Schwann cell cytoplasm. VEGF binds to its receptors, VEGR-1 and VEGR-2, which are located on vascular endothelial cells. VEGF and its receptors have been shown to promote the proliferation of endothelial cells, to increase the cells' vascular permeability, and to induce the production of plasminogen activator (Uesaka et al. 2007).

Furthermore, VEGF supports the survival of the pre-existing tumor vasculature. More recently, VEGF has also emerged in mobilization of endothelial progenitor cells from the bone marrow to distant sites of neovascularization. Expression of VEGF in tumors of the peripheral nerve, such as neurofibroma and benign as well as malignant peripheral nerve sheath tumor (BPNST/MPNST), has been demonstrated (Wasa et al. 2008). Large tumors, recurrent tumors and tumors with a high growth rate had higher levels of VEGFR-1 (Cayè-Thomasen et al. 2005). A clinical trial with humanized monoclonal IgG1 antibody against VEGF antibody for vestibular schwannomas revealed a significantly decrease of tumor growth rate from 62% before treatment to 26% after treatment, as well as hearing improvement in most patients in selected patients with NF2-associated vestibular schwannomas (Plotkin et al. 2009). Furthermore, radiographic regression of vestibular schwannomas following anti-VEGF therapy (Bevacizumab) was described (Mautner et al. 2010).

Angiogenesis and tumor invasion require controlled degradation of the extracellular matrix (ECM) components in order to allow cell migration and new tissue formation. Tumor invasion involves always degradation of the ECM, involving enzymes such as heparinase, serine proteinases, cathepsins, and matrix metalloproteinases (MMPs). MMPs are a large family of over 20 zinc-dependent endopeptidases that proteolytically degrade most components of the extracellular matrix. MMPs are known to facilitate repair mechanisms by promoting cell differentiation and axonal growth during axonal regeneration and nerve remyelination (Lehmann et al. 2009). It was supposed, that neoplastic Schwann cells express MMPs abnormally, and degrade the ECM sheaths in the basal lamina of peripheral nerve, which is formed by collagenous matrices. MMPs are known to be able to proliferate, migrate and remodel basal lamina. The exact process, however, how Schwann cells express MMPs and invade peripheral nerve ECM remains unclear. Among the MMPs, MMP-2 and MMP-9 are most frequently involved in human tumor invasion and metastasis. MMP-2 and MMP-9 are also known to support tumor growth by neovascularization. They allow endothelial cell migration and invasion into the surrounding tissue during angiogenesis (Møller et al. 2010). Interestingly, MMP-2 and MMP-9 are expressed in diverse tumors of peripheral nerves, such as schwannoma, malignant peripheral nerve sheath tumor, and neurofibroma. In vestibular schwannoma, tumor concentration of MMP-9 correlates significantly with absolute tumor growth rate (Møller et al. 2010).

COX-2 generated prostaglandins may also enhance bFGF-induced angiogenesis through the induction of VEGF. bFGF is found in basement membranes and sub-ECM (Majima et al. 2000). Previous studies suggested that overexpression of bFGF in Schwann cells stimulates mitosis and increases the differentiation and proliferation of various neuronal populations (Grothe and Wewetzer 1996). Furthermore, bFGF has been shown to significantly increase the production of TXB_2, the active metabolite of TXA_2, which supports endothelial migration. In sporadic vestibular schwannomas, bFGF and VEGF correlated positively with microvascular density and tumor growth index (Koutsimpelas et al. 2007; Blair et al. 2011).

Other studies have suggested also that apoptosis resistance of tumor cells correlates with overexpression of COX-2 (Tsujii and DuBois 1995; Arico et al. 2002; Elder et al. 2002). The overexpression of COX-2 can lead to the increased production of the protein Bcl-2, which increases the survival of vascular endothelial cells. This antiapoptotic process may also support tumor growth of human schwannomas. Utermark et al. (2005) demonstrated the reduction of basal apoptosis rate of primary human schwannoma cells in comparison to that of normal Schwann cell. Finally, decreased expression of integrin $\alpha_v\beta_3$ may be involved in the changes in Schwann cell morphology, loss of extracellular matrix adhesion, and increased migration (Eliceiri and Cheresh 2006). Integrin $\alpha_v\beta_3$ and its receptors are involved in cell adhesion, migration, survival, morphology, and angiogenesis in neoplasms.

In conclusion, current data suggest that the COX-2 pathway is involved in the development and growth of human schwannomas by different mechanisms. Overexpression of COX-2 correlated with high proliferation rates of schwannomas. The increase of COX-2 products like prostaglandins resulting in stimulation of tumor proangiogenic factors is supposed to be the main mechanism. Furthermore, prostaglandin products support the release of various neurotrophic factors, which contribute additionally to tumor growth. COX-2 may present an interesting new target for medical therapy in recurrent or difficult-to-operate schwannomas. Further studies with selective COX-2 inhibitors are required to underline such concepts.

References

Arico S, Pattingre S, Bauvy C, Gane P, Barbat A, Codogno P, Ogier-Denis E (2002) Celecoxib induces apoptosis by inhibiting 3-phosphoinositide-dependent protein kinase-1 activity in the human colon cancer cell line HT-29. J Biol Chem 277:27613–27621

Bennet A (1986) The production of prostanoids in human cancers, and their implications for tumor progression. Prog Lipid Res 25:539–542

Blair KJ, Kiang A, Wang-Rodriguez J, Yu MA, Doherty JK, Ongkeko WM (2011) EGF and bFGF promote invasion that is modulated by PI3/Akt Kinase and Erk in vestibular schwannoma. Otol Neurotol 32:308–314

Buccoliero AM, Caldarella A, Gheri CF, Taddei A, Paglierani M, Pepi M, Mennonna P, Taddei GL (2006) Inducible cyclooxygenase (COX-2) in glioblastoma – clinical and immunohistochemical (COX-2-VEGF) correlations. Clin Neuropathol 25:59–66

Cardillo MR, Filipo R, Monini S, Aliotta N, Barbara M (1999) Transforming growth factor-beta1 expression in human acoustic neuroma. Am J Otol 20:65–68

Cayè-Thomasen P, Werther K, Nalla A, Bøg-Hansen TC, Nielsen HJ, Stangerup SE, Thomsen J (2005) VEGF and VEGF receptor-1 concentration in vestibular schwannoma homogenates correlates to tumor growth rate. Otol Neurotol 26:98–101

Charabi S, Simonsen K, Charabi B, Jacobsen GK, Moos T, Rygaard J, Tos M, Thomsen J (1996) Nerve growth factor receptor expression in heterotransplanted vestibular schwannoma in athymic nude mice. Acta Otolaryngol 116:59–63

Dillard DG, Venkatraman G, Cohen C, Delgaudio J, Gal AA, Mattox DE (2001) Immunolocalization of erythropoietin and erythropoietin receptor in vestibular schwannoma. Acta Otolaryngol 121:149–152

Durrenberger PF, Facer P, Casula MA, Yiangou Y, Gray RA, Chessell IP, Day NC, Collins SD, Bingham S, Wilson AW, Elliot D, Birch R, Anand P (2006) Prostanoid receptor EP1 and Cox-2 in injured human nerves and a rat model of nerve injury: a time-course study. BMC Neurol 6:1

Elder DJ, Halton DE, Playle LC, Paraskeva C (2002) The MEK/ERK pathway mediates COX-2 selective NSAID-induced apoptosis and induced COX-2 protein expression in colorectal carcinoma cells. Int J Cancer 99:323–327

Eliceiri BP, Cheresh DA (2006) Role of alpha v integrins during angiogenesis. Cancer J 6(Suppl 3):S245–S249

Gately S, Li WW (2004) Multiple roles of COX-2 in tumor angiogenesis: a target for antiangiogenic therapy. Semin Oncol 31(2 Suppl 7):2–11

Grothe C, Wewetzer K (1996) Fibroblast growth factor and its implications for developing and regenerating neurons. Int J Dev Biol 40:403–410

Hansen MR, Linthicum FH Jr (2004) Expression of neuregulin and activation of erbB receptors in vestibular schwannomas: possible autocrine loop stimulation. Otol Neurotol 25:155–159

Hong B, Krusche CA, Schwabe K, Friedrich S, Klein R, Krauss JK, Nakamura M (2011) Cyclooxygenase-2 supports tumor proliferation in vestibular schwannomas. Neurosurgery 68:1112–1117

Hu W, Mathey E, Hartung HP, Kieseier BC (2003) Cyclo-oxygenases and prostaglandins in acute inflammatory demyelination of the peripheral nerve. Neurology 61:1774–1779

Kökoğlu E, Tüter Y, Sandıkçi KS, Yazıcı Z, Ulakoğlu EZ, Sönmez H, Ozyurt E (1998) Prostaglandin E2 levels in human brain tumor tissues and arachidonic acid levels in the plasma membrane of human brain tumors. Cancer Lett 132:17–21

Kolev Y, Uetake H, Iida S, Ishikawa T, Kawano T, Sugihara K (2007) Prognostic significance of VEGF expression in correlation with COX-2, microvessel density, and clinicopathological characteristics in human gastric carcinoma. Ann Surg Oncol 14:2738–2747

Komiya I, Yamaguchi K, Miyake Y, Honda S, Tsuchihashi T, Shimizu T, Fukushima N, Saito Y (1991) Retroperitoneal neurilemoma presenting with humoral hypercalcemia associated with markedly elevated plasma prostaglandin levels. Cancer 68:1086–1091

Koutsimpelas D, Stripf T, Heinrich UR, Mann WJ, Brieger J (2007) Expression of vascular endothelial growth factor and basic fibroblast growth factor in sporadic vestibular schwannomas correlates to growth characteristics. Otol Neurotol 28:1094–1099

Lehmann HC, Köhne A, Bernal F, Jangouk P, Meyer Zu Hörste G, Dehmel T, Hartung HP, Previtali SC, Kieseier BC (2009) Matrix metalloproteinase-2 is involved in myelination of dorsal root ganglia neurons. Glia 57:479–489

Liu XH, Yao S, Kirschenbaum A, Levine AC (1998) NS398, a selective cyclooxygenase-2 inhibitor, induces apoptosis and down-regulates bcl-2 expression in LNCaP cells. Cancer Res 58:4245–4249

Majima M, Hayashi I, Muramatsu M, Katada J, Yamashina S, Katori M (2000) Cyclo-oxygenase-2 enhances basic fibroblast growth factor-induced angiogenesis through induction of vascular endothelial growth factor in rat sponge implants. Br J Pharmacol 130:641–649

Masferrer JL, Leahy KM, Koki AT, Zweifel BS, Settle SL, Woerner BM, Edwards DA, Flickinger AG, Moore RJ, Seibert K (2000) Antiangiogenic and antitumor activities of cyclooxygenase-2 inhibitors. Cancer Res 60:1306–1311

Mautner VF, Nguyen R, Knecht R, Bokemeyer C (2010) Radiographic regression of vestibular schwannomas induced by bevacizumab treatment: sustain under continuous drug application and rebound after drug discontinuation. Ann Oncol 21:2294–2295

Møller MN, Werther K, Nalla A, Stangerup SE, Thomsen J, Bøg-Hansen TC, Nielsen HJ, Cayè-Thomasen P (2010) Angiogenesis in vestibular schwannomas: expression of extracellular matrix factors MMP-2, MMP-9, and TIMP-1. Laryngoscope 120:657–662

Murphy PR, Myal Y, Sato Y, Sato R, West M, Friesen HG (1989) Elevated expression of basic fibroblast growth factor messenger ribonucleic acid in acoustic neuromas. Mol Endocrinol 3:225–231

Neufeld AH, Hernandez MR, Gonzalez M, Geller A (1997) Cyclooxygenase-1 and cyclooxygenase-2 in the human optic nerve head. Exp Eye Res 65:739–745

Perdiki M, Korkolopoulou P, Thymara I, Agrogiannis G, Piperi C, Boviatsis E, Kotsiakis X, Angelidakis D, Diamantopoulou K, Thomas-Tsagli E, Patsouris E (2007) Cyclooxygenase-2 expression in astrocytomas. Relationship with microvascular parameters, angiogenic factors expression and survival. Mol Cell Biochem 295:75–83

Perrone G, Santini D, Vincenzi B, Zagami M, La Cesa A, Bianchi A, Altomare V, Primavera A, Battista C, Vetrani A, Tonini G, Rabitti C (2005) COX-2 expression in DCIS: correlation with VEGF, HER-2/neu, prognostic molecular markers and clinicopathological features. Histopathology 46:561–568

Pistolesi S, Boldrini L, Gisfredi S, Ursino S, Ali G, Nuti S, De Ieso K, Pieracci N, Parenti G, Fontanini G (2007) Expression of cyclooxygenase-2 and its correlation with vasogenic brain edema in human intracranial meningiomas. Cancer Invest 25:555–562

Plotkin SR, Stemmer-Rachamimov AO, Barker FG 2nd, Halpin C, Padera TP, Tyrrell A, Sorensen AG, Jain RK, di Tomaso E (2009) Hearing improvement after bevacizumab in patients with neurofibromatosis type 2. N Engl J Med 361:358–367

Schmedtje JF Jr, Ji YS, Liu WL, DuBois RN, Runge MS (1997) Hypoxia induces cyclooxygenase-2 via the NF-kappaB p65 transcription factor in human vascular endothelial cells. J Biol Chem 272:601–608

Sturgis EM, Woll SS, Aydin F, Marrogi AJ, Amedee RG (1996) Epidermal growth factor receptor expression by acoustic neuromas. Laryngoscope 106:457–462

Subbaramaiah K, Altorki N, Chung WJ, Mestre JR, Sampat A, Dannenberg AJ (1999) Inhibition of cyclooxygenase-2 gene expression by p53. J Biol Chem 274:10911–10915

Tsujii M, DuBois RN (1995) Alterations in cellular adhesion and apoptosis in epithelial cells overexpressing prostaglandin endoperoxide synthase 2. Cell 83:493–501

Tsujii M, Kawano S, Tsujii S, Sawaoka H, Hori M, DuBois RN (1998) Cyclooxygenase regulates angiogenesis induced by colon cancer cells. Cell 93:705–716

Uesaka T, Shono T, Suzuki SO, Nakamizo A, Niiro H, Mizoguchi M, Iwaki T, Sasaki T (2007) Expression of VEGF and its receptor genes in intracranial schwannomas. J Neurooncol 83:259–266

Utermark T, Kaempchen K, Antoniadis G, Hanemann CO (2005) Reduced apoptosis rates in human schwannomas. Brain Pathol 15:17–22

Wasa J, Nishida Y, Suzuki Y, Tsukushi S, Shido Y, Hosono K, Shimoyama Y, Nakamura S, Ishiquro N (2008) Differential expression of angiogenic factors in peripheral nerve sheath tumors. Clin Exp Metastasis 25:819–825

Zhi YH, Liu RS, Song MM, Tian Y, Long J, Tu W, Guo RX (2005) Cyclooxygenase-2 promotes angiogenesis by increasing vascular endothelial growth factor and predicts prognosis in gallbladder carcinoma. World J Gastroenterol 11:3724–3728

Adult Primary Gliosarcoma: Epidemiology

6

Maryam Hamidi, John Moody, and Kevin Kozak

Contents

M. Hamidi (✉) • J. Moody • K. Kozak
Department of Human Oncology and Medical Physics, School of Medical and Public Health, University of Wisconsin-Madison, 600 Highland Avenue, Madison, WI 53792, USA
e-mail: mah244@musc.edu; kevin_kozak@yahoo.com

Abstract

Comprising approximately 2% of all glioblastoma cases, adult primary gliosarcoma is a rare aggressive tumor, composed of a mixture of malignant glial and sarcomatous elements, with dismal outcomes. While the general epidemiology and presentation reflect that of other glioblastomas, gliosarcoma has a much higher rate of metastases, may carry a worse prognosis, and has not been found to carry the hallmark characteristic of EGFR overexpression generally found in glioblastomas. The histogenesis of the disease remains unclear but there is increasing molecular and genetic evidence that the mixed components of the tumor have a monoclonal origin. In part because of a lack of information on the disease, patients with gliosarcoma are generally managed in the same manner as patients with other glioblastomas.

Introduction

Adult primary gliosarcoma is a central nervous system tumor composed of a mixture of malignant glial and sarcomatous elements. Stroebe (1885) described the first reported case in 1895 but it was not a widely accepted diagnosis until 60 years later when Feigin and Gross (1954) described three cases of gliosarcoma in detail. The malignancy is exceedingly rare, and epidemiological studies have been limited to small retrospective studies and case reports. The largest

M. Hayat (ed.), *Tumors of the Central Nervous System, Volume 13*,
DOI 10.1007/978-94-007-7602-9_6, © Springer Science+Business Media Dordrecht 2014

study to date has been a SEER database study comparing about 350 gliosarcoma patients to more than 16,000 patients with glioblastoma (Kozak et al. 2009). Currently, the malignancy is considered a variant of glioblastoma but there is growing evidence suggesting that it may be a separate entity. However, the underlying pathogenesis has not been definitively clarified and the disease continues to be managed in the same manner as glioblastoma.

Clinical Presentation and Prognosis

Gliosarcomas account for about 2% (Meis et al. 1990; Galanis et al. 1998; Kozak et al. 2009) of all glioblastoma cases. The annual incidence in North America and Europe is approximately one new case per 1,000,000 (Lantos et al. 1996). Otherwise, the general epidemiology follows that of glioblastoma. The median age at presentation is between 50 and 70 years of age, with the largest study finding a median age of 63 years (Kozak et al. 2009). Like glioblastoma, gliosarcoma patients are slightly more likely to be male (M:F, 1.4–1.8: 1.0) (Morantz et al. 1976; Kozak et al. 2009). Also, similar to other glioblastomas, gliosarcomas are almost always located supratentorially but gliosarcomas specifically have a predilection for the temporal lobe. Kozak et al. (2009) found that approximately 34.0% of gliosarcomas were located in the temporal lobe compared to 23.0% of other glioblastomas.

The clinical presentation of gliosarcoma is similar to that of other rapidly expanding cerebral tumors. Reflecting their location, presenting symptoms of gliosarcomas include headache, weakness, nausea, personality changes and confusion, seizures, and aphasia (Morantz et al. 1976; Perry et al. 1995). Salient signs consist of papilledema, hemiparesis, homonymous hemianopsia, and dysphasia (Morantz et al. 1976). These signs and symptoms correlate strongly to the clinical features of glioblastoma.

Interestingly, in contrast to the majority of glioblastomas, which rarely metastasize outside the cranium, gliosarcomas have a much higher likelihood of hematogenous dissemination to other organs. Han et al. (2009) found that 11% of the reported cases in literature were associated with extracranial metastases. The lungs (72%), liver (41%), and lymph nodes (18%) were most often involved (Beaumont et al. 2007), but gliosarcoma metastases have been reported in a variety of anatomic locations, including adrenal glands, kidneys, skin, oral mucosa, spleen, bone marrow, ribs and spine. Beaumont et al. (2007) reported on one patient who had almost universal spread of disease including to previously unreported organs such as the thyroid, pericardium, myocardium, diaphragm, pancreas and stomach. The past several decades have witnessed an increase in the number of gliosarcoma distant metastasis reports, potentially attributable to both an increased awareness of the diagnosis and the modestly prolonged survival offered by better treatments.

Unfortunately, gliosarcoma is an aggressive disease with dismal outcomes. In fact, overall survival of untreated patients has been reported to be a mere 4 months (Morantz et al. 1976; Kozak et al. 2009). Kozak et al. found that the median survival of all gliosarcoma patients was only 9 months, which is comparable to the median survival found in the smaller retrospective studies. They also found that the prognosis of patients with gliosarcoma was slightly, but significantly, worse than patients with glioblastoma. In addition to receiving treatment, younger age at diagnosis is also associated with a longer median survival. Patients diagnosed before the age of 50 had a median survival of 15 months compared to 7 months for those diagnosed after the age of 50. Furthermore, tumor location has an impact on overall survival; ventricle involvement predicts worse outcomes.

Interestingly, two distinct morphological types of gliosarcomas have been described on gross appearance, suggesting that there may be two variants of gliosarcomas. On surgical resection, some gliosarcomas are firm, well-circumscribed, easily resectable tumors, often attached to the dura, resembling meningiomas. Other tumors are necrotic and diffusely infiltrating and thus difficult to resect, resembling glioblastomas (Salvati et al. 2005). On histological analysis of 11 patients, Salvati et al. (2005) found that the meningioma-like tumors had a more prevalent sarcomatous component, while the glioblastoma-like tumors had a more prevalent gliomatous component. Han et al. (2010a) found that patients with meningioma-like tumors had a median survival of 16 months compared to 9.5 months for patients with glioblastoma-like tumors. Furthermore, the time to progression after treatment (resection and radiotherapy with concurrent tomozolamide) was also longer in the former group, supporting the findings of previous studies (Salvati et al. 2005; Maiuri et al. 1990). The more favorable prognosis of the meningioma-like tumors may simply reflect the greater ease of achieving macroscopic total resection rather than an inherent difference in disease natural history.

Histology

The 2007 World Health Organization classification defines gliosarcoma as "a malignant grade IV neoplasm with both glial and phenotypically mesenchymal components" (Louis et al. 2007). The gliomatous component follows the histologic criteria for glioblastoma while the mesenchymal component displays a variety of morphologies, including fibroblastic, cartilaginous, osseous, smooth muscle and adipocytic differentiation. In order to differentiate from other sarcomas and gliomas, the 2007 WHO criteria recommends that the glial areas stain positive for glial fibrillary acidic protein (GFAP) while the sarcomatous areas stain negative for GFAP but positive for reticulin.

There has been much controversy concerning the pathogenesis of gliosarcomas. Feigin and Gross (1954) supported the "collision" theory, which purports that the hyperplastic blood vessels of high grade gliomas undergo a neoplastic transformation. The theory suggested that excessive stimulation of endothelial cells by neoplastic glial cells induced a malignant transformation. This early theory has been supported by the finding that the sarcomatous areas of some tumors are reactive to vascular endothelial markers such as factor VIII, von Willebrand factor, and *Ulex europaeus* agglutinin-1, a lectin that binds to surface glycoproteins on endothelium (Perry et al. 1995). However, these markers are not a ubiquitous finding, shedding doubt on the veracity of this explanation.

More recently, the idea of a monoclonal origin of both the sarcomatous and glial components has become popular. The monoclonal hypothesis suggests that the neoplastic glial cells undergo a metaplastic transformation, acquiring sarcomatous phenotypes (Meis et al. 1990). Numerous genetic studies support this hypothesis. Biernat et al. (1995) and Reis et al. (2000) found that in many gliosarcoma tumors both the glial and sarcomatous cells have identical *TP53* mutations. Reis et al. (2000) also found identical PTEN mutations, p16 deletion, and CDK4 amplification. In fact, 57% of all chromosomal imbalances detected in a gliosarcoma tumor are shared by both components of the tumor (Actor et al. 2002). Kleihues et al. (2000) also reported *TP53* and PTEN mutations in approximately 25 and 35%, respectively, of glioblastomas while Reifenberger et al. (1999) reported CDK4 amplification in glioblastomas. Further studies have demonstrated many other glioblastoma-like genetic alterations common to both areas.

While gliosarcomas generally arise *de novo*, there have been 11 reported cases following whole brain irradiation for either another primary brain tumor or acute lymphoblastic leukemia. In their retrospective study of 32 cases of gliosarcoma, Perry et al. (1995) reported that seven of the cases consisted of tumor recurrence in patients who received 50-Gy whole-brain irradiation for a primary glioblastoma. Thus, in these patients, the radiation apparently induced gliosarcomatous transformation of the initial glioblastoma. The median time to tumor recurrence, which correlates with the time from irradiation of the primary malignancy, was 36 weeks. In a review of radiation induced primary gliomas and brain sarcomas, Kaschten et al. (1995) found that in contrast to gliomas, radiation-induced sarcomas did not occur at doses below 20 Gy. The four cases of radiation-induced primary gliosarcomas occurred following a mean dose of

37 Gy. The time from irradiation of the primary disease to presentation of the secondary malignancy varied significantly, from 1 to 12 years.

More recently, there has been some debate about the possibility of gliosarcomas being a separate entity from glioblastoma. As previously mentioned, gliosarcomas are strikingly different in their tendency to metastasize and possibly have a worse prognosis than other glioblastomas. Though gliosarcomas do share some of the genetic alterations found in other glioblastomas, significant differences have also been found. Actor et al. (2002) discovered that although the number of imbalances per tumor did not vary significantly between gliosarcomas and glioblastomas, the number of chromosomes affected by an imbalance was lower for gliosarcomas. That is, they found that gliosarcomas have a greater degree of chromosomal stability than glioblastomas. Their finding was corroborated by cell culture studies showing that a gliosarcoma-derived cell line had a far lower number of chromosomal imbalances compared to nine glioblastoma-derived cell lines.

Importantly, no case of gliosarcoma has demonstrated EGFR overexpression/amplification, which is considered one of the molecular hallmarks of primary glioblastomas (Reifenberger et al. 1999; Actor et al. 2002). The rarity of gliosarcoma and the modest utility of EGFR inhibitors in glioblastoma management suggest the therapeutic implications of this observation remain undefined. In light of the paucity of data, patients with gliosarcomas will continue to be managed with the same trimodal approach as patients with other primary glioblastomas.

References

Actor B, Cobbers JM, Buschges R, Wolter M, Knobbe CB, Lichter P, Reifenberger G, Weber RG (2002) Comprehensive analysis of genomic alterations in gliosarcoma and its two tissue components. Genes Chromosomes Cancer 34:416–427

Beaumont T, Kuspky WJ, Barger GR, Sloan AE (2007) Gliosarcoma with multiple extracranial metastases: case report and review of the literature. J Neuro Oncol 83:39–46

Biernat W, Aguzzi A, Sure U, Grant JW, Kleihues P, Hegi ME (1995) Identical mutations of the p53 tumor suppressor gene in the gliomatous and the sarcomatous components of gliosarcoma suggest a common origin from glial cells. J Neuropathol Exp Neurol 54:651656

Feigin I, Gross SW (1954) Sarcoma arising in glioblastoma of the brain. Am J Pathol 31:633–653

Galanis E, Buckner JC, Dinapoli RP, Scheithauer BW, Jenkins RB, Wang CH, O'Fallon JR, Farr G Jr (1998) Clinical outcome of gliosarcoma compared with glioblastoma multiforme: North Central Cancer Treatment Group results. J Neurosurg 89:425–430

Han SJ, Isaac Y, Tihan T, Prados MD, Parsa AT (2010a) Primary gliosarcoma: key clinical and pathologic distinctions from glioblastoma with implications as a unique oncologic entity. J Neurooncol 96:313–320

Han S, Yang I, Ahn BJ, Otero JJ, Tihan T, McDermott MW, Berger MS, Prados MD, Parasa AT (2010b) Clinical characteristics and outcomes for a modern series of primary gliosarcoma patients. Cancer 116:1358–1366

Kaschten B, Flandrov P, Reznik M, Hainaut H, Stevenaert A (1995) Radiation-induced gliosarcoma case report and review of the literature. J Neurosurg 83:154–162

Kleihues P, Burge PC, Collins VP, Newcomb EW, Ohgaki H, Cavenee WK (2000) Pathology and genetics of tumours of the nervous system. IARC Press, Lyon, pp 29–39

Kozak KR, Mahadevan A, Moody JS (2009) Adult gliosarcoma: epidemiology, natural history, and factors associated with outcome. Neuro Oncol 11:183–191

Lantos PL, VandenBerg SR, Kleihues P (1996) Tumours of the nervous system. Greenfield's neuropathology. Edward Arnold, London, pp 583–879

Louis D, Ohgaki H, Wiestler O, Cavenee W, Burger P, Jouvet A, Scheithauer B, The KP (2007) WHO classification of tumours of the central nervous system. IARC Press, Lyon, pp 196–198

Maiuri F, Stella L, Benvenuti D, Giamundo A, Pettinato G (1990) Cerebral gliosarcomas: correlation of computed tomographic findings, surgical aspect, pathological features, and prognosis. Neurosurgery 26:261–267

Meis J, Martz KL, Nelson JS (1990) Mixed glioblastoma multiforme and sarcoma: a clinicopathologic study of 26 radiation therapy oncology group cases. Cancer 67:2342–2349

Morantz R, Feigin I, Ransohoff J (1976) Clinical and pathological study of 24 cases of gliosarcoma. J Neurosurg 45:398–408

Perry J, Ang LC, Bilbao JM, Muller PJ (1995) Clinicopathologic features of primary and post-irradiation cerebral gliosarcoma. Cancer 75:2910–2918

Reifenberger G, Weber T, Weber RG, Wolter M, Brandis A, Kuchelmeister K, Pilz P, Reusche E, Lichter P, Wiestler OD (1999) Chordoid glioma of the third ventricle: immunohistochemical and molecular genetic characterization of a novel tumor entity. Brain Pathol 9:617–626

Reis R, Konu-Leblecliciogl u D, Lopes JM, Kleihues P, Ohgaki H (2000) Genetic profile of gliosarcomas. Am J Pathol 156:425–432

Salvati M, Caroli E, Raco A, Giangaspero F, Delfini R, Ferrante L (2005) Gliosarcomas: analysis of 11 cases do two subtypes exist? J Neurooncol 74:59–63

Stroebe H (1885) 195 Ueber Entstehung und Bau der Gehirnglima. Beitr Pathol Anat Allg Pathol 19:405–486

Mesenchymal Chondrosarcoma in the Central Nervous System: Histological Diagnosis

7

Lisa Lin, Winny Varikatt, and Thomas Ng

Contents

L. Lin (✉) • W. Varikatt • T. Ng
Department of Tissue Pathology, Level 3, ICPMR, Westmead Hospital, Hawkesbury Rd, Westmead, NSW 2145, Australia
e-mail: lisalin10@gmail.com

Abstract

Mesenchymal chondrosarcoma is a rare malignant neoplasm typically arising in the bones of young adults. It may also arise in somatic soft tissue, central nervous system and other organs. There are no specific clinical or radiological characteristics, and histological assessment remains the key to diagnosis. Histological features are similar regardless of site, and display a characteristic biphasic pattern composed of highly undifferentiated small round cells and islands of well differentiated hyaline cartilage. In this chapter, we discuss the clinical, radiological and histological features of mesenchymal chondrosarcoma arising in the central nervous system (CNS), the important differential diagnoses of small round cell tumour within the CNS, and the differentiating features of mesenchymal chondrosarcoma from Ewing sarcoma/PNET, medulloblastoma, haemangiopericytoma, monophasic synovial sarcoma and atypical teratoid/rhabdoid tumour.

Introduction

Mesenchymal chondrosarcoma is an aggressive tumour mostly, but not exclusively, involving the skeletal system of adolescents and young adults. Involvement of the central nervous system is extremely rare, and has only been acknowledged by isolated case reports. Primary cartilaginous tumours account for 0.16% of all intracranial

M. Hayat (ed.), *Tumors of the Central Nervous System, Volume 13*,
DOI 10.1007/978-94-007-7602-9_7, © Springer Science+Business Media Dordrecht 2014

neoplasms, and these include chondromas, chondrosarcomas, myxoid chondrosarcomas, and mesenchymal chondrosarcomas. Mesenchymal chondrosarcoma has a propensity for local aggressiveness and frequent recurrence, and is considered one of the most malignant subtypes of chondrosarcoma (Bingaman et al. 2000).

The characteristic biphasic histological appearance has been widely described, composing of an undifferentiated small round cell proliferation and islands of cartilaginous differentiation. Diagnostic complexities arise when only one of the two characteristic components is sampled for histological examination. The anatomical and radiological features, histological characteristics, the phenomenon of post-chemotherapy cytomaturation, as well as differential diagnoses of small round cell tumours within the central nervous system are discussed.

Epidemiology/Sites of Involvement

Mesenchymal chondrosarcoma is a rare aggressive malignancy which usually affects young adults in their second to third decades with equal gender distribution (Fletcher et al. 2002; Unni et al. 2005). It was first described by Lichtenstein and Bernstein (1959), and has traditionally been regarded as a neoplasm of bone. However, extraosseous examples involving areas such as somatic soft tissues, mediastinum, orbit, and meninges have been reported in the literature in recent years (Unni and Inwards 2010; Bingaman et al. 2000), and the central nervous system (CNS) is now considered the most common site of extraosseous mesenchymal chondrosacroma (Louis et al. 2007). The CNS can be involved by direct extension from a nearby osseous primary (cranial or spinal), a lesion of dural origin, or by arising directly within the brain parenchyma (Burger and Scheithauer 2007). The most common location of mesenchymal chondrosarcoma within the central nervous system is the craniospinal meninges (Fig. 7.1), and most cases are supratentorial, located in the frontoparietal region (Chen et al. 2004).

Clinical Presentation

The clinical presentation of intracranial mesenchymal chondrosarcoma is similar to those of other mass lesions within the CNS.

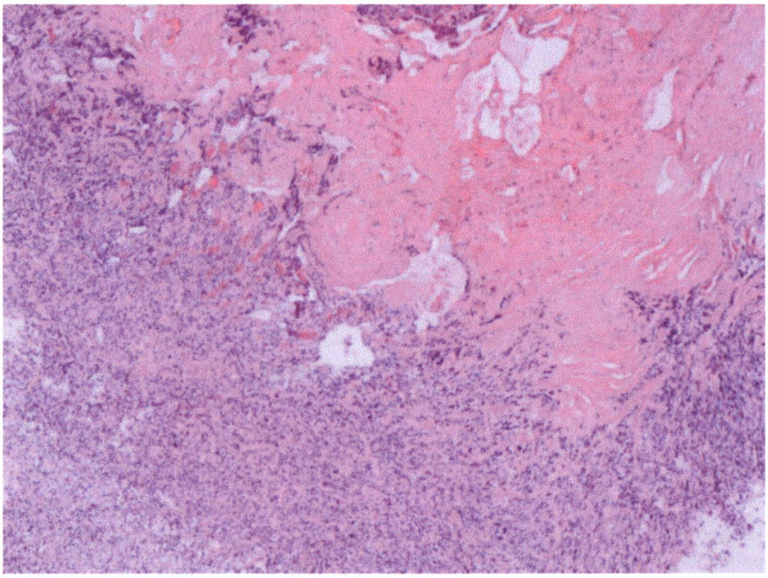

Fig. 7.1 Low power view showed the *small round blue* cell component of the mesenchymal chondrosarcoma attached to the dura mater (H&E, ×20)

Symptoms and signs are dependent on the location of the tumour, and frequently reflected the compressive effects of involved neurological structures, with examples including cranial nerve palsies in tumours involving the base of skull. Most patients have a tendency to present with headaches and other symptoms related to increased intracranial pressure (Scheithauer and Rubinstein 1978), and isolated cases may only be detected after the discovery of metastatic disease.

Radiological Features

Imaging of intracranial mesenchymal chondrosarcoma show inconsistent features, however, the majority of cases are hypointense to normointense on T1-weighted MRI, with strong enhancement after administration of gadolinium (Huang et al. 2004). It can be extremely hypervascular on angiography, and embolisation may be required prior to surgery. Due to its common involvement of the meninges and strong enhancement on MRI, these tumours may resemble malignant meningioma or haemangiopericytoma on radiological imaging (Chen et al. 2004).

Histopathology

Pathological examination of both extraosseous and osseous examples shows the same characteristic biphasic appearance (Scheithauer and Rubinstein 1978). The two distinct elements include islands of hyaline cartilage and a proliferation of undifferentiated small round cells (Fig. 7.2). The cartilaginous component is well differentiated and may appear as benign hyaline cartilage, low-grade chondrosarcoma, and rarely, as an intermediate-grade chondrosarcoma (Unni et al. 2005). The undifferentiated areas show sheets or alveolar arrangements of small round cells with relatively uniform nuclei, dense chromatin and sparse cytoplasm (Fletcher et al. 2002; Unni et al. 2005; Louis et al. 2007). Staghorn vascular spaces are commonly seen. The two characteristic components have a tendency to show an abrupt transition, however, occasional cases may show a gradual merging of the two elements. More importantly, the proportion of each element is highly variable and diagnostic problems arise when a limited biopsy only shows one component (Bingaman et al. 2000; Lin et al. 2012).

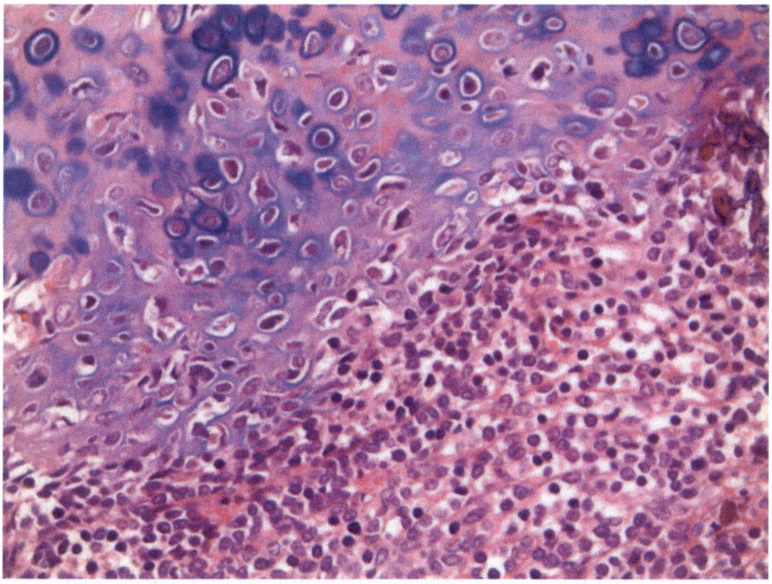

Fig. 7.2 Section showed the characteristic biphasic histological appearance, with islands of hyaline cartilage juxtaposed against a *small round blue* cell population (H&E, ×200)

Histochemical and Immunoperoxidase Studies

Histochemical and immunoperoxidase studies of mesenchymal chondrosarcoma are not specifically helpful in distinguishing it from other differential diagnoses. The small cells are glycogen rich (Burger and Scheithauer 2007), and stain positively for CD99, vimentin and Leu7. The chondroid cells are positive for S100 (Fletcher et al. 2002). Sox9, a master regulator of chondrogenesis, has been shown in a study by Wehrli et al. (2003) to reliably differentiate mesenchymal chondrosarcoma from other small round blue cell tumours, based on the hypothesis that both small cell and cartilaginous components are derived from primitive chondroprogenitor cells. In their study, 21 of 22 cases of mesenchymal chondrosarcoma showed positive nuclear staining in both components, and the remaining 73 small round blue cell tumours (including rhabdomyosarcoma, neuroblastoma, Ewing sarcoma/PNET, lymphoma, small cell carcinoma, merkel cell carcinoma, small cell osteosarcoma, extraskeletal myxoid chondrosarcoma and small cell desmoplastic tumour) displayed negative staining. Hence, Sox9 may be an useful ancillary stain for diagnosing mesenchymal chondrosarcoma when an undifferentiated small round cell component is present with a lack of chondroid elements.

Ancillary Tests

Electron microscopy of the undifferentiated small cells show large nuclei and little organelles, similar to primitive mesenchymal cells, and the cartilaginous areas show a usual chondrocyte appearance (Fletcher et al. 2002; Unni et al. 2005). No specific molecular aberration has been identified, although an identical Robertsonian translocation involving chromosomes 13 and 21 [der(13;21)(q10;q10)] has been detected in two cases (Fletcher et al. 2002; Unni et al. 2005).

Diagnostic Complexities

As mentioned previously, the proportion of the two components is highly unpredictable and the predominance of either component may result in diagnostic difficulties. The dominance of cartilaginous component may lead to a misdiagnosis of a pure chondroid lesion, and the sole presence of the small round cell component may result in the erroneous diagnosis of other small round cell tumours. Even if the tumour contained both elements in significant amounts, diagnostic dilemma may arise if only one element is sampled in a limited biopsy or debulking procedure.

Furthermore, a recent case report by Lin et al. (2012) highlighted a diagnostic pitfall in the diagnosis of mesenchymal chondrosarcoma arising from the tentorium cerebelli of a 21 year old woman. Histological examination of the tissue obtained at the initial debulking procedure demonstrated a dural-based pure small round cell population with hyperchromatic ovoid cells, finely to coarsely granular chromatin, inconspicuous nucleoli and scanty cytoplasm. The cells were Periodic acid Schiff (PAS) positive and diastase sensitive, and immunohistochemically showed focal strong membranous immunoreactivity with CD99, and negative staining for epithelial markers. The diagnosis of Ewing sarcoma/primitive neuroectodermal tumour (PNET) was made and the patient underwent chemotherapy and radiotherapy according to a PNET protocol. However, the tissue obtained during a definitive complete macroscopic removal several months after the initial procedure showed prominent hypercellular lobules of atypical chondroid cells, prompting the re-diagnosis as a mesenchymal chondrosarcoma. This case highlighted diagnostic complexities of mesenchymal chondrosarcomas in an intracranial location where limited biopsy material is often obtained, and potential misdiagnosis if only one component is sampled. The authors also raised a second hypothesis of post-chemotherapy cellular maturation in an attempt to explain the differences in histological appearances seen in the tissue obtained from the first and second procedure.

Post-chemotherapy cellular maturation is a well known phenomenon described in a range of pediatric tumours, including pediatric sarcomas and embryonal tumours, Wilms' tumour and germ cell tumours (McCartney et al. 1984; Omar et al. 1986; Coffin et al. 2005; Smith et al. 2002; Nozza et al. 2010). This phenomenon is thought to be a secondary event to anti-cancer agents selectively destroying immature and more anaplastic clones, and hence, offering relatively benign and mature cells a survival advantage (McCartney et al. 1984; Omar et al. 1986). The post-chemotherapy cyto-maturation phenomenon may potentially explain the marked differentiation of the cartilaginous component in this case.

Differential Diagnoses

A small round cell tumour arising in the brain raises several differential diagnoses. These include Ewing sarcoma/PNET, medulloblastoma, haemangiopericytoma, monophasic synovial sarcoma and atypical teratoid/rhabdoid tumour. Firstly, Ewing sarcoma/PNET has a tendency to affect a younger population, with the peak incidence in the second decade (Louis et al. 2007). Secondly, most reported primary CNS Ewing sarcoma/PNET are extra-axial, dural based masses (Theeler et al. 2009). Histologically, the small round cell appearance in Ewing sarcoma/PNET show a striking histological similarity to mesenchymal chondrosarcoma, both exhibiting sheets of small round cells with PAS-positive, diastase sensitive, glycogen rich cytoplasm (Louis et al. 2007; Burger and Scheithauer 2007). Homer-Wright rosettes may be helpful in distinguishing Ewing sarcoma/PNET, however it is only present in occasional cases. Furthermore, CD99, a characteristic immunoperoxidase stain positive in Ewing sarcoma/PNET, may also be seen in other tumours such as mesenchymal chondrosarcoma, haemangiopericytoma and synovial sarcoma (Louis et al. 2007; Kazmi et al. 2007). As a result of these non-specific features, there is an increasing reliance on the utilisation of molecular testing, and these include reverse transcription polymerase chain reaction (rt-PCR) looking for EWS-FLI1 and EWS-ERG (or other EWS variant) fusion transcript, and fluorescence in-situ hybridisation studies (FISH), looking for characteristic chromosomal translocation t(11;22)(q24;q12) or other variant translocations, such as t(21;22)(q22;q12) (Louis et al. 2007; Kazmi et al. 2007). Sensitivities and specificities of 91 and 100% respectively have been reported with FISH, and this is considered a confirmatory test if positive (Kazmi et al. 2007). However, it is noted that 5% of these tumours have no detectable chromosomal translocation and in these cases, making the correct diagnosis may be challenging (Navarro et al. 2007).

Medulloblastoma is also another important differential diagnosis, as it shows sheets of similar round blue cells, and Homer Wright rosettes are only present in less than 40% of cases. However, it rarely shows a dural-based location, and has an absence of glycogen rich cytoplasm and CD99 positivity (Theeler et al. 2009; Kazmi et al. 2007). Haemangiopericytoma is similarly meningeal-based, and typically displays characteristic staghorn vessels, which may be seen in mesenchymal chondrosarcoma (Kazmi et al. 2007). However, they have different immunophenotype, with haemangiopericytoma being Factor XIIIa and CD34 positive. Monophasic synovial sarcoma may also display a prominent haemangiopericytomatous vascular pattern, cytokeratin and EMA positivity, and SYT locus rearrangement on FISH (Fletcher et al. 2002). Atypical teratoid/rhabdoid tumour shows presence of rhabdoid cells, EMA positivity and distinctive loss of INI-1 nuclear expression (Louis et al. 2007).

In conclusion, mesenchymal chondrosarcoma is a malignant neoplasm rarely encountered in the central nervous system. Clinical experience with cranial or spinal mesenchymal chondrosarcoma is limited, but frequent local recurrence and distant metastasis is the norm. Mesenchymal chondrosarcoma should be considered in the differential diagnosis of any small round blue cell lesion encountered in a dural-based or parenchymal lesion, and correlation with the clinical and radiological information is mandatory, especially to ascertain whether the entire lesion has been removed. Obtaining adequate material displaying both components, performing the appropriate immunoperoxidase panel and molecular testing to exclude other conditions, and a high index of suspicion is the mainstay to diagnosis.

References

Bingaman KD, Alleyne CH, Olson J (2000) Intracranial extraskeletal mesenchymal chondrosarcoma: case report. Neurosurgery 46:207–211

Burger PC, Scheithauer BW (2007) AFIP atlas of tumor pathology fourth series fascicle 7: tumors of the central nervous system. Am Regist Pathol, Washington, DC, pp 390–391

Chen JY, Hsu SS, Ho JT (2004) Extraskeletal intracranial mesenchymal chondrosarcoma: case report and literature review. Kaohsiung J Med Sci 20:240–245

Coffin CM, Lowichik A, Zhou H (2005) Treatment effects in pediatric soft tissue and bone tumours. Am J Clin Pathol 123:75–90

Fletcher CD, Unni KK, Mertens F (eds) (2002) WHO classification of pathology and genetics of tumours of soft tissue and bone. IARC Press, Lyon, pp 200–204, 255–256

Huang YL, Lai PH, Lin SL, Yuan MK, Pan HB (2004) Mesenchymal chondrosarcoma at the falx cerebri. Chin J Radiol 29:337–341

Kazmi SA, Perry A, Pressey JG, Wellons JC, Hammers Y, Palmer CA (2007) Primary Ewing sarcoma of the brain: a case report and literature review. Diagn Mol Pathol 16:108–111

Lichtenstein L, Bernstein D (1959) Unusual benign and malignant chondroid tumors of the bone: a survey of some mesenchymal cartilage tumors and a malignant chondroblastic tumor including a few multicentric ones, as well as many atypical benign chondroblastomatas and chondromyxoid fibromas. Cancer 12:1142–1157

Lin L, Varikatt W, Dexter M, Ng T (2012) Diagnostic pitfall in the diagnosis of mesenchymal chondrosarcoma arising in the central nervous system. Neuropathology 32:82–90

Louis DN, Ohgaki H, Wiestler OD, Cavenee WK (eds) (2007) WHO classification of tumours of the central nervous system, 4th edn. IARC Press, Lyon, p 176

McCartney AC, Paradinas FJ, Newlands ES (1984) Significance of the 'maturation' of metastases from germ cell tumours after intensive chemotherapy. Histopathology 8:457–467

Navarro R, Laguna A, De Torres C, Cigudosa JC, Suñol M, Cruz O, Mora J (2007) Primary Ewing sarcoma of the tentorium presenting with intracranial hemorrhage in a child. J Neurosurg 107:411–415

Nozza P, Casciana ML, Rossi A, Cama A, Milanaccio C, Raso A, Ravegnani M, Morreale G, Garrè ML (2010) Post-chemotherapy maturation of a pineoblastoma. Acta Neuropathol 119:651–653

Omar R, Davidian MM, Marcus JR, Rose J (1986) Significance of the "maturation" of metastases from Wilms' tumor after therapy. J Surg Oncol 33:239–242

Scheithauer BW, Rubinstein LJ (1978) Meningeal mesenchymal chondrosarcoma: report of 8 cases with review of the literature. Cancer 42:2745–2752

Smith LM, Anderson JR, Coffin CM (2002) Cytodifferentiation and clinical outcome after chemotherapy and radiation therapy for rhabdomyosarcoma (RMS). Med Pediatr Oncol 38:398–404

Theeler BJ, Keylock J, Yoest S, Forouhar M (2009) Ewing's sarcoma family tumors mimicking primary central nervous system neoplasms. J Neurol Sci 284:186–189

Unni KK, Inwards CY (2010) Dahlin's bone tumors: general aspects and data on 10165 cases, 6th edn. Lippincott Williams and Wilkins, Philadelphia, pp 92–96

Unni KK, Inwards CY, Bridge JA, Kindblom LF, Wold LE (2005) AFIP atlas of tumor pathology 4th series fascicle 2: tumors of the bones and joints. Am Regist Pathol, Washington, DC, pp 99–104

Wehrli BM, Huang W, De Crombrugghe B, Ayala AG, Czerniak B (2003) Sox9, a master regulator of chondrogenesis, distinguishes mesenchymal chondrosarcoma from other small blue round cell tumors. Hum Pathol 34:263–269

Supratentorial Primitive Neuroectodermal Tumors (PNETs)

8

Kourosh Karimi Yarandi and Abbas Amirjamshidi

Contents

Abstract

Supratentorial PNETs are a group of heterogeneous neoplasms which are rarely encountered especially in adulthood. Because of the rarity of these tumors, articles describing the clinical features and their response to treatment are mostly limited to case reports and retrospective studies of relatively small series. In Addition, many of the investigators have analyzed mixed varieties of these heterogeneous tumors found in different age groups. Therefore, it is very hard to assess the clinical features, true effect of various treatment modalities, and prognosis of these tumors. Nowadays, supratentorial PNETs are distinct from medulloblastomas and peripheral PNETs.

One of the most crucial steps in treatment of these neoplasms is total surgical evacuation; however, this cannot be achieved in the majority of the cases. Accordingly, radiotherapy, especially in pineoblastomas, and chemotherapy are used as useful adjuvant therapies to extend the survival. Multiple recent reports emphasize on the therapeutic role of high dose chemotherapy with autologous stem cell rescue in children and adults. Various prognostic factors are suggested to predict the outcome. Unfortunately, the true effect of many of these factors has remained controversial and accurate evaluation of the prognostic significance of those variants requires larger well-designed studies and higher level of evidence.

K. Karimi Yarandi (✉) • A. Amirjamshidi
Department of Neurosurgery, Sina Hospital, Tehran University of Medical Sciences, Tehran, Iran
e-mail: Kourosh@ymail.com; abamirjamshidi@yahoo.com

M. Hayat (ed.), *Tumors of the Central Nervous System, Volume 13*,
DOI 10.1007/978-94-007-7602-9_8, © Springer Science+Business Media Dordrecht 2014

Introduction

Primitive neuroectodermal tumors (PNETs) include a heterogeneous group of highly aggressive and malignant embryonal neoplasms thought to originate from primitive or undifferentiated neuroepithelial cells. These tumors occur usually in children and may arise in any place along the central nervous system (CNS). Traditionally, several terms have been applied to these tumors. In 1973, Hart and Earle described a group of 23 poorly differentiated supratentorial tumors and introduced the term *Primitive neuroectodermal tumors* to refer to *supratentorial* malignant small cell neoplasms that were believed to arise from primitive embryonal cells. Rorke (1983) expanded the term "PNET" to include infratentorial tumors as well. There has been considerable debate in the literature concerning the most appropriate classification of PNETs, and some have postulated that all PNETs, including medulloblastomas, arise from similar cells and should be classified as PNETs and then sub-classified based on evidence of cellular differentiation and possibly the tumor location. Others have suggested that the term medulloblastoma should be maintained for those tumors arising in the posterior fossa and, similarly, "pineoblastoma" for tumors arising in the pineal region.

According to the fourth edition of the World Health Organization *Classification of Tumours of the Central Nervous System*, published in 2007, the term *central nervous system (CNS) PNET* should be applied to the neoplasms histologically indistinguishable from medulloblastoma but occurring outside the cerebellum. This classification endorses the theory that medulloblastoma is a distinct entity from other PNETs, as recent molecular studies and reports of the outcome analysis and responses to therapies have suggested. In addition, the incidence of medulloblastomas is much higher than that of all the other PNETs, further distinguishing it among this type of tumors.

In 1986, Dehner proposed the terms "central" and "peripheral" PNETs. This classification is widely accepted and PNETs may occur outside the CNS throughout the body, as peripheral neuroblastomas and Ewing sarcomas. It should be reminded that our discussion is limited to tumors within the brain. Supratentorial PNETs are uncommon and highly heterogeneous tumors. Because of the rarity of supratentorial PNETs, articles on clinical features and their response to treatment are mostly limited to case reports and retrospective studies of relatively small series including various pathologies in different age groups. Accordingly, it should be emphasized that there are numerous unknown aspects about these tumors which are yet to be revealed by well-designed larger series in the future.

Epidemiology

Supratentorial PNETs are rare, accounting for less than 1% of all primary CNS neoplasms and 1.9–7% of childhood brain tumors (Dai et al. 2003; Gaffney et al. 1985; Johnston et al. 2008; Rickert and Paulus 2001; Yang et al. 1999). The annual incidence of these tumors is estimated to be two per million (Yang et al. 1999). Supratentorial PNETs are found in both children and adults, although they are more commonly discovered in pediatric age groups and tend to occur in young children. More than half of these tumors manifest during the first 5 years of life. The mean age at diagnosis is 3–6.8 years in children (Johnston et al. 2008; Yang et al. 1999). These tumors have also been discovered in adults of different age groups (Kim et al. 2002). No sex predilection exists either in children or adults (Jakacki 1999; Johnston et al. 2008; Kim et al. 2002), whereas medulloblastomas are more frequently found in males and young adults.

Pathology and Etiology

Historically, the terminology used to describe medulloblastomas or PNETs has been confusing.

Upon gross inspection, supratentorial PNETs are usually pinkish-gray lobulated tumors with well-demarcated margins. In microscopic evaluation, medulloblastoma and other PNETs are densely cellular tumors composed predominantly of small cells with little ill-defined cytoplasm and hyperchromatic nuclei. The tumor cells are arranged in cords or nests surrounded by loose or sometimes desmoplastic connective tissue. Cell clusters can also aggregate around fibrinoid matrix. In this case, the nuclei surround a central clear area made up of cell processes rather than a vessel forming Homer-Wright rosettes. Numerous mitoses are usually present, however, the rate of mitotic activity can be quite variable (Kim et al. 2002). These tumors are thought to originate from primitive or undifferentiated neuro-epithelial cells but they frequently display histological heterogeneity demonstrating glial or neuronal differentiation or both in some regions of the neoplasm. In other words, PNETS are composed of undifferentiated neuro-epithelial cells, but frequently contain one or more populations of differentiated neoplastic cells. Other types of cellular differentiation such as mesenchymal differentiation can also be expected in these tumors (Kim et al. 2002). Cystic change, necrosis, and calcification can be detected as well (Kim et al. 2002).

In Immunohistochemical analysis, synaptophysin is expressed in 58% of the cases as well as GFAP which is also expressed in 58% of these tumors (Kim et al. 2002). Neuron-specific enolase can be positive in such neoplasms as well. PNETs with ependymal differentiation frequently show perivascular pseudorosetts with GFAP expression and glial differentiation (Kim et al. 2002). In different series, the mean value of KI-67 labeling index (LI) in supratentorial PNETs is reported to be 6.4–20.6% ranging from 0.4 to 38% (Kim et al. 2002; Yang et al. 1999). Moreover, the mean value of p53 LI is 5.6–7.6% ranging from 0 to 19% (Kim et al. 2002; Yang et al. 1999).

Regardless of location, PNETs share some morphologic similarities and have the propensity to spread within the CNS, via CSF dissemination, and beyond. Similar tumors found in the pineal gland are termed *pineoblastomas*, whereas those in the supratentorial space are termed PNETs, *ependymoblastomas* or *neuroblastomas*. According to Dehner's classification (1986) ependymoblastomas are PNETs with ependymal differentiations, while cerebral neuroblastomas contain neuroblastic-neuronal differentiation and medulloepitheliomas contain embryonic neural canal. Dehner also classified *retinoblastomas* of the eye and *olfactory neuroblastomas* of the intranasal cavity as central neuroectodermal tumors. Nowadays, retinoblastomas are not classically recognized as supratentorial PNETs and recent data basically does not support the idea that olfactory neuroblastomas can be regarded as typical PNETs. Supratentorial PNETs may also be seen in conjunction with bilateral retinoblastomas (altogether called *trilateral* retinoblastoma). This situation is caused by germline mutations in the *Rb* gene. The supratentorial tumors often occur in pineal region and thus are classified as pineoblastomas. Suprasellar or parasellar PNETs may also accompany retionoblastomas.

Medulloblastoma has been the term applied to PNETs arising in the posterior fossa. But recently, several reports have shown different biological features including transcriptional and cytogenetic profiles which have changed the whole previous concept. Large scale protein expression studies demonstrate distinct gene activation patterns between supratentorial PNETs and medulloblastomas. An isochromosome 17q has been recognized as an important abnormality in the medulloblastomas, whereas only a limited number of supratentorial PNETs have been found to have such a phenomenon (Raffel et al. 1993). Furthermore, it is shown that cerebellar tumors with similar pathologic features with supratentorial PNETs (i.e., medulloblastomas) have different expression of neurofilament proteins, vimentin, GFAP and platelet derived growth factor (PDGF) receptors (Smits et al. 1996). Spatial organization of tumor vessels is also different among medulloblastomas and supratentorial PNETs (Goldbrunner et al. 1999).

According to the fourth edition of the World Health Organization (WHO) classification, three separate subtypes of embryonal central nervous system tumors are: medulloblastomas, central nervous system (CNS) primitive neuroectodermal tumors (PNETs), and atypical teratoid rhabdoid tumors. Accordingly, medulloblastomas are now classified into a distinct tumor entity. Supratentorial PNETs are also subcategorized into: CNS neuroblastoma, CNS ganglioneuroblastoma, medulloepithelioma, and ependymoblastoma. As mentioned before similar histologic subtype is also found in pineal region which is known as pineoblastoma. Despite some morphological similarities, supratentorial PNETs are shown to be clinicopathologically and genetically distinct lesions from peripheral embryonal tumors as well (Gyure et al. 1999; Vogel and Fuller 2003). Although peripheral PNETs are also categorized among the poorly differentiated neoplasms, they seem to be a completely separate entity. The cells of origin, pathophysiology and natural history, and molecular biology of these two classes of PNETs are totally different. In addition, evidence suggests that PNETs of the central nervous system (CNS) differ in their genetic aberrations. Peripheral PNETs are believed to be the more differentiated end of the spectrum including skeletal and extra-skeletal Ewing's sarcoma (Vogel and Fuller 2003).

A number of individuals with known Li-Fraumeni syndrome and documented germline *p53* mutations have been shown to develop supratentorial PNETs (Taylor et al. 2000). As discussed before, pineoblastomas or supratentorial PNETs can also occur in the setting of a germline *Rb* mutation (Taylor et al. 2000). In addition, theoretical relationship between Rubinstein-Taybi syndrome, due to germline mutations in the CREB-binding protein CBP, and development of supratentorial PNETs has also been described (Taylor et al. 2000). PNETs can also arise as a consequence of cranial irradiation (Amirjamshidi and Abbassioun 2000; Chen et al. 2005; Hader et al. 2003).

Clinical Evaluation: Signs and Symptoms

The majority of symptoms are nonspecific referable to tumor's mass effect. Regardless of the age group, the most frequent presenting symptoms are those related to increased intracranial pressure including nausea or vomiting and headache (Johnston et al. 2008; Kim et al. 2002; Yang et al. 1999). Seizures and focal neurological signs are also common (Yang et al. 1999). Irritability, visual difficulties, increased head circumference, lethargy, and ataxia are some other reported symptoms of supratentorial PNETs (Dai et al. 2003; Johnston et al. 2008; Kim et al. 2002). Seizures are frequent in children younger than 12 months (Dai et al. 2003). In physical examination, papilledema, visual field defects (homonymous or unilateral hemianopsia), cranial nerves palsy, dysphasia, and focal neurological deficits can be found in these patients (Kim et al. 2002). The diagnosis is usually made over a brief period of time after appearance of the first symptoms. The median duration between onset of symptoms and diagnosis is reported to be as short as 1 month in children (Johnston et al. 2008). Furthermore, the mean duration of symptoms before diagnosis is 26 days in infants (Dai et al. 2003), 3–17 weeks in children (Yang et al. 1999), and 4 months in adults (Kim et al. 2002).

Paraclinical Evaluation

Radiologic Evaluation

PNETs or medulloblastomas are usually diagnosed by means of computed tomography (CT) or magnetic resonance imaging (MRI). Nonetheless, preoperative radiological diagnosis is difficult and demands high level of suspicion, due to nonspecific CT and MRI characteristics and low incidence in adults. Supratentorial PNETs can mimic other high-grade tumors on MR images.

On CT scans, PNETs are typically large and hyperdense (Fig. 8.1a), showing homogeneous contrast enhancement, and may be partially cystic. Calcification can also be found in 41–44% of the cases (Dai et al. 2003; Johnston et al. 2008; Kim et al. 2002). They usually have small areas of calcification (Fig. 8.1a) and small cysts; rarely, they may be extensively calcified. They are mostly hemispheric (81%) and the remaining 19% of these tumors are reported to occur in pineal region (Johnston et al. 2008). The hemispheric

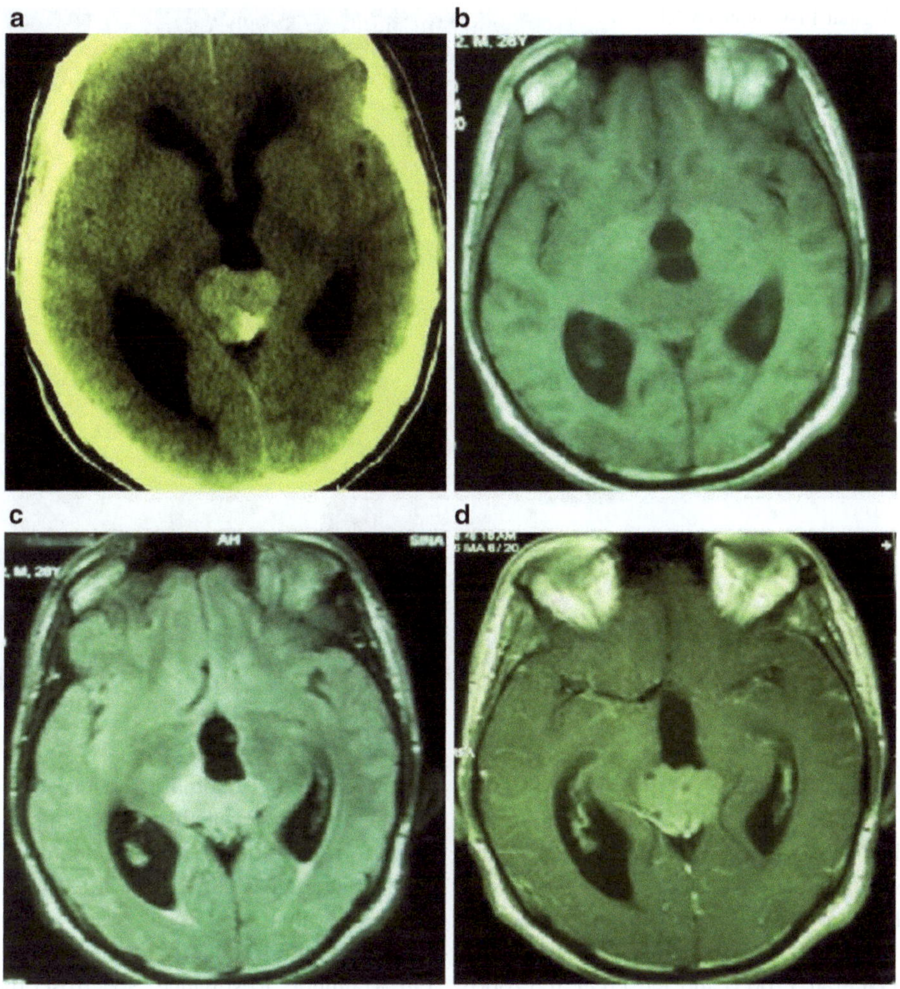

Fig. 8.1 MRI study of a 28-year old male with pineoblastoma. (**a**) Axial CT scan without contrast: The sharply demarcated tumor is hyperdense in CT images. Hydrocephalus is also evident. A small portion of the tumor is strongly hyperdense representing calcification. (**b**) and (**c**) Axial T1-weighted without contrast and FLAIR MRI: The tumor is isointense to cortical parenchyma in T1 and hyperintense in FLAIR. Again signs of hydrocephalus are visible. (**d**) Axial T1-weighted with contrast: The lesion is homogenously enhanced with gadolinium administration

tumors can occur in frontal, parietal, temporal, and occipital lobes of the brain (Dai et al. 2003; Jakacki 1999; Kim et al. 2002). They can also involve lateral ventricles and third ventricle or thalamus (Dai et al. 2003; Klisch et al. 2000).

On MRI scans, the tumors are usually isointense or hypointense on T1-weighted images, isointense or hyperintense on T2-weighted images, and strongly enhancing after gadolinium injection (Figs. 8.1, 8.2 and 8.3). Usually, there is a relatively good demarcation between the borders of the tumor and the adjacent brain tissue (Dai et al. 2003; Kim et al. 2002). Approximately 11% of supratentorial PNETs are not contrast enhancing on an MRI scan (Johnston et al. 2008). This makes postoperative assessment of residual disease more difficult. Intra- and peri-tumoral hemorrhage can be noticed in 12–37% of the cases (Dai et al. 2003; Johnston et al. 2008; Kim et al. 2002). Radiologic signs of necrosis can also be visualized in some cases (Kim et al. 2002). Lepto-meningeal involvement does not preclude

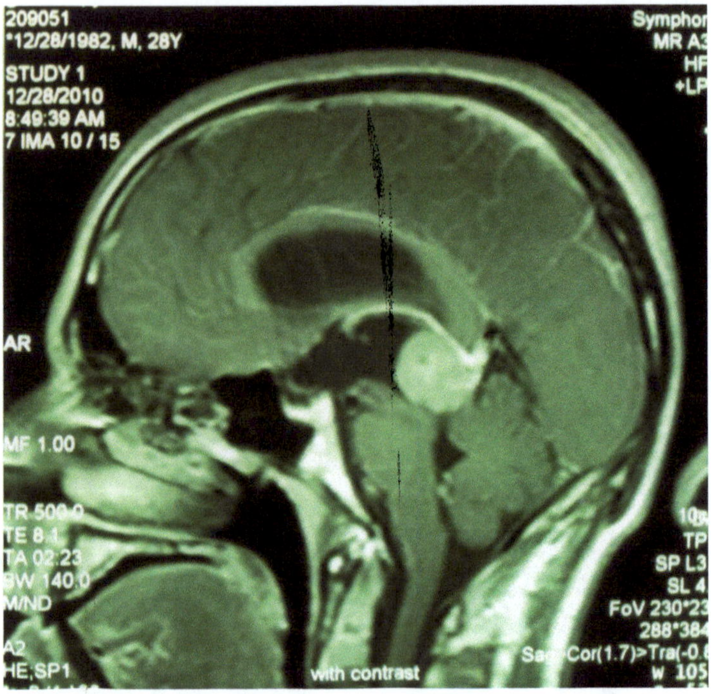

Fig. 8.2 Sagittal T1-weighted MRI with contrast of the same patient discussed in Fig. 8.1: The tumor is located at the posterior border of third ventricle, superior to the quadrigeminal plate proposing a lesion arising from pineal gland and enlarging this structure. Tri-ventricular hydrocephalus is evident

the diagnosis and can be discovered in 18% of cases (Johnston et al. 2008). Non-pineal supratentorial PNETs may be firmly attached to the dura misleading to the diagnosis of an extra-axial tumor (Fig. 8.3). The evidence seems to be conflicting regarding the amount of peri-tumoral edema. Dai et al. (2003) and Klisch et al. (2000) believe that presence of significant vasogenic peri-tumoral edema is very unusual in these tumors, while Kim et al. (2002) report it to be present in almost every case in adults. Solid portions of the supratentorial PNETs are usually hyperintense in diffusion-weighted images (DWI) of MRI mainly due to high cell density (Klisch et al. 2000).

In summary, supratentorial PNETs should be kept in the list of differential diagnosis when a large hyperdense supratentorial tumor with sharp margins is discovered.

Because PNETs can spread throughout the CNS, many physicians obtain MRI scan of the spine as soon as the diagnosis of PNET is entertained. Postsurgical blood and protein may confound spinal imaging for weeks after surgery, therefore preoperative spinal MRI may give the best assessment of spinal metastasis. At the time of diagnosis, MRI of the spine is reported to be normal in 88–100% of supratentorial PNETs (Johnston et al. 2008; Kim et al. 2002). In angiographic evaluations, focal areas of prominent vascularity are frequently seen (Kim et al. 2002). On Magnetic Resonance Spectroscopy (MRS) examinations, choline increases in these tumors. The amount of choline is significantly more than low grade and anaplastic astrocytomas, glioblastomas, and metastases (Majos et al. 2002).

Cerebrospinal Fluid (CSF) Analysis

PNETs often seed within the central nervous system (CNS) via cerebrospinal fluid (CSF). These tumors may be multicentric at the time of diagnosis. In the Children's Cancer Group study, metastatic spread at the time of diagnosis was found in 19% of supratentorial PNETs (Cohen et al. 1995). Dirks et al. (1996) reported a 38.9% rate of intracranial or spinal dissemination at the time of diagnosis. Other investigators reported a rate ranging from 7 to 12% for spinal

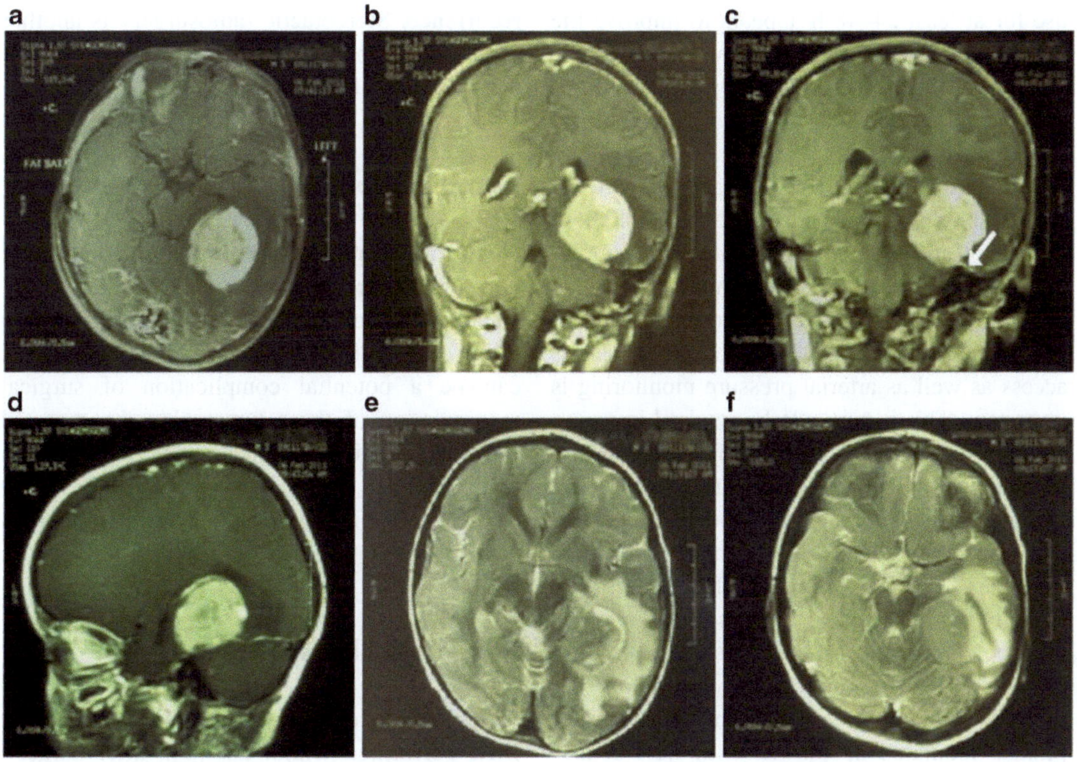

Fig. 8.3 MRI study of a 3-year old boy with ependymoblastoma. (**a**) Axial Fat Sat image with contrast: The supratentorial tumor has a well- defined border with the adjacent parenchyma and is homogenously enhanced with gadolinium administration. (**b**), (**c**), and (**d**) Sagittal and coronal T1-weighted MRI with contrast: The tumor seems to have a base on the tentorium and a dural tail along its surface (shown with the *white arrow*) resembling a tentorial meningioma. This dural tail is actually believed to be a sign of tumor aggression and malignancy. (**e**) and (**f**) Axial T2-weighted MRI: Significant surrounding edema exists around the borders of the tumor

seeding in supratentorial PNETs (Yang et al. 1999). Accordingly, CSF analysis can be helpful, if not mandatory, in the primary assessment; in order to discover any metastatic dissemination of these tumors into CNS. CSF analysis is reported to be positive for malignant cells in 9% of the cases (Johnston et al. 2008).

Bone Marrow Analysis and Bone Scan Analysis

Supratentorial PNETs may even metastasize beyond neuraxis. However, systemic metastases are not frequent, and usually occur at the time of recurrence (Jakacki 1999; Yang et al. 1999). Thus, for the newly diagnosed cases of supratentorial PNETs, preoperative bone marrow analysis and bone scans may be unnecessary and are no longer recommended as part of the staging process, unless clear symptoms and signs of involvement of other locations exist (Johnston et al. 2008).

Surgical Treatment

Surgery is a crucial part of treatment in these tumors and is beneficial for both diagnosis and decompression. Nevertheless, total resection is not always possible in these cases. The preoperative administration of corticosteroids appears to help the surgeon by decreasing the amount of peritumoral edema and ICP. Their preoperative use for at least 24–48 h appears to improve the whole clinical and neurological status of the patient. Similarly, pre-and intraoperative administration of anticonvulsant agents can be helpful for these tumors.

Standard neurosurgical and anesthetic techniques should be used in the removal of these tumors. Careful monitoring of urine output through a Foley catheter is recommended. Since these can be vascularized tumors, transfusion may be necessary. Accordingly, adequate venous access as well as arterial pressure monitoring is important. This is particularly required in young children in whom the amount of blood loss can be significant and may limit the extent of resection. Doppler cardiac monitoring is also required when the head is positioned significantly higher than the heart. Hence, placement of a CV line is strongly proposed in such cases. Air embolism must be considered if there is a sudden decrease of end-tidal PCO_2 or a drop in O_2 saturation.

These tumors are generally purplish gray or pinkish lobulated neoplasms with well demarcated borders along the adjacent normal brain in the majority of cases (Kim et al. 2002). Debulking of the tumor can allow the surgeon to bring in the edges of the lesion to better identify the margins of it. Gross total resection of the tumor is the optimal goal of surgery, but it must be balanced against potential neurological deficits particularly when there is adjacency to eloquent locations. Complete resection can be achieved in only 33–58% of the cases (Johnston et al. 2008; Kim et al. 2002; Timmermann et al. 2002) and as usual, surgeons most familiar with these tumors achieve the greatest amount of resection. The wide involvement of the brain and increased vascularity of these tumors are the main reasons precluding total or radical excision. Use of an ultrasonic aspirator greatly enhances the speed of tumor removal and assists with reducing blood loss. When the tumors are in the vicinity of eloquent regions, performing awake surgeries is another option. Also, surgical approach can be tailored according to preoperative diffusion tensor (DTI) images and concomitant use of neuronavigation. Additional surgical interventions may be needed when there is local recurrence. Stereotactic radiosurgery is another option when the local recurrence is not voluminous (Kim et al. 2002).

Complications

Hemorrhage in the tumor bed, brain edema, and new neurological deficits such as hemiparesis are some of the reported complications of surgery (Kim et al. 2002). Cerebrospinal dissemination can be a potential complication of surgical manipulation of these tumors but the reported incidence of this phenomenon is negligible (Kim et al. 2002). In the modern era, the rate of operative mortality is very low (Jakacki 1999; Kim et al. 2002).

Postsurgical Assessments

Determination of residual disease is best done by MRI performed within 24–48 h after surgery, before any enhancement attributable to postoperative inflammation or gliosis can cloud the imaging of residual neoplasm. Cytologic examination of CSF can also be performed by lumbar puncture about 2 weeks after surgery, after sufficient time has passed to avoid contamination by operative debris. Similarly, intraoperative samples may be taken at the beginning of the procedure. A CSF cytology determination that is positive for tumor cells, either preoperatively or postoperatively, predicts a poor outcome. A negative cytology test, however, does not preclude advanced disease. Spinal MRI should be done after surgery if it was not already performed. Postoperative imaging can occasionally be difficult because of the presence of postoperative debris and blood.

Outcomes and Adjuvant Therapy

Despite the concomitant use of hyperfractionated craniospinal radiation and adjuvant chemotherapy in the postoperative period, survival is generally poor in supratentorial PNETs particularly in children (Johnston et al. 2008). Local recurrence and CSF dissemination are chief causes of such a poor outcome. Current data suggest that children diagnosed with supratentorial PNETs have a poorer median survival interval than children with infratentorial medulloblastomas (Nishio et al. 1998a; Packer and Finlay 1996; Paulino and Melian 1999; Reddy et al. 2000). It is possibly because of younger age at time of diagnosis and special considerations for radiotherapy or frequent dissemination of such tumors (especially pineoblastomas) in the earlier phase of the disease. The 4-year survival of these tumors is also reported to be 37.7% in children (Johnston et al. 2008). Accordingly, most of the management protocols have included these patients within treatment regimens designed for children with poor-risk medulloblastoma. The prognosis is usually better for adults (Paulino and Melian 1999; Terheggen et al. 2007), in whom the mean survival is as high as 86 months and 3-years survival is 75% (Kim et al. 2002). Another important related issue in the management of PNETs is the quality of life among long-term survivors. It is now well recognized that a significant number of long-term surviving children have noticeable neurocognitive, endocrinologic, and psychological sequelae (Packer and Finlay 1996). However, this does not seem to be true among adults. The karnofsky performance scales (KPS) in the adult survivors is generally more than 70 with the mean follow-up duration of 49 months (Kim et al. 2002).

There is a paucity of articles discussing the prognostic factors of supratentorial PNETs.

Some factors have been proposed as being consistently important for outcome including: adjuvant therapy, age at the time of diagnosis, extent of initial resection, evidence of metastatic disease, histopathology, proliferation markers, and site of tumor. The impact of most of these factors on the outcome is highly controversial and the true prognostic effect of each is yet to be recognized:

Adjuvant therapy: Although the survival is poor in general, it can be prolonged with chemotherapy and radiation therapy in children (Jakacki 1999; Johnston et al. 2008; Yang et al. 1999) and adults (Terheggen et al. 2007). According to some reports radiotherapy is even more advantageous compared to chemotherapy (Nishio et al. 1998a; Paulino and Melian 1999; Timmermann et al. 2002). Radiotherapy is especially beneficial when it can be tolerated in pineal PNETs, which are somewhat resistant to chemotherapy in infants and young children (Hinkes et al. 2007; Jakacki 1999). Nevertheless, contradictory results can also be found in literature. There are several investigators who could not find any statistically significant benefit attributed to chemotherapy in children (Tomita et al. 1988) or adults (Kim et al. 2002) suffering from supratentorial PNETs. It seems that long-term survival rates with adjuvant chemotherapy, however, is much higher in recent reports than the previous ones, especially when high-dose chemotherapy with autologous stem-cell rescue is used (Broniscer et al. 2004; Butturini et al. 2009; Fangusaro et al. 2008; Gururangan et al. 2003; Perez-Martinez et al. 2004, 2005; Sung et al. 2007).

Age at the time of diagnosis: Largest series on PNETs have reported improved survival rates when these tumors are seen in older children and adults. Young age (less than 3 years of age) has been consistently shown to have an adverse effect on prognosis of these patients (Albright et al. 1995; Dirks et al. 1996; Hinkes et al. 2007; Jakacki 1999; Johnston et al. 2008; Kim et al. 2002). But among adults, age at the time of diagnosis does not seem to have any significant relation to outcome (Kim et al. 2002).

Extent of initial resection: Although radical resection, if possible, is suggested by the majority of the investigators, the effect of gross total or radical resection on the outcome is not clear. Nishio et al. (1998a) proposed such a beneficial effect in their review of literature. Albright et al. (1995) reported improved survival, albeit not statistically significant due to small sample size, for the patients in whom postoperative residual tumor was measured less than 1.5 cm^2. Yang et al. (1999) found a statistically significant relationship between the extent of surgery and the outcome in univariate analysis but they failed to show such a relationship in the multivariate analysis. However, more recent reports could not find any relationship between the extent of surgery and outcome either in children (Johnston et al. 2008) or adults (Kim et al. 2002).

Microscopic features and proliferation markers: Yang et al. (1999) reported a strong relationship between presence of tumor necrosis and worse outcome, whereas according to another report, presence of intratumoral necrosis, cyst, and hemorrhage does not affect the survival (Kim et al. 2002). The reports on the influence of histological differentiation on progression free survival in supratentorial PNETs have been inconsistent, and conclusions cannot be drawn since patients were not treated uniformly (Jakacki 1999). It is claimed that glial differentiation in PNETs predicts a poor clinical outcome but more recent studies have rejected this theory (Kim et al. 2002; Yang et al. 1999). Yang et al. (1999) did not find any prognostic significance for KI-67 LI. In contrast, some researchers believe that when KI-67 LI is greater than 30% worse outcomes are expected (Kim et al. 2002). Also, there are some reports emphasizing on the value of MIB1 staining index in prediction of clinical behavior and survival in these tumors (Nishio et al. 1998b). According to several reports, p53 LI does not have any relationship to the outcome (Kim et al. 2002; Yang et al. 1999).

Presence of metastatic disease at diagnosis: Most studies concerning M stage have demonstrated a statistically significant influence on outcome (Paulino and Melian 1999; Yang et al. 1999). Albright et al. (1995) found M stage as the most important prognostic factor for survival in children with non-pineal supratentorial PNETs. But Johnston et al. (2008) did not find any prognostic significance for presence of metastatic disease.

Presenting symptoms: No difference was found between various presenting symptoms (the ones associated with increased intracranial pressure versus others) in terms of outcome (Kim et al. 2002).

Race or sex: There is no difference in survival time according to race or sex (Johnston et al. 2008).

Tumor site: According to some reports, pineal location is accompanied with more favorable outcome (Gilheeney et al. 2008; Jakacki 1999; Timmermann et al. 2002). Again, Johnston et al. (2008) reported the opposite. According to their results the tumor site does not affect the whole survival. Meticulous analysis of different reports shows the importance of age and radiotherapy in these tumors. When pineal region tumors are seen in older children and adults, in whom radiotherapy can be performed, the prognosis seems to significantly improve (Gilheeney et al. 2008; Hinkes et al. 2007; Jakacki 1999). In contrast, in very young children and infants, in whom radiotherapy is not advised, the worst prognosis is attributed to pineal tumors (Jakacki 1999; Johnston et al. 2008). This seems to be due to the combined effect of radiotherapy, tumor site, presence of metastasis, and age.

Tumor size: Kim et al. (2002) did not find any relationship between size of these tumors and survival.

Currently, no established standard of care for patients with supratentorial PNETs exists. Nowadays, a variety of therapeutic strategies ranging from radiation therapy with or without chemotherapy, to high-dose chemotherapy with autologous stem cell transplantation are adopted by different treatment centers.

Radiation therapy: Standard treatment has usually included radiation therapy. Cushing was the first to use craniospinal irradiation to treat these tumors, recognizing their tendency for CSF propagation. Whole neuraxis radiation with additional boost to the tumor region is strongly proposed by most of the investigators, regardless of the metastatic stage. The exact effect of such a protocol on survival is not easy to assess. Whether it is required in every case is still controversial. The dose of irradiation is also another matter of debate. The more accepted suggested dose is 54 Gys at primary site, 30–36 Gys at whole brain, and

27–36 Gys at whole spine (Kim et al. 2002; Timmermann et al. 2002).

Attempts have been made to decrease the morbidity associated with craniospinal irradiation especially in young children. Hyperfractionated irradiation and dose reduction of craniospinal irradiation have been used to try to reduce the global effects of irradiation. Radiotherapy is especially effective in pineal region tumors but is generally not advised in children below the age of 2 or 3 years. The reported survival is significantly better in pineal PNETs in comparison with the non-pineal tumors when both radiotherapy and chemotherapy are used as adjuvant treatments. Interestingly, in young children and infants, where radiotherapy is not allowed, the survival is much better in the non-pineal tumors comparing with the pineal ones. Furthermore, in pineoblastomas the median overall and progression free survival in older children (more than years of age) is 8.8 and 7.9 years respectively, while these figures are only 0.9 and 0.6 years in children under 3 years of age (Hinkes et al. 2007). This data suggests radiotherapy as an important determinant of survival for pineal region PNETs (Jakacki 1999). It should be emphasized again that such a significant difference can be due to combined effect of age, radiotherapy, and presence of metastasis (which is more common among young children).

Chemotherapy: Chemotherapy is now an integral part of the management of the majority, if not all, of childhood PNETs. The results are encouraging especially in infants and young children with nonpineal PNETs (Jakacki 1999). It is also successfully used as a complementary treatment along with radiotherapy in older children (Chintagumpala et al. 2009; Jakacki 1999; Yang et al. 1999) and adults (Kim et al. 2002). There are several investigators who tried preirradiation chemotherapy as an adjunct to the treatment but some of the results have been disappointing. Any delay in radiotherapy according to this type of treatment can be associated with worse outcome (Timmermann et al. 2002). Multiple recent reports emphasize on the dramatic therapeutic role of high dose chemotherapy with autologous stem cell rescue in children (Broniscer et al. 2004; Butturini et al. 2009; Fangusaro et al. 2008; Gururangan et al. 2003; Perez-Martinez et al. 2004, 2005; Sung et al. 2007) and adults (Gururangan et al. 2003).

In conclusion, multimodality therapy including maximal surgical resection, aggressive chemotherapy, and craniospinal radiotherapy (if allowed) is strongly suggested in supratentorial PNETs.

Follow-Up

Again, no unanimous protocol for follow-up can be found throughout the literature. Evidence of metastasis during recurrence of the disease justifies frequent examination of CSF cytology or spinal imaging during the follow-up period (Yang et al. 1999).

References

Albright AL, Wisoff JH, Zeltzer P, Boyett J, Rorke LB, Stanley P, Geyer JR, Milstein JM (1995) Prognostic factors in children with supratentorial (nonpineal) primitive neuroectodermal tumors. Pediatr Neurosurg 22:1–7

Amirjamshidi A, Abbassioun K (2000) Radiation-induced tumors of the central nervous system occurring in childhood and adolescence. Four unusual lesions in three patients and a review of the literature. Child Nerv Syst 16(7):390–397

Broniscer A, Nicolaides TP, Dunkel IJ, Gardner SL, Johnson J Jr, Allen JC, Sposto R, Finlay JL (2004) High-dose chemotherapy with autologous stem-cell rescue in the treatment of patients with recurrent non-cerebellar primitive neuroectodermal tumors. Pediatr Blood Cancer 42:261–267

Butturini AM, Jacob M, Aguajo J, Vander-Walde NA, Villablanca J, Jubran R, Erdreich-Epstein A, Marachelian A, Dhall G, Finlay JL (2009) High-dose chemotherapy and autologous hematopoietic progenitor cell rescue in children with recurrent medulloblastoma and supratentorial primitive neuroectodermal tumors. The impact of prior radiotherapy on outcome. Cancer 115(13):2956–2963

Chen AY, Lee H, Hartman J, Greco C, Ryu JK, O'Donnell R, Boggan J (2005) Secondary supratentorial primitive neuroectodermal tumor following irradiation in a patient with low-grade astrocytoma. Am J Neuroradiol 26:160–162

Chintagumpala M, Hassall T, Palmer S, Ashley D, Wallace D, Kasow K, Merchant TE, Krasin MJ, Dauser R, Boop F, Krance R, Woo S, Cheuk R, Lau C, Gilbertson R, Gajjar A (2009) A pilot study of risk-adapted radiotherapy and chemotherapy in patients with supratentorial PNET. Neuro Oncol 11:33–40

Cohen BH, Zeltzer PM, Boyett JM, Geyer JR, Allen JC, Finlay JL, McGuire-Cullen P, Milstein JM, Rorke LB, Stanley P (1995) Prognostic factors and treatment results for supratentorial primitive neuroectodermal tumors in children using radiation and chemotherapy: a Children's Cancer Group randomized trial. J Clin Oncol 13:1687–1696

Dai A, Backstrom JW, Burger PC, Duffner PK (2003) Supratentorial primitive neuroectodermal tumors of infancy: clinical and radiologic findings. Pediatr Neurol 29:430–434

Dehner LP (1986) Peripheral and central primitive neuroectodermal tumors. A nosologic concept seeking a consensus. Arch Pathol Lab Med 110:997–1005

Dirks PB, Harris L, Hoffman HJ, Humphreys RP, Drake JM, Rutka JT (1996) Supratentorial primitive neuroectodermal tumors in children. J Neurooncol 29:75–84

Fangusaro J, Finlay J, Sposto R, Ji L, Saly M, Zacharoulis S, Asgharzadeh S, Abromowitch M, Olshefski R, Halpern S, Dubowy R, Comito M, Diez B, Kellie S, Hukin J, Rosenblum M, Dunkel I, Miller DC, Allen J, Gardner S (2008) Intensive chemotherapy followed by consolidative myeloablative chemotherapy with autologous hematopoietic cell rescue (AuHCR) in young children with newly diagnosed supratentorial primitive neuroectodermal tumors (sPNETs): report of the head start I and II experience. Pediatr Blood Cancer 50:312–318

Gaffney CC, Sloane JP, Bradley NJ, Bloom HJG (1985) Primitive neuroectodermal tumors of the cerebrum. J Neurooncol 3:23–33

Gilheeney SW, Saad A, Chi S, Turner C, Ullrich NJ, Goumnerova L, Scott RM, Marcus K, Lehman L, De Girolami U, Kieran MW (2008) Outcome of pediatric pineoblastoma after surgery, radiation and chemotherapy. J Neurooncol 89:89–95

Goldbrunner RH, Pietsch T, Vince GH, Bernstein JJ, Wagner S, Hageman H, Selby DM, Krauss J, Soerensen N, Tonn J (1999) Different vascular patterns of medulloblastoma and supratentorial primitive neuroectodermal tumors. Int J Dev Neurosci 17:593–599

Gururangan S, McLaughlin C, Quinn J, Rich J, Reardon D, Halperin EC, Herndon J II, Fuchs H, George T, Provenzale J, Watral M, McLendon RE, Friedman A, Friedman HS, Kurtzberg J, Vredenbergh J, Martin PL (2003) High-dose chemotherapy with autologous stem-cell rescue in children and adults with newly diagnosed pineoblastomas. J Clin Oncol 21:2187–2191

Gyure KA, Prayson RA, Estes ML (1999) Extracerebellar primitive neuroectodermal tumors: a clinicopathologic study with *bcl-2* and cd99 immunohistochemistry. Ann Diagn Pathol 3:276–280

Hader WJ, Drovini-Zis K, Maguire JA (2003) Primitive neuroectodermal tumors in the central nervous system following cranial irradiation. A report of four cases. Cancer 97:1072–1076

Hart MN, Earle KM (1973) Primitive neuroectodermal tumors of the brain in children. Cancer 32:890–897

Hinkes BG, von Hoff K, Deinlein F, Warmuth-Metz M, Soerensen N, Timmermann B, Mittler U, Urban C, Bode U, Pietsch T, Schlegel PG, Kortmann RD, Kuehl J, Rutkowski S (2007) Childhood pineoblastoma: experiences from the prospective multicenter trials HIT-SKK87, HIT-SKK92 and HIT91. J Neurooncol 81:217–223

Jakacki RI (1999) Pineal and nonpineal supratentorial primitive neuroectodermal tumors. Child Nerv Syst 15:586–591

Johnston DL, Keene DL, Lafay-Cousin L, Steinbok P, Sung L, Carret A, Crooks B, Strother D, Wilson B, Odame I, Eisenstat DD, Mpofu C, Zelcer S, Huang A, Bouffet E (2008) Supratentorial primitive neuroectodermal tumors: a Canadian pediatric brain tumor consortium report. J Neurooncol 86:101–108

Kim DG, Lee DY, Paek SH, Chi JG, Choe G, Jung H (2002) Supratentorial primitive neuroectodermal tumors in adults. J Neurooncol 60:43–52

Klisch J, Husstedt H, Hennings S, Velthoven V, Pagenstecher A, Schumacher M (2000) Supratentorial primitive neuroectodermal tumours: diffusion-weighted MRI. Neuroradiology 42:393–398

Majos C, Alonso J, Aguilera C, Serrallonga M, Acebes JJ, Arus C, Gili J (2002) Adult primitive neuroectodermal tumor: proton MR spectroscopic findings with possible application for differential diagnosis. Radiology 225:556–566

Nishio S, Morioka T, Fukui M (1998a) Primitive neuroectodermal tumors. Crit Rev Neurosurg 8:261–268

Nishio S, Morioka T, Suzuki S, Hamada Y, Kaneko Y, Fukui M (1998b) Supratentorial primitive neuroectodermal tumours: a report of four cases with an unusual clinical course in one patient. Acta Neurochir (Wien) 140:207–213

Packer RJ, Finlay JL (1996) Chemotherapy for childhood medulloblastoma and primitive neuroectodermal tumors. Oncologist 1:381–393

Paulino AC, Melian E (1999) Medulloblastoma and supratentorial primitive neuroectodermal tumors. An institutional experience. Cancer 86:142–148

Perez-Martinez A, Quintero V, Vicent MG, Sevilla J, Diaz MA, Madero L (2004) High-dose chemotherapy with autologous stem cell rescue as first line of treatment in young children with medulloblastoma and supratentorial primitive neuroectodermal tumors. J Neuro Oncol 67:101–106

Perez-Martinez A, Alvaro L, Vicent MG, Sevilla J, Diaz MA, Madero L (2005) High-dose chemotherapy with autologous stem cell rescue for children with high risk and recurrent medulloblastoma and supratentorial primitive neuroectodermal tumors. J Neuro Oncol 71:33–38

Raffel C, Thomas GA, Tishler DM, Lassoff S, Allen JC (1993) Absence of p53 mutations in childhood central nervous system primitive neuroectodermal tumors. Neurosurgery 33:301–305

Reddy AT, Janss AJ, Phillips PC, Weiss HL, Packer RJ (2000) Outcome for children with supratentorial primitive neuroectodermal tumors treated with surgery, radiation, and chemotherapy. Cancer 88(9):2189–2193

Rickert CH, Paulus W (2001) Epidemiology of central nervous system tumors in childhood and adolescence based on the new WHO classification. Child Nerv Syst 17:503–511

Rorke LB (1983) The cerebellar medulloblastoma and its relationship to primitive neuroectodermal tumors. J Neuropathol Exp Neurol 42:1–15

Smits A, van Grieken D, Hartman M, Lendahl U, Funa K, Nistér N (1996) Coexpression of platelet-derived growth factor a and b receptors on medulloblastomas and other primitive neuroectodermal tumors is consistent with an immature stem cell and neuronal derivation. Lab Invest 74:188–198

Sung KW, Yoo KH, Cho EJ, Koo HH, Lim DH, Shin HJ, Ahn SD, Ra YS, Choi ES, Ghim TT (2007) High-dose chemotherapy and autologous stem cell rescue in children with newly diagnosed high-risk or relapsed medulloblastoma or supratentorial primitive neuroectodermal tumor. Pediatr Blood Cancer 48:408–415

Taylor MD, Mainprize TG, Rutka JT (2000) Molecular insight into medulloblastoma and central nervous system primitive neuroectodermal tumor biology from hereditary syndromes: a review. Neurosurgery 47(4):888–901

Terheggen F, Troost D, Majoie CB, Leenstra S, Richel DJ (2007) Local recurrence and distant metastasis of supratentorial primitive neuro-ectodermal tumor in an adult patient successfully treated with intensive induction chemotherapy and maintenance temozolomide. J Neurooncol 82:113–116

Timmermann B, Kortmann R, Kuhl J, Meisner C, Dieckmann K, Pietsch T, Bamberg M (2002) Role of radiotherapy in the treatment of supratentorial primitive neuroectodermal tumors in childhood: results of the prospective German brain tumor trials HIT 88/89 and 91. J Clin Oncol 20(3):842–849

Tomita T, McLone DG, Yasue M (1988) Cerebral primitive neuroepithelial tumors in childhood. J Neurooncol 6:233–243

Vogel H, Fuller GN (2003) Primitive neuroectodermal tumors, embryonal tumors, and other small cell and poorly differentiated malignant neoplasms of the central and peripheral nervous systems. Ann Diagn Pathol 7:387–398

Yang HJ, Nam DH, Wang K, Kim YM, Chi JG, Cho B (1999) Supratentorial primitive neuroectodermal tumor in children: clinical features, treatment outcome and prognostic factors. Child Nerv Syst 15:377–383

Retinoblastoma and Reproductive Decision-Making

9

Charlotte J. Dommering and Lidewij Henneman

Contents

C.J. Dommering (✉) • L. Henneman
Department of Clinical Genetics,
VU University Medical Center, P.O. Box 7057,
1007 MB Amsterdam, The Netherlands
e-mail: CJ.Dommering@vumc.nl

Abstract

Studies on reproductive decision-making of couples at increased risk of a child with retinoblastoma (Rb) are scarce. The few studies that have been done, however, clearly show that Rb influences childbearing. Many couples decide against having children, or, if they already have a child, decide against having more children. Some choose prenatal diagnosis. Several factors were shown to be of influence on these decisions: the perceived burden of the disease (perceived consequences for the child and parents, type of treatment and sequellae of treatment) and the perceived impact of ophthalmological screening for children at risk. In a recent study, the most important factor of influence appeared couples' perceived risk more than objective risk. It is important to explore these factors of influence with couples at increased risk when discussing family planning. Since couples' decisions and considerations on childbearing may change over time, continued access to genetic counseling should be offered, even when the increased risk for offspring is relatively small.

Introduction

Retinoblastoma (Rb) is a childhood cancer of the eye, estimated to affect between 1:15,000 and 1:20,000 live births (Moll et al. 1997). The initiating event in Rb is inactivation of both alleles of the retinoblastoma (*RB1*) gene (Knudson 1971).

M. Hayat (ed.), *Tumors of the Central Nervous System, Volume 13*,
DOI 10.1007/978-94-007-7602-9_9, © Springer Science+Business Media Dordrecht 2014

About 40% of Rb cases are hereditary in an autosomal dominant way based on a positive family history, bilateral disease and/or a germline mutation in the retinoblastoma tumor suppressor gene *RB1*. Of all non-familial unilateral cases, around 15% are caused by a *de novo* germline *RB1* mutation (Rushlow et al. 2009). The other 85% of unilateral non-familial cases is assumed to be caused by two somatic *RB1* mutations. In 1987 Lee et al. cloned the *RB1* gene (Lee et al. 1987). Current molecular screening techniques detect around 90% of *RB1* mutations in familial or bilateral cases (Rushlow et al. 2009).

Rb can be treated by enucleation of the affected eye, by local therapy (laser photocoagulation or cryotherapy), by external beam radiation therapy or radioactive plaque brachytherapy or chemotherapy (Shields and Shields 2010). Chemotherapy can be administered systemically or can be delivered locally as superselective ophthalmic artery chemotherapy, via microcatheterization. Choice of therapy may depend on laterality, seize and location of the tumor, as well as the age of the child, among other factors.

Management of children with Rb and their families is complex and relies on close cooperation between members of a multidisciplinary team including a specialized ophthalmologist, pediatric oncologist, radiologist, interventional neuroradiologist, pathologist and clinical geneticist, as well as specialized nurses and sometimes a psychologist. Children at increased risk for Rb are offered ophthalmological screening from birth. Starting at the age of 3 months these exams are performed under anesthesia. When no retinoblastoma has developed, screening is discontinued at the age of 4 years (Moll et al. 2000).

Healthy parents of a child with Rb may have an increased risk of having another child with Rb, as do Rb patients. This increased risk varies between less than 1 and 50%, and depends on the results of DNA testing and family history. Four possible situations for a counselee with an increased risk are displayed in Table 9.1. When a child is affected by unilateral non-familial Rb, the eye is enucleated and *RB1* mutation testing on tumor material detects two *RB1* mutations and these mutations are not detected in DNA from leucocytes of the child, there is no increased recurrence risk for the parents (Lohmann and Gallie 2010). Because of the small chance (1–2%) of low-grade mosaicism in the child for one of the *RB1* mutations found in the tumor, there will be a small recurrence risk, however, for future children of the affected child (0.6%) (Lohmann and Gallie 2010).

Table 9.1 Possible family situations leading to an increased risk of developing retinoblastoma in offspring of counselees

Family situation	Rb risk for (future) offspring (%)[a]
Counselee is a carrier of a germline *RB1* mutation	50
Counselee has had unilateral non-familial Rb, without a detectable germline *RB1* mutation	0.5–1
Counselees are healthy parents of a child with a de novo *RB1* mutation	2–3
Counselees are healthy parents of a child with unilateral non-familial Rb without a detectable germline *RB1* mutation	<1

Rb retinoblastoma
[a]Based on GeneReviews (http://www.ncbi.nlm.nih.gov/books/NBK1452/) and a *RB1* mutation detection rate of 90%

Couples with an increased risk have several reproductive options. They may decide to remain childless, or, if they already have an affected child, have no more children.

They may adopt a child or choose gamete donation. If a germline *RB1* mutation is detected, couples can choose prenatal diagnosis (PND). If the child is a carrier of the mutation, they could opt for termination of the pregnancy, although some may find termination of a pregnancy of a potentially treatable disease debatable.

If one of the parents is a *RB1* mutation carrier, preimplantation genetic diagnosis (PGD) may also be an option to avoid the birth of an affected child. PGD involves DNA testing of the parental *RB1* mutation in two cells of a 3-day old embryo during in vitro fertilization. Only embryos without the mutation are transferred to the uterus. Alternatively, couples can decide to accept the risk and have biological children without using the above mentioned options. Genetic counseling may facilitate informed reproductive decision-making by couples at increased risk of a child with Rb. To accommodate the needs of these couples, it is important to obtain more insight into their reproductive behavior and into difficulties they may experience. In this chapter the current knowledge about reproductive decision-making of couples at increased risk of a child with retinoblastoma is discussed.

Reproductive Decision-Making Process

Little is known about the reproductive decision-making process of affected individuals with Rb or of healthy parents with a child with Rb. Several studies, however, have examined reproductive decision-making of couples at increased risk of having a child with another hereditary disease. One study in the Netherlands, for example, did a follow-up study of 164 couples who had visited a Clinical Genetics Department for a variety of disorders concerning childbearing (Frets et al. 1991) during the 1980s. Forty-three percent of the couples experienced the reproductive decision-making process as difficult, had doubts about the decision that was made, or were unable to make a decision. Factors associated with problems in the decision-making process in these couples were: no post-counseling relief, anticipation of a high risk of affected children prior to counseling, relatives' disapproval of the decision, a decision against having children and the presence of an affected child (Frets et al. 1991). A more recent study by (Kelly 2009) showed that parents of children with a genetic condition or impairment may not pursue further childbearing or decline the use of prenatal diagnostics, in order to avoid making a difficult decision (Kelly 2009), i.e., parents chose not to choose. As was already shown in 1979, in the study by Lippman-Hand: parents tend to make a "non-decision" if they are not able to process the "facts" to provide a sense of coping (Lippman-Hand and Fraser 1979a).

Dommering et al. published a qualitative interview study of 14 couples of childbearing age at increased risk for a child with Rb, exploring the impact of prospective risk on reproductive decisions and the needs of couples with regard to reproductive counseling (Dommering et al. 2010). This study was followed by a cross-sectional questionnaire survey among 81 individuals with an increased risk of a child with Rb (Dommering et al. 2012). The results show that 44% of the 81 respondents from all four risk groups (see Table 9.1) had had doubts about their reproductive decisions as a result of Rb, while 38% had changed their minds about their decision whether or not to have any (or more) children. Some respondents changed their opinion and decided not to have another child, after having a child with Rb. This can be illustrated with quotes from the qualitative study (Dommering et al. 2010), like this couple with a child with a de novo *RB1* mutation:

> That second pregnancy was emotionally heavy. It was a hell in a way. [...] I really wanted three children, but now I am hesitating. I mean, I have two healthy children [one with bilateral Rb], all is going well. I don't want to tempt fate. I feel Rb should not rule your life, but basically it does.

Others changed their mind the other way, deciding to have children (or more children), for different reasons: e.g., some time having passed since treatment of their child, after testing negative for a germline *RB1* mutation, after talking to other parents of affected children. This is illustrated by the following comment of another healthy couple with a child affected by hereditary Rb:

> We always wanted to have two children. But after we had X [child with Rb] we never wanted to go through this process ever again. [...] But after you get out of the hospital, you forget what you have been through. And your child is playing again and everything is all right and you start to think: Well, it doesn't have to happen again...

These results clearly show that for many couples at increased risk of having a child with Rb, having children is not self-evident.

Reproductive Behavior

Again, few studies have addressed the actual reproductive behavior of couples at increased risk of a child with Rb. Based on research primarily aimed at investigating other aspects of living with Rb (Byrne et al. 1995; van Dijk et al. 2010; Cohen et al. 2001), it appears that for many individuals the risk of Rb influences childbearing. One study from the USA described the long-term effects of a genetic testing service for families with a child with unilateral, non-familial Rb (Cohen et al. 2001). It was found that *RB1* testing of these children influenced parents' decision to have more children in 20% (10/49) of the families. In the early 1980s, Byrne et al. interviewed 56 Rb survivors, diagnosed before 1962, to assess long-term consequences of Rb (Byrne et al. 1995). Fifteen of these Rb patients had a 50% risk for a child with Rb. One of their main findings was that fewer married Rb survivors than controls reported a pregnancy. Moreover, of the female survivors who married and became pregnant, 42% had only one pregnancy compared to 16% of female controls. The authors concluded that these differences reflected lifestyle or personal choices more than impaired fertility (Byrne et al. 1995). However, these patients were interviewed in the pre-molecular era, making it difficult to use this data in current-day practice. Van Dijk et al. interviewed Dutch Rb survivors in 2005/2006 and found that 68% of the 38 adult hereditary Rb survivors and 32% of the 54 non-hereditary Rb survivors indicated that Rb had an impact on their desire to have children (van Dijk et al. 2010). Overall, 12% of all 92 adult Rb survivors in their study decided not to have children due to the increased risk of having a child with Rb.

In the questionnaire study by Dommering et al. (2012) 43 of the 81 respondents were considering children after becoming aware of their increased recurrence risk (Dommering et al. 2012). Twenty-five (58%) of these 43 respondents reported that they had changed their reproductive behavior because of Rb; 20 (80%) decided against having any (or more) children, including 11 respondents with a recurrence risk of less than 3%. A healthy couple with a child with unilateral Rb, without a detectable *RB1* mutation put it this way:

> We would have considered another child. But now we decided against it. [...] We didn't want to go through this another time. The agony! They say it's not hereditary, but still...

A couple with a 50% risk decided not to have any children at all:

> I think Rb is a very serious disease and looking at it that way, you know, that decides I don't want to try it. I know for sure I don't want a child with Rb.

In many centers, prenatal diagnosis (PND) is available for couples at risk. In PND, DNA-testing of the familial *RB1* mutation is performed during early pregnancy, with the option to terminate the pregnancy if the fetus is affected. Five couples out of the 43 respondents considering children from the study of Dommering et al. (2012) had used chorionic villi sampling to determine whether the fetus was affected, including three healthy couples with a child affected by hereditary Rb and a recurrence risk of 2–3% (Dommering et al. 2012). Like this healthy couple who became pregnant again shortly after their first child started intensive treatment for hereditary Rb and who chose to do PND.

> We were barely thinking about the pregnancy, only in a way like: checking, checking, and checking. We were 180% busy with X [child with Rb] and I kept thinking: what if X's treatment is not finished and we will have to go to the hospital with the new baby? We will get a problem who to give our attention to. So to get more certainty, we did the test.

Eighteen of the forty-three respondents who had considered having children after becoming aware of their increased risk, decided to have a (subsequent) child and did not choose any of the alternative reproductive options (Dommering et al. 2012). A woman affected by hereditary Rb, who did not feel like using PND or PGD said:

> It feels for me like going to the market and buying yourself a baby. [...] You don't take a child, you get one and you get it like it is. Actually, that is how it's meant to be. [...] You know life after Rb is a good and complete life, it's what you make of it.

Several studies have also documented preimplantation genetic diagnosis (PGD) for Rb (Xu et al. 2004; Dommering et al. 2004; Rechitsky et al. 2002; Girardet et al. 2003). Since 1999, ten couples have been referred for PGD in the Netherlands and four couples in effect decided to go through with the procedure (Dommering et al. 2012).

Several studies about reproductive decisions of individuals at increased risk for children with other genetic diseases have also found that many decide against having more children to avoid having an affected child. For example, 41% of 230 parents of children with metabolic diseases (Read 2002) and 32% of 181 parents of children with cystic fibrosis had no further children (Henneman et al. 2001). Two studies on reproductive decisions of hemophilia carriers showed that carriers not choosing PND often decided not to have any (or more) children (Tedgard et al. 1999; Kadir et al. 2000).

In studies assessing attitudes towards PND and/or PGD for other hereditary cancers (Levy and Richard 2000; Kastrinos et al. 2007; Lammens et al. 2009; Douma et al. 2009), between 33 and 71% considered the use of these methods to avoid the birth of an affected child. The recurrence risk in these studies was 50%. In the study by Dommering et al. (2012) some of the couples who decided to have PND for Rb were healthy couples with a child with a de novo *RB1* mutation, with a recurrence risk of 2–3% (Dommering et al. 2012). So in the case of Rb, even couples with a relatively small recurrence risk may decide to opt for PND.

Factors of Influence on Reproductive Decision-Making

Several factors, such as the perceived burden and familiarity with the disorder, can be of influence on reproductive decisions of couples at increased risk of a child with a genetic disorder (Frets et al. 1990; Henneman et al. 2002). With regard to Rb, type of treatment was associated with reproductive decisions, such as whether or not to have (further) children or use assisted reproductive technologies. Treatment with bilateral enucleation, chemotherapy and/or radiotherapy was more of influence on reproductive decisions than treatment with just unilateral enucleation and/or local therapy (Dommering et al. 2012):

> You can't explain to your child what's happening. He is suffering but doesn't know what's happening at all. [...] Due to the chemotherapy his skin was ruined and changing diapers became very painful. So with a second child you risk having to go through this again. No, we couldn't face it.

A study of the quality of life of adult Rb survivors concluded that bullying in childhood was one of the predictors of a worse quality of life (van Dijk et al. 2007). For some Rb survivors the sequellae of treatment of Rb (impaired vision, cosmetic deformities due to treatment) have an impact on their social life and sometimes these negative experiences also influence reproductive decision-making. Like this person, who had been treated for bilateral Rb by external beam radiotherapy, which had lead to orbital deformity and who chose PND:

> Yes, my childhood has been unpleasant. I was bullied for being different, until halfway through high school, day in and day out: they beat me up, they ruined my glasses, and they nicked my stuff. [...] I don't want this for my child, such a life.

Other aspects that were shown to be of influence on reproductive behavior were the perceived consequences of Rb (Dommering et al. 2012). This included factors such as the risk of passing Rb on to offspring, the risk of a child with impaired vision or blindness and fear or worries about developing second primary tumors later in life. A couple at 50% risk said:

> We decided that we would terminate the pregnancy if the child turned out to be a carrier. [...] They can't predict the severity of the disease. We were afraid it [the child] would be blind at birth or that it would not survive cancer. I know someone who has lost a child to retinoblastoma. I absolutely would not want to live through that.

Children at increased risk for Rb are advised to undergo frequent ophthalmological screening under anesthesia during the first 4 years of life. Although most parents feel that the screening program is needed because it leads to detection and treatment at an early stage, at the same time they describe it as a burden, which for some affects reproduction:

> We understood that our next child would have to be screened. X [affected child] hated going to the hospital, she had to be held down before going under anesthesia and when she woke up, she was always sick, so I really felt: No, never again.

Earlier it has been shown that the actual magnitude of the risk for offspring seems to be of relative importance on reproductive decision-making with regard to genetic disorders (Frets et al.

1990). People tend to take a high risk of having an affected child, even if they perceive the disorder as severe. Moreover, it has been shown that people's perception of the genetic risk, more than the magnitude of the actual risk, is of influence on reproductive decisions (Marteau et al. 1991; Sivell et al. 2008; Helbig et al. 2010). One of the aims of genetic counseling is to correct misperceptions of risk and increase understanding of the genetic risk information (Fraser 1974). However, risk perception is a complex process and is influenced by many factors. Personal beliefs and expectations about risk prior to counseling, psychological impact of the family history and emotional aspects also influence risk perception (Sivell et al. 2008; Shiloh and Saxe 1989). Some parents tend to relate to their risk as a two-way option: it will or will not happen, rather than a probabilistic figure, provided by genetic counseling (Lippman-Hand and Fraser 1979b; Beeson and Golbus 1985). In the study by Dommering et al. (2012) the only factor significantly associated with influence of Rb on reproductive behavior in multiple logistic regression was perceived risk (Dommering et al. 2012). For example this healthy couple with a child affected by hereditary Rb, who perceived their recurrence risk of 2–3% as high:

> Yes, the risk is very small, but for X [child with Rb] the risk had also been very small, so very small is of no value for me. [..] During my second pregnancy, that two percent really started to feel as something much bigger.

Others may interpret the same risk as low:

> We did all the genetic tests. [...] If we had had the mutation in our blood, it would have been a 50% chance, and then the decision would have been difficult for us. But it turned out to be negligible really. So we went ahead and got pregnant again.

Table 9.2 shows findings from the 81 respondents of the questionnaire study of Dommering et al. (2012) on risk perception. Displayed are the difference between objective risk of having a child with Rb, the respondents' recollection of their risk and whether respondents perceived their risk to be low, medium or high. Overall, most respondents recalled their objective numerical risk correctly, although three respondents in the 50% risk group remembered their risk to be less than 50% and five thought their risk was more than 50%. Of all respondents, 41% perceived their risk as medium or high, including 13 respondents (22%) from the three lowest risk categories (i.e., risk <3%).

Table 9.2 Recollection of risk and perceived risk of having a child with retinoblastoma as compared to objective risk (*n*=81)

	Objective risk			
	50%	2–3%	0.5–1%	<1%
	n=21	*n*=28	*n*=11	*n*=21
Risk recollection (*n*)				
<1%	0	15	2	12
2–5%	0	8	5	3
5–15%	1	4	2	0
15–50%	2	0	0	1
50%	12	0	0	1
>50%	5	0	0	0
No risk	0	0	0	2
Don't know	0	1	1	2
Missing	1		1	
Perceived risk (*scale 1–7*)				
Low (1,2,3)	1	21	6	18
Medium (4)	1	6	1	2
High (5,6,7)	18	1	2	1
Missing	1		2	

In summary, reproductive decision-making in Rb is impacted by the perceived burden of the disease (perceived consequences, type of treatment and sequellae of treatment) and the need for ophthalmological screening for children at risk. However, the recent study by Dommering et al. suggests that the most important factor of influence is perceived risk, more than objective risk (Dommering et al. 2012).

Recommendations

The following issues are important when discussing family planning with couples at increased risk of a child with Rb:

- Perceived Rb risk for offspring, as in the interpretation of the objective risk by the counselee.
- Perceived consequences of Rb for child and parents, including the risk of passing Rb on to offspring, the risk of a child with impaired

vision or blindness and fear or worries about developing second primary tumors later in life.

- The perceived impact of extensive treatment of Rb patients, including the negative experience with the sequellae of Rb treatment of affected individuals.
- The perceived burden of ophthalmological screening under anesthesia for children at risk.

Attitudes may change over time. Offering easy access to follow-up genetic counseling sessions, and/or extra support from a psychologist to discuss the reproductive options in the light of new views of counselees is therefore important.

References

Beeson D, Golbus MS (1985) Decision making: whether or not to have prenatal diagnosis and abortion for X-linked conditions. Am J Med Genet 20:107–114

Byrne J, Fears TR, Whitney C, Parry DM (1995) Survival after retinoblastoma: long-term consequences and family history of cancer. Med Pediatr Oncol 24: 160–165

Cohen JG, Dryja TP, Davis KB, Diller LR, Li FP (2001) RB1 genetic testing as a clinical service: a follow-up study. Med Pediatr Oncol 37:372–378

Dommering CJ, Moll AC, Imhof SM, de Die-Smulders CE, Coonen E (2004) Another liveborn after preimplantation genetic diagnosis for retinoblastoma. Am J Ophthalmol 138:1088–1089

Dommering CJ, van den Heuvel MR, Moll AC, Imhof SM, Meijers-Heijboer H, Henneman L (2010) Reproductive decision-making: a qualitative study among couples at increased risk of having a child with retinoblastoma. Clin Genet 78:334–341

Dommering C, Garvelink M, Moll A, Van DJ, Imhof S, Meijers-Heijboer H, Henneman L (2012) Reproductive behavior of individuals with increased risk of having a child with retinoblastoma. Clin Genet 81:216–223

Douma KF, Aaronson NK, Vasen HF, Verhoef S, Gundy CM, Bleiker EM (2009) Attitudes toward genetic testing in childhood and reproductive decision-making for familial adenomatous polyposis. Eur J Hum Genet 18:186–193

Fraser FC (1974) Genetic counseling. Am J Hum Genet 26:636–659

Frets PG, Duivenvoorden HJ, Verhage F, Niermeijer MF, van de Berge SM, Galjaard H (1990) Factors influencing the reproductive decision after genetic counseling. Am J Med Genet 35:496–502

Frets PG, Duivenvoorden HJ, Verhage F, Peters-Romeyn BM, Niermeijer MF (1991) Analysis of problems in making the reproductive decision after genetic counselling. J Med Genet 28:194–200

Girardet A, Hamamah S, Anahory T, Dechaud H, Sarda P, Hedon B, Demaille J, Claustres M (2003) First preimplantation genetic diagnosis of hereditary retinoblastoma using informative microsatellite markers5. Mol Hum Reprod 9:111–116

Helbig KL, Bernhardt BA, Conway LJ, Valverde KD, Helbig I, Sperling MR (2010) Genetic risk perception and reproductive decision making among people with epilepsy. Epilepsia 51:1874–1877

Henneman L, Bramsen I, Van Os TA, Reuling IE, Heyerman HG, van Der LJ, van der Ploeg HM, ten Kate LP (2001) Attitudes towards reproductive issues and carrier testing among adult patients and parents of children with cystic fibrosis (CF). Prenat Diagn 21:1–9

Henneman L, Kooij L, Bouman K, ten Kate LP (2002) Personal experiences of cystic fibrosis (CF) carrier couples prospectively identified in CF families. Am J Med Genet 110:324–331

Kadir RA, Sabin CA, Goldman E, Pollard D, Economides DL, Lee CA (2000) Reproductive choices of women in families with haemophilia. Haemophilia 6:33–40

Kastrinos F, Stoffel EM, Balmana J, Syngal S (2007) Attitudes toward prenatal genetic testing in patients with familial adenomatous polyposis. Am J Gastroenterol 102:1284–1290

Kelly SE (2009) Choosing not to choose: reproductive responses of parents of children with genetic conditions or impairments. Soc Health Illn 31:81–97

Knudson AG Jr (1971) Mutation and cancer: statistical study of retinoblastoma. Proc Natl Acad Sci U S A 68:820–823

Lammens C, Bleiker E, Aaronson N, Vriends A, Ausems M, Jansweijer M, Wagner A, Sijmons R, van Den OA, van Der LR, Spruijt L, Gomez GE, Ruijs M, Verhoef S (2009) Attitude towards pre-implantation genetic diagnosis for hereditary cancer. Fam Cancer 8: 457–464

Lee WH, Bookstein R, Hong F, Young LJ, Shew JY, Lee EY (1987) Human retinoblastoma susceptibility gene: cloning, identification, and sequence. Science 235:1394–1399

Levy M, Richard S (2000) Attitudes of von Hippel-Lindau disease patients towards presymptomatic genetic diagnosis in children and prenatal diagnosis. J Med Genet 37:476–478

Lippman-Hand A, Fraser FC (1979a) Genetic counseling – the postcounseling period: II Making reproductive choices. Am J Med Genet 4:73–87

Lippman-Hand A, Fraser FC (1979b) Genetic counseling – the postcounseling period: I. Parents' perceptions of uncertainty. Am J Med Genet 4:51–71

Lohmann DR, Gallie BL (2010) Retinoblastoma – GeneReviews. http://www.ncbi.nlm.nih.gov/books/NBK1452/

Marteau TM, Kidd J, Cook R, Michie S, Johnston M, Slack J, Shaw RW (1991) Perceived risk not actual risk predicts uptake of amniocentesis. Br J Obstet Gynaecol 98:282–286

Moll AC, Kuik DJ, Bouter LM, Den OW, Bezemer PD, Koten JW, Imhof SM, Kuyt BP, Tan KE (1997) Incidence and survival of retinoblastoma in the Netherlands: a register based study 1862–1995. Br J Ophthalmol 81:559–562

Moll AC, Imhof SM, Meeteren AY, Boers M (2000) At what age could screening for familial retinoblastoma be stopped? A register based study 1945–98. Br J Ophthalmol 84:1170–1172

Read CY (2002) Reproductive decisions of parents of children with metabolic disorders. Clin Genet 61:268–276

Rechitsky S, Verlinsky O, Chistokhina A, Sharapova T, Ozen S, Masciangelo C, Kuliev A, Verlinsky Y (2002) Preimplantation genetic diagnosis for cancer predisposition. Reprod Biomed Online 5:148–155

Rushlow D, Piovesan B, Zhang K, Prigoda-Lee NL, Marchong MN, Clark RD, Gallie BL (2009) Detection of mosaic RB1 mutations in families with retinoblastoma. Hum Mutat 30:842–851

Shields CL, Shields JA (2010) Retinoblastoma management: advances in enucleation, intravenous chemoreduction, and intra-arterial chemotherapy. Curr Opin Ophthalmol 21:203–212

Shiloh S, Saxe L (1989) Perception of risk in genetic counseling. Psychol Health 3:45–61

Sivell S, Elwyn G, Gaff CL, Clarke AJ, Iredale R, Shaw C, Dundon J, Thornton H, Edwards A (2008) How risk is perceived, constructed and interpreted by clients in clinical genetics, and the effects on decision making: systematic review. J Genet Couns 17:30–63

Tedgard U, Ljung R, McNeil TF (1999) Reproductive choices of haemophilia carriers. Br J Haematol 106: 421–426

van Dijk J, Huisman J, Moll AC, Schouten-van Meeteren AY, Bezemer PD, Ringens PJ, Cohen-Kettenis PT, Imhof SM (2007) Health-related quality of life of child and adolescent retinoblastoma survivors in the Netherlands. Health Qual Life Outcome 5:65

van Dijk J, Oostrom KJ, Huisman J, Moll AC, Cohen-Kettenis PT, Ringens PJ, Imhof SM (2010) Restrictions in daily life after retinoblastoma from the perspective of the survivors. Pediatr Blood Cancer 54:110–115

Xu K, Rosenwaks Z, Beaverson K, Cholst I, Veeck L, Abramson DH (2004) Preimplantation genetic diagnosis for retinoblastoma: the first reported liveborn. Am J Ophthalmol 137:18–23

Trigeminal Nerualgia with Cerebellopontine Angle Tumors

10

Joseph Hobbs, Gregory D. Arnone, and Raymond F. Sekula Jr.

Contents

J. Hobbs • R.F. Sekula Jr.
Department of Neurosurgery, Center for Cranial Nerve Disorders, Allegheny General Hospital, 420 East North Avenue, Suite 302, Pittsburgh, PA 15212, USA
e-mail: rsekula@wpahs.org

G.D. Arnone (✉)
Department of Neurosurgery, University of Illinois at Chicago, 912 S. Wood Street (MC-799), Chicago, IL 60612, USA
e-mail: garnone@uic.edu

Introduction

Trigeminal neuralgia (TN) is a debilitating facial pain disorder with an incidence of 4.3 per 100,000 patients (Katusic et al. 1990). It is characterized by the sudden onset of lancinating, "electric shock-like" pain in the unilateral distribution of the trigeminal nerve, usually lasting for a matter of seconds, followed by complete resolution of pain. These paroxysms occur infrequently at first, instigated by activities such as chewing, talking, and the stimulation of so-called "trigger points" on the face by light touch, breezes, or cold temperatures. Recently, trigeminal neuralgia has been further classified based upon the presentation and nature of the pain: type I TN, for which pains are paroxysmal and stabbing, is the classic form of the disease previously referred to as "tic douloureux" or typical TN; type IIa TN is described by constant pain more than 50% of the time, in addition to paroxysms; type IIb is described by constant pains in the trigeminal distribution with no paroxysms (Eller et al. 2005).

The etiology of trigeminal neuralgia is incompletely understood, though it is generally accepted that vascular compression at the root entry zone (REZ) of the trigeminal nerve by an artery (usually the anterior inferior cerebellar or superior cerebellar artery, sometimes in addition to veins) is the inciting event. Chronic pulsatile arterial compression about the root entry zone of the trigeminal nerve causes demyelination or dysmyelination of the oligodendrocyte-derived myelin at the REZ, allowing for ephaptic

M. Hayat (ed.), *Tumors of the Central Nervous System, Volume 13*,
DOI 10.1007/978-94-007-7602-9_10, © Springer Science+Business Media Dordrecht 2014

transmission to occur. Resultantly, the patient experiences pain in response to trigger stimuli. Nuclear hyperactivity and trigeminal ganglion "ignition" (self-sustaining neural discharge) have also been suggested in the pathophysiology of TN (Moller 1991; Rappaport and Devor 1994). Prior to the work by Dandy, Gardner, and Jannetta in the recognition of vascular compression in the etiology of TN, many surgeons recognized a small segment of TN being caused by posterior fossa tumors. The presence of posterior fossa tumors in cases of trigeminal neuralgia has critical implications for the diagnosis and treatment of facial pain.

Incidence of Tumors in Patients with Facial Pain

In his personal cases dating back as early as 1905, Oppenheim recognized instances of trigeminal neuralgia associated with tumors (Oppenheim 1905). In 1910, Weisenburg published his report of a case of a cranial neuralgia (in his case, glosspharyngeal neuralgia) being caused by a tumor of the cerebellopontine angle (Weisenburg 1910). Interestingly, this tumor did not cause additional "general pressure" symptoms that would implicate a mass lesion of the cerebellopontine angle. In 1934, Dandy published the first series of operations for TN, noting that in 10.7% of cases, the neuralgia was associated with a mass lesion, including cerebellopontine angle tumors (5.6%) and vascular lesions (5.1%) (Dandy 1934). In a series of 1,211 patients undergoing their first operation for trigeminal neuralgia, Barker et al. (1996a, b) noted a 2.1% incidence of posterior fossa tumors. In the largest series to date of 5,058 patients evaluated for facial pain, Cheng et al. (1993) found facial pain due to tumors in 5.85% of cases, 19.6% of which cases described pain characteristic of type I TN. Overall, pooled data from numerous series suggest that tumors accompany the presentation of trigeminal neuralgia in 0.8–9.5% of cases of TN (Barker et al. 1996a, b; Bullitt et al. 1986; Cheng et al. 1993; Dandy 1934).

Clinical Presentation and Examination

Some practitioners argued that upon exam, evidence of trigeminal nerve involvement, usually in the form of sensory loss in one or more distributions, would be sufficient in excluding mass lesions in patients with TN, particularly with lesions in the area of the gasserian ganglion (Love and Woltman 1942). Diminished corneal reflexes or weakness of the masticators has also been suggested to rule out secondary causes of trigeminal neuralgia. Additionally, the nature of facial pain may differ according to the location of the pathology along the trigeminal nerve. Hamby and Love both noted that mass lesions on the gasserian ganglion tend to produce a constant, burning background pain not influenced by eating or speaking (type IIb TN), possibly associated with added paroxysms (type IIa TN) (Hamby 1943; Love and Woltman 1942). In his series of operations, Dandy concluded that in order for tumors to cause pains characteristic of type I TN, they must invade the sensory root at the junction between the pons and Meckel's cave, the so-called "root entry zone" of the trigeminal nerve complex (Dandy 1934). Despite the convenience of such theories, many instances of tumors involving the trigeminal nerve fail to follow a pattern in terms of natural history and presentation. Various trigeminal nerve symptoms and signs, in addition to variable involvement of the vestibular, cochlear, and lower cranial nerves, have been described in patients with tumors and concomitant facial pains. In rare cases, the facial nerve may also be affected and cause tic convulsive, either ipsilateral or contralateral to the side of the tumor (Ogasawara et al. 1995). Hamby (1946) documented two cases of cerebellopontine angle meningiomas causing contralateral TN, after which he concluded that the cause of TN cannot be distinguished by the character of the pains. In fact, lesions compressing all parts of the extramedullary trigeminal pathway have been reported to cause TN pains, and large tumors can present as trigeminal pains with no evidence of focal neurological signs.

In an effort to identify secondary causes of TN in patients, trigeminal evoked potentials and brainstem auditory evoked potentials have been studied. Leandri et al. (1988) described the various changes in evoked potentials of the trigeminal nerve in the setting of tumors, finding that the potentials were altered in all patients with tumors of the skull base (parasellar and cp angle), even in lesions with no clinical evidence of involvement of the trigeminal nerve. Out of 38 patients studied with TN, only 9 (23.7%) manifested altered evoked potentials. In a separate review, Metzer (1991) noted that 4% of patients with TN caused by vascular compression have an abnormal R1 response of the blink reflex, while 60% of patients with demonstrable causes of facial pain have significant prolongation of R1 on blink reflex testing. Though these modalities were originally proposed as providing useful information in discriminating secondary causes of TN, the advent of magnetic resonance imaging has emerged as the screening test of choice due to its convenience and cost-effectiveness. Subtle hints to secondary etiology, such as presentation at a young age, high association with V1 pain, and diminished corneal reflexes may be helpful findings in cases of TN caused by tumors; however, due to the variability of symptoms caused by tumors, high resolution MRI is now indicated to rule out mass lesions in all patients presenting with trigeminal neuralgia (Nomura et al. 1994).

Tumors Associated with Facial Pain

Numerous types of neoplasms have been reported in association with facial pain or trigeminal neuralgia, including meningioma, vestibular and trigeminal schwannoma, cholesteatoma, lipoma, lymphoma, metastases, angioma, epidermal cyst, basal cell carcinoma, choroidal epithelial cyst, high cervical neurinoma, embryonal rhabdomyosarcoma, mandibular lipoma, glomus tumor, pontine cyst, tuberculoma, traumatic neuroma, pituitary adenoma, pontine glioma, glioblastoma, angiolipoma, ependymoma, dermoid cyst, medulloblastoma, and osteoma (Cheng et al. 1993). Additionally, tumors arising anywhere along the course of the trigeminal nerve and its branches may cause facial pain. Dandy (1934) noted in his original series that the most common tumor type associated with TN was vestibular schwannoma, though varying reports since have cited either vestibular schwannoma or meningioma as the most prevalent tumor in series of TN associated with mass lesions. In the posterior fossa, vestibular schwannomas, meningiomas and cholesteatomas (epidermoids) are the most common lesions associated with TN (Barker et al. 1996a, b; Dandy 1934; Revilla 1947a, b). Although epidermoids account for less cases of TN than other posterior fossa tumors, when present, they more often cause facial pain than meningiomas or neurinomas. As much as 19.4–40% of cp angle epidermoids may be associated with TN (Meng et al. 2005; Rappaport 1985), while 4/53 (7.5%) patients in a separate series of acoustic schwannomas complained of face pain (Parker 1937). In two large series of cp angle meningiomas, 15% of cases presented with TN (Granick et al. 1985). Overall, the presence of facial pain with posterior fossa tumors may be even less, with 12.3% of such tumors manifesting TN symptoms (Puca et al. 1995).

A rare but interesting cause of TN, trigeminal schwannomas are of particular interest in examining the causative effect of tumors in TN. Tumors of the trigeminal nerve constitute 0.08–0.28% of intracranial tumors, and only 0.8–8% of neurinomas (Tancioni et al. 1995). In a study of 160 cases of cp angle neurinomas (schwannomas), Revilla et al. (1947a, b) noted 3 (1.8%) schwannomas arising from the trigeminal nerve. A review of 22 cases of trigeminal root schwannoma revealed that only 50% presented with TN, either type IIa (5 cases) or type IIb (6 cases).

Treatment Options

Microvascular decompression (MVD) surgery is the treatment of choice for "idiopathic" type I TN (Barker et al. 1996a, b). Antiepileptic medications, however, are frequently initially prescribed. Carbamazepine is the most effective AED in treating TN, particularly for patients with Type I TN; however, medications often lose efficacy over a period of time and produce unwanted side-effects (most commonly impaired cognition) for patients. Bullitt et al. (1986) studied a series of 200 patients evaluated for facial pain, 16 of which were found to have tumors involving the trigeminal nerve. Of these cases of TN caused by tumors, Carbamazepine and Phenytoin were effective in relieving pain in patients with symptoms characteristic of type I TN. The results of Bullitt's study suggest that the efficacy of AEDs in cases of TN caused by tumors is similar to that in cases of "idiopathic" TN, again, provided that the nature of the pain is that of type I TN and not an atypical pain or type II TN. Unfortunately, TN pains secondary to tumors more commonly and more rapidly become refractory to medications, within 1 year of initial treatment.

Procedures which are damaging to the trigeminal nerve are also utilized to treat TN, including sectioning of the trigeminal root via a temporal or subcerebellar approach, glycerol rhizolysis, alcohol rhizolysis, radiofrequency ablation, balloon compression, thermocoagulation, and stereotactic radiosurgery. Such so-called "destructive" procedures have also been partially effective in cases of secondary trigeminal neuralgia. In a small series of tumors presenting as TN, Cheng et al. (1993) noted that alcohol blocks were the most effective of the destructive procedures in bringing total pain relief, with a 33% failure rate (only 3 total pts). For recurrent TN due to tumors, eventual failure rates were 40% for nerve sectioning, 55% for alcohol blocks, and 66% for glycerol and radiofrequency ablation.

Tumor Resection

Tumor resection is successful in relieving TN pains in 82.6–92% of overall cases, and outcomes are more favorable for patients with type I TN pain (Barker et al. 1996a, b; Cheng et al. 1993). A common cause of long-term treatment failure is tumor recurrence, especially common in cases of malignant neoplasms. Indeed, direct compression on the trigeminal nerve by a mass lesion seems a viable cause of demyelination and resulting ephaptic transmission. Additionally, some physicians argue that distortion of the brainstem may stretch the trigeminal nerve or compress the contralateral trigeminal nerve against the petrous ridge severely enough to cause symptoms. An optimal treatment plan, however, is dependent on the type of tumor and the ramifications of total resection, as tumors that infiltrate the cranial nerves or invade critical structures may be associated with an unacceptable operative morbidity. For example, cp angle lipomas are often adherent to cranial nerves, and, even in cases where compression of the nerve by the tumor was not apparent on gross examination, histological sections of the trigeminal root revealed subpial infiltration of the trigeminal rootlets by the lipoma (Kato et al. 1995). Resultingly, outcomes of lipoma resection are unpredictable, and total resection of these lesions is associated with considerable damage to the cranial nerves and permanent morbidity for patients. In such cases, options of subtotal resection or destructive procedures should be considered.

Some practitioners argue that the mere presence of a cp angle tumor does not mandate its removal, even when indirectly symptomatic. Removal of certain tumors could leave long lasting sensory deficits with no added benefit for pain relief. Even after complete tumor resection, treatment failure may occur. Barker et al. (1996a, b) studied 26 cases of TN associated with tumors, inspecting the root entry zone of the trigeminal nerve in 21/26 cases. Vascular compression was noted in all 21 instances and the researchers hypothesized that tumors cause displacement of vessels or distortion of anatomy that brings the root of the trigeminal nerve into contact with blood vessels. Hence, the etiology of TN in cases of posterior fossa tumors may, in fact, be arterial compression. In support of this notion is the fact that the majority of cp angle tumors and tumors in contact with the trigeminal nerve do not cause facial pain; however, Barker's hypothesis fails to explain the resolution of symptoms after tumor removal in numerous cases where compressive blood vessels were not identified during resection (Celik et al. 2000). Due to the possibility of vascular compression in the setting of tumors, surgeons should generally be prepared to perform MVD regardless of the amount of planned tumor removal. An important consideration, however, is the increased operative mortality during MVD in the setting of posterior fossa tumors. Due to this risk, complete removal of a posterior fossa tumor is preferable prior to performing vascular decompression.

Radiosurgery

As more data regarding the long-term efficacy of stereotactic radiosurgery for tumors emerges, investigators are more frequently using radiosurgery for treatment of secondary trigeminal neuralgia. While MVD is always the treatment of choice in "idiopathic" TN (provided that the patient can tolerate the procedure), standards are less well-defined for TN caused by tumors. In situations where radiosurgery may achieve favorable tumor control/shrinkage, those procedures may also be viable options in treating the facial pain associated with the tumor (whether this is due to a decrease in the bulk of the tumor itself pressing on the nerve or a change in the vascular relationships of the PF due to tumor shrinkage, or other unknown mechanisms is still unknown). After a single gamma knife (GK) radiosurgery treatment directed at a posterior fossa tumor associated with facial pain (the majority of which were either meningiomas or schwannomas), Huang et al. (2008) found an excellent outcome (complete pain relief without medication) in 57% of patients. Additionally, 50% of initial failures in their series achieved complete pain relief following a second GK procedure, this time directed at the TN root. Overall, they reported pain relief without the aid of AEDs in 23.7–78.4% of cases of posterior fossa tumors associated with facial pain treated with GK. Despite the added efficacy of a second GK procedure, 50% of such patients developed hypesthesia after the repeat procedure, compared to a 0–3.8% risk of facial hypesthesia following a first time radiosurgical procedure (Huang et al. 2008; Regis et al. 2001).

With radiosurgery, pain improvement often occurs over a prolonged time interval, and pain recurrence frequently occurs. Chang et al. (1999) noted an 85.7% initial pain response to GK radiation directed at tumors over a mean time interval of 6.3 months in a series of 27 patients. Half of those patients experienced a recurrence of pain at a mean of 10.3 months post-radiation. For patients who were pain free after radiosurgery, the latency from the procedure to complete relief of pain was 11.3 months. In total, 42.9% of patients experienced more than 50% relief of their pre-procedure pain at 32.1 months. Interestingly, they found no relationship between tumor volume change and pain relief, or correlation of tumor type, maximum dose administered, preoperative sensory change, or extent of root involvement to outcomes.

In a more detailed study of radiosurgery treatment for secondary TN, Regis et al. (2001) proposed separating patients into groups depending upon the amenability of the tumor to radiation and the visualization of the trigeminal REZ on imaging. For tumors that are amenable to radiation, the dosing was directed at the tumor with the intent to shrink the tumor and debulk, thereby secondarily decompressing the trigeminal nerve. For tumors that are not amenable to radiosurgery, if the REZ was not visualized, dosing was directed at the supposed location of the REZ obscured by the tumor, with doses similar to those used for tumor control. If the REZ was visualized on imaging and the tumor not amenable to radiation, treatment was directed toward the retrogasserian root in doses intermediate to those used for tumor control and those used for idiopathic trigeminal neuralgia. In doing so, pain free outcomes (without medications) were achieved in 79.5, 66.7, and 75% of patients at mean follow-ups of 4.5, 5.5, and 3.8 years, respectively. Similarly, in a series of 53 patients with type I or type IIa TN secondary to tumors treated with GK directed at the tumor, in addition to augmented treatment to the trigeminal REZ (when it was visualized), 46.9% of patients maintained greater than 50% reduction in pain over 26 months (Chang et al. 2000).

Considering the recurrence rates of pain in patients treated with radiosurgery, results seem to be less robust than those of tumor resection. Although Chang et al. (1999) reported no difference in outcomes for various tumor types, the durability of radiosurgery for tumor control is not well established for all tumors. For vestibular schwannomas, however, where long-term efficacy is established for small tumors (<3 cm) (Pollock et al. 2006), GK may be considered a primary treatment option, given the low risk of morbidity from a first time procedure compared to the risks of surgical resection.

In the case of trigeminal schwannomas, two cases treated initially with GK failed to give long-term pain relief (4 months at the most) (Miller et al. 2008). Some practitioners would argue that a second GK could be performed in lieu of tumor resection; however, the higher risk of cranial neuropathy must be weighed against the added benefit of additional radiation.

Conclusion

TN pain may be caused by tumors along any location of the trigeminal nerve. The presentation of tumors causing TN is variable, and no clinical features exist which are sensitive or specific enough to differentiate tumors from idiopathic TN; hence, an MRI of the brain is indicated for every patient with TN. Medications and destructive procedures (not including radiosurgery) are not durable treatments and complete tumor resection is preferable when possible. Preparation for vascular decompression should be included in any surgical approach and subtotal resection is a viable option for certain tumors, though MVD risk is probably higher in the setting of PF tumors. Finally, the classifications of outcomes from various studies are not consistent enough to properly compare microsurgery to GK. Longer follow-up intervals are also needed to make further conclusions. Posterior fossa exploration may be necessary in cases refractory to radiosurgery. Both modalities probably have valid uses and consideration of tumor type and location, patient comorbidities, and procedural risks should be discussed with patients.

References

Barker FG 2nd, Jannetta PJ, Babu RP, Pomonis S, Bissonette DJ, Jho HD (1996a) Long-term outcome after operation for trigeminal neuralgia in patients with posterior fossa tumors. J Neurosurg 84:818–825

Barker FG 2nd, Jannetta PJ, Bissonette DJ, Larkins MV, Jho HD (1996b) The long-term outcome of microvascular decompression for trigeminal neuralgia. N Engl J Med 334:1077–1083

Bullitt E, Tew JM, Boyd J (1986) Intracranial tumors in patients with facial pain. J Neurosurg 64:865–871

Celik SE, Kocaeli H, Cordan T, Bekar A (2000) Trigeminal neuralgia due to cerebellopontine angle lipoma. Case illustration. J Neurosurg 92:889

Chang JW, Kim SH, Huh R, Park YG, Chung SS (1999) The effects of stereotactic radiosurgery on secondary facial pain. Stereotact Funct Neurosurg 72(Suppl 1):29–37

Chang JW, Chang JH, Park YG, Chung SS (2000) Gamma knife radiosurgery for idiopathic and secondary trigeminal neuralgia. J Neurosurg 93(Suppl 3):147–151

Cheng TM, Cascino TL, Onofrio BM (1993) Comprehensive study of diagnosis and treatment of trigeminal neuralgia secondary to tumors. Neurology 43:2298–2302

Dandy WE (1934) Concerning the cause of trigeminal neuralgia. Am J Surg 24:9

Eller JL, Raslan AM, Burchiel KJ (2005) Trigeminal neuralgia: definition and classification. Neurosurg Focus 18:E3

Granick MS, Martuza RL, Parker SW, Ojemann RG, Montgomery WW (1985) Cerebellopontine angle meningiomas: clinical manifestations and diagnosis. Ann Otol Rhinol Laryngol 94:34–38

Hamby WB (1943) Trigeminal neuralgia due to radicular lesions. Arch Surg 46:9

Hamby WB (1946) Trigeminal neuralgia due to contralateral tumors of the posterior cranial fossa; report of 2 cases. J Neurosurg 4:4

Huang CF, Tu HT, Liu WS, Lin LY (2008) Gamma Knife surgery for trigeminal pain caused by benign brain tumors. J Neurosurg 109(Suppl):154–159

Kato T, Sawamura Y, Abe H (1995) Trigeminal neuralgia caused by a cerebellopontine-angle lipoma: case report. Surg Neurol 44:33–35

Katusic S, Beard CM, Bergstralh E, Kurland LT (1990) Incidence and clinical features of trigeminal neuralgia, Rochester, Minnesota, 1945–1984. Ann Neurol 27:89–95

Leandri M, Parodi CI, Favale E (1988) Early trigeminal evoked potentials in tumours of the base of the skull and trigeminal neuralgia. Electroencephalogr Clin Neurophysiol 71:114–124

Love JG, Woltman HW (1942) Trigeminal neuralgia and tumors of the gasserian ganglion. Proc Staff Meet Mayo Clin 17:7

Meng L, Yuguang L, Feng L, Wandong S, Shugan Z, Chengyuan W (2005) Cerebellopontine angle epidermoids presenting with trigeminal neuralgia. J Clin Neurosci 12:784–786

Metzer WS (1991) Trigeminal neuralgia secondary to tumor with normal exam, responsive to carbamazepine. Headache 31:164–166

Miller JP, Acar F, Burchiel KJ (2008) Trigeminal neuralgia and vascular compression in patients with trigeminal schwannomas: case report. Neurosurgery 62:E974–E975; discussion E975

Moller AR (1991) The cranial nerve vascular compression syndrome: II. A review of pathophysiology. Acta Neurochir (Wien) 113:24–30

Nomura T, Ikezaki K, Matsushima T, Fukui M (1994) Trigeminal neuralgia: differentiation between intracranial mass lesions and ordinary vascular compression as causative lesions. Neurosurg Rev 17:51–57

Ogasawara H, Oki S, Kohno H, Hibino S, Ito Y (1995) Tentorial meningioma and painful tic convulsif. Case report. J Neurosurg 82:895–897

Oppenheim H (1905) Lehrbuch der Nervenkrankheiten. S Karger, Berlin

Parker HL (1937) Paroxysmal trigeminal pain with tumours of the nervus acusticus. J Neurol Psychopathol s1-17:6

Pollock BE, Driscoll CL, Foote RL, Link MJ, Gorman DA, Bauch CD, Mandrekar JN, Krecke KN, Johnson CH (2006) Patient outcomes after vestibular schwannoma management: a prospective comparison of microsurgical resection and stereotactic radiosurgery. Neurosurgery 59:77–85; discussion 77–85

Puca A, Meglio M, Vari R, Tamburrini G, Tancredi A (1995) Evaluation of fifth nerve dysfunction in 136 patients with middle and posterior cranial fossae tumors. Eur Neurol 35:33–37

Rappaport ZH (1985) Epidermoid tumour of the cerebellopontine angle as a cause of trigeminal neuralgia. Neurochirurgia (Stuttg) 28:211–212

Rappaport ZH, Devor M (1994) Trigeminal neuralgia: the role of self-sustaining discharge in the trigeminal ganglion. Pain 56:127–138

Regis J, Metellus P, Dufour H, Roche PH, Muracciole X, Pellet W, Grisoli F, Peragut JC (2001) Long-term outcome after gamma knife surgery for secondary trigeminal neuralgia. J Neurosurg 95:199–205

Revilla AG (1947a) Neurinomas of the cerebellopontine recess: a clinical study of one hundred and sixty cases including operative mortality and end results. Bull Johns Hopkins Hosp 80:43

Revilla AG (1947b) Tic douloureux and its relationship to tumors of the posterior fossa; analysis of twenty-four cases. J Neurosurg 4:7

Tancioni F, Gaetani P, Villani L, Zappoli F, Rodriguez Y, Baena R (1995) Neurinoma of the trigeminal root and atypical trigeminal neuralgia: case report and review of the literature. Surg Neurol 44:36–42

Weisenburg TH (1910) Cerebellopontine tumor diagnosed for six years as tic douloureux: the symptoms of irritation of the 9th and 12th cranial nerves. JAMA 54:5

Part II

Diagnosis

The Concept of a Preniche for Localization of Future Metastases

11

Vladimir M. Perelmuter and Vasiliy N. Manskikh

Contents

V.M. Perelmuter
Pathological Anatomy and Cytology Department, Cancer Research Institute, Siberian Branch of the Russian Academy of Medical Sciences, 634028 Tomsk, Russia

V.N. Manskikh (✉)
Faculty of Bioengineering and Bioinformatics, Lomonosov Moscow State University, 119991 Moscow, Russia
e-mail: manskikh@mail.ru

Abstract

The hypothesis of metastatic niches (advanced kind of "seed and soil" hypothesis) is very promising concept. It has been proposed to supplement the metastatic niche concept with a stage of "preniche" that determines the site of development of a premetastatic niche and of a subsequent metastasis. The "preniche" includes all cellular and molecular events in the site of a prospective metastasis preceding the entrance of myeloid progenitor cells. The preniche integrates an activation of vascular endothelium of the microcirculatory vessels of target organs in the site of a future metastasis under conditions of chronic persistent productive inflammation that can be induced by cytokines from the primary tumor and independently of it. The endothelium activation is responsible for adhesion and clustering of the recruited myeloid progenitor cells and also for the retention of cells of malignant tumors. The preniche easily arises in organs enriched with organ-specific macrophages (lungs, liver, brain, etc.) where the endothelium is predisposed for intensive recruiting of myeloid progenitor cells of macrophages, especially under conditions of inflammation. The feature of CNS is especial population of macrophage cells (microglia) which could be activated and to form metastatic niches without recruiting myeloid progenitor cells and preniche formation as well as bone morrow. Nevertheless, inflammatory prenicha seems to be the factor enhancing brain metastases by the recruiting

M. Hayat (ed.), *Tumors of the Central Nervous System, Volume 13*,
DOI 10.1007/978-94-007-7602-9_11, © Springer Science+Business Media Dordrecht 2014

of additional niche cells and cells of a tumor. Introduction of the preniche concept allows us to avoid difficulties associated with the development of the metastatic niche concept, especially concerning the problem of organ-preferential localization of metastases, and to make potential approaches for preventing metastasizing in some oncologic patients.

Introduction

Metastases are most common tumor lesions in brain. To summarize, about 25% of patients who die of cancer have CNS metastases detected at autopsy. Of these, about 15% are in the brain and for about 10% of them the brain is the only site of CNS metastases (Gavrilovic and Posner 2005). Numerous researches have elucidated many primary malignant tumors which most often metastasize to CNS. The list of these cancers is similar in all publications although relative incidences of brain metastases are different for every nosological form and vary from paper to paper. These tumors are (in decreasing order of incidence): lung carcinomas (especially small cell carcinomas and adenocarcinomas), breast cancer, carcinomas of kidneys, colorectal cancer, melanomas (in adult patients), leucosis, lymphomas and sarcomas (in children). It has been noted that border between white and gray matter is preferential place of brain metastases development. Metastatic lesions of spinal cord have less frequency than brain. Usually, it is epidural metastases which have incidence 5–10% of all cases of malignant tumors.

Obviously, CNS is one of the main sites for hematogenous metastases development as well as lung, liver and bone marrow. It is essential that brain metastatic lesions have high incidence despite CNS is not a venous blood collector as opposed to liver and lung. Causes of more frequent development of metastases in brain in comparison with some other organs such as kidney and skeletal muscles remain unknown. There are only few researches of the problem why tumors of some histological types and localization metastasize preferentially to CNS. It has been found that DCUN1D1, also known as squamous cell carcinoma-related oncogene, expression may play a role in development of brain metastasis in patients with Non-small cell lung carcinoma (NSCLC). DCUN1D1-positive tumor cells may have the ability to disrupt the blood-brain-barrier and colonize the brain. Those findings suggest that DCUN1D1 may play a role in the brain parenchyma invasion (Yoo et al. 2012).

The other known example of relation between molecular marker and brain metastatic lesions is breast cancer with amplification or over-expression of the human epidermal growth factor receptor 2 (HER2/neu). Patients with HER2/neu gene aberrations have more aggressive disease, frequent disease recurrence and a shorter survival. The addition of trastuzumab to chemotherapy in HER2/neu-positive advanced breast cancer patients has increased complete and partial response rates, and prolonged time to progression and overall survival. However, a relatively common failure site in patients administered trastuzumab is CNS. CNS metastases in these patients seem to develop despite responses achieved in extracerebral sites. It was postulated that HER2/neu over-expression and/or amplification might predispose to brain metastases (Duchnowska and Szczylik 2005). ER-negative and PgR-negative breast cancers are known to be more likely to develop CNS metastases. Also, the high risk of subsequent CNS recurrence was elevated in patients with breast cancer experiencing lung metastases. It has been suggested that some steps of metastases development are the same in lungs and brain (but not the same in bone marrow) (Pestalozzi et al. 2006).

Recently, a promising hypothesis of metastatic niches has been proposed. This concept is able to explain a lot of unclear questions of metastasis development. We supplemented the metastatic niche concept with a stage of "preniche" (Perelmuter and Manskikh 2012) that determines the site of development of a premetastatic niche and of a subsequent metastasis. Introduction of the preniche concept allows to avoid some difficulties associated with the development of the metastatic niche concept, especially concerning the problem of organ-preferential localization of metastases which includes brain metastatic lesions. In this chapter, we shall expound the concept of metastatic niches and preniches with emphasis on analysis of brain niche and preniche features which seem to determine preferential metastases development in this localization.

Concept of Metastatic Niches and Preniches

Physiological Reactions as a Basis of Preniche and Niche Concepts

It seems that the abundance of special information about mechanisms of tumor progression and the role of various molecules in metastasis obtained with different model systems of human blastomas rather prevents than promotes understanding of carcinogenesis and especially the control of tumor growth. There are different approaches to systematizing such information. We think that searching for prototypes of physiological reactions among pathological processes can be a rather promising approach. This approach is now not very popular in the case of metastasis of malignant tumors, which often appears to be a cascade of molecular processes as if created by Nature purposefully to generalize malignancies. But the "physiological approach" allows us not only to remove this apparent uniqueness of processes associated with tumor progression but also to subordinate different mechanisms involved in metastasis; this approach can also reveal yet unknown aspects of this process and pathways to control tumor dissemination.

The recently proposed concept of metastatic niches (Psaila and Lyden 2009) can be very helpful in searching for physiological prototypes of metastasis (Perelmuter and Manskikh 2012). This concept allowed us to quite otherwise elucidate many problems associated with metastasis and explain experimental data that could not be interpreted earlier. Although the concept of metastatic niches still has many blank spots, its development (especially on searching for a probable physiological prototype of the metastatic niche) can be very promising for comprehension of such problem of oncology as organ-preferential localization of metastases. But it must be stipulated beforehand that in the present work, first, it is admitted that metastasis of all, or at least the majority, of carcinomas and melanomas can be described by the concept of metastatic niches (although the available experimental and clinical data concern only a limited range of studied tumors) and, second, mesenchymal tumors will be deliberately not considered because by now about them there are no data necessary for the theory of niches. A clear subordination of the metastasizing stages is also emphasized – in the present paper we shall speak only about processes preceding formation of micrometastases leaving aside the problem of formation of a macroscopic node of a secondary tumor.

Selectiveness of Metastases Localization

Although there is no organ which would be absolutely protected against development of metastases of malignant tumors, such metastases are relatively often developed in a rather limited number of "typical" localizations: regional (with respectively to the primary tumor location); lymphatic nodes (lymphogenous metastases); lungs, liver, bone marrow, and brain (hematogenous metastases); peritoneum and pleura. Much less often hematogenous metastases are found in kidneys, gonads, spleen, subcutaneous fat tissue, and extremely seldom in the walls of the gastrointestinal tract, uterus, heart, and skeletal muscles. It should be noted that in the overwhelming majority of cases metastasis into atypical locations is associated with the generalization of the process affecting many organs and tissues. However, the spleen is an interesting exception. This organ is rarely damaged by macrometastases, except for the cases of generalized tumors (especially melanomas), but it has been shown earlier that micrometastases in the spleen occur rather often, whereas is muscles micrometastases are virtually not found.

There is no doubt that localization of metastases is partially associated with specific features of the lymph and venous blood outflow from the region of the primary tumor location. Just this determines the development of lymphogenous metastases into the regional lymph nodes and of hematogenous metastases of abdominal cavity organ tumors (stomach and pancreas carcinomas, colorectal cancer) into the liver. However, it is impossible to explain the localization of metastases only by specific features of the vascular system responsible for delivery of tumor cells to the site of metastasis. Thus, the bone marrow and liver are usual sites for hematogenous metastasis of kidney tumors, although these organs are not located on the pathway of venous outcome from the liver. Mechanisms responsible for differences in organ vulnerability are intensively discussed in the literature, but there is still no integral concept describing the causes of organ-preferential metastasis.

Concept of Metastatic Niches

The presence in blood of circulating tumor cells not always leads to development of macro- and micrometastases in target organs (Alix-Panabieres et al. 2008) and experimental works have shown the absence of a direct and constant correlation between the ability of endothelial cells to constitutively express selectins halting tumor cells and the adhesion of these cells on the endothelium, on one hand, and the sites of preferential localization of metastases, on the other hand (Wong et al. 1997). Even more interesting is the discovery of a phenomenon of "inefficient metastasis" when immigrated tumor elements are present in the target organs but fail to produce metastases (Bidard et al. 2008). These findings clearly suggest that formation and localization of micrometastases are more likely determined not by the presence of tumor elements in the blood flow but rather by some specific features of target organs (including those arising under the influence of the primary tumor) that are responsible for occupation of a suitable organ by blastoma cells and formation from them of a micrometastasis. The same findings also show that the halting of tumor cells in the target organs without some additional conditions is yet insufficient for development in them of metastases. This so-called "seed and soil" hypothesis was proposed by Stephen Paget in the beginning of the last century (Psaila and Lyden 2009), but only recent data filled it with concrete content. Researchers of D. Lyden's group have established that the development of micrometastases in target organs is preceded by the accumulation in them of cells immigrated from the bone marrow and creating a stromal microenvironment that is adequate for the tumor and determines the development of metastases (Kaplan et al. 2005; Peinado et al. 2008). To describe this process, the concept of "niche" was proposed, borrowed from hematology where it was used for description of microenvironment that regulates the proliferation, homing, and differentiation of stem cells (Wilson and Trumpp 2006).

Lyden's concept suggests the formation and step-by-step changes in the site of a future metastasis of the following forms of microenvironment: a premetastatic niche with bone marrow precursor cells without tumor elements; a micrometastatic niche characterized by the presence of a cluster of immature bone marrow and tumor cells; a macrometastatic niche with angiogenesis added to the preceding processes (Wels et al. 2008). According to scheme of Peinado et al. (2011), the formation of a micrometastatic niche is determined by several processes:

- first, the primary tumor cells capable of secreting the vascular endothelium growth factor A (VEGFA) mobilize the myeloid bone marrow derived cells (BMDC) (vascular endothelium growth factor receptor 1-positive, VEGFR1$^+$) into the peripheral blood flow; on the surface of these cells there is an integrin "very late antigen 4" (VLA4) interacting with fibronectin and thus promoting the homing of BMDC (Scott et al. 2003);
- second, fibronectin is accumulated in the sites of future metastases. This fibronectin is synthesized in situ by fibroblasts and seems to be also produced in the primary tumors, released in the blood flow, and accumulated in the target organ (Scott et al. 2003);
- third, VEGFR1$^+$ BMDCs due to the VLA4$^+$ migrate into the sites of fibronectin accumulation where they form a cluster of immature cells.

Note that the phenotype of these cells is significantly overlapped with the phenotype of macrophage series cells with different maturity. Thus, a premetastatic niche is formed. Formation of the premetastatic niche is also promoted by other factors that are secreted by the primary tumor and in situ (lysyl oxidase (LOX), macrophage inflammatory protein 2 (MIP2), matrix metalloproteinase 9 (MMP9), KIT-ligand, transforming growth factor β (TGFβ), tumor necrosis factor α, (TNFα));

– fourth, tumor cells and macrophages are recruited into the produced cell cluster (into the premetastatic niche) due to chemokines (serum amyloid component A3 (SAA3), chemokines S100A8, S100A9, and stromal cell-derived factor, SDF-1) synthesized by these cells. These cells are also supplemented with elements of the fibroblast series, and this results in formation of a full micrometastatic niche capable of providing for survival and proliferation of tumor cells.

Macrometastatic formation niche and clinical manifestation of metastases request activation of angiogenesis and migration of bone marrow endothelial precursors to metastatic niches (Kaplan et al. 2005). Besides angiogenesis, a lot of other factors have significance for macrometastatic niche development. Using mouse models of spontaneous breast cancer, Gao et al. (2012) has shown enhanced recruitment of bone marrow–derived $CD11b^{+}Gr1^{+}$ myeloid progenitor cells in the premetastatic lungs. Various protumorigenic activities were associated with $Gr1^{+}$ myeloid cells, including expression of proangiogenic factor BV8, metastasis-promoting LOX and MMP9, contribution to TGF-β–mediated metastasis, and immune tolerance. Gene expression profiling revealed that the myeloid cells from metastatic lungs express versican, a large extracellular matrix chondroitin sulfate proteoglycan. Notably, versican in metastatic lungs was mainly contributed by the $CD11b^{+}Ly6C^{high}$ monocytic fraction of the myeloid cells and not the tumor cells or other stromal cells. Versican attenuated the Smad-mediated epithelial-mesenchymal transition (EMT) signaling pathway as determined by the reduction of p-Smad2 levels and suppression of transcription factor Snail. The suppression of Smad2 pathway–induced mesenchymal-epithelial transition (MET) increased cell proliferation. In addition, versican did not impact apoptosis, suggesting that versican-mediated stimulation of MET and enhanced proliferation may be the main mechanisms of increased tumor outgrowth and formation of focal macrometastases. The contribution of fibroblasts and tumor cells to the metastatic lung was about tenfold lower compared with $CD11b^{+}Ly6C^{high}$ cells. Furthermore, no significant increase in fibroblast numbers was observed in the metastatic lungs compared with controls. Analysis of a cohort of patients with breast cancer from whom distal metastases were available showed clusters of myeloid cells expressing versican in the metastatic lungs. Immunohistochemistry and RT-PCR analysis showing enhanced versican expression in metastatic lungs, brain, and liver in patients with breast cancer compared with control normal lungs. Gao et al. (2012) has supposed that selectively targeting tumor-elicited myeloid cells or versican-mediated proliferation pathways, perhaps in combination with conventional chemotherapeutics, may represent a potential therapeutic strategy for combating metastatic disease.

Due to introduction of BMDCs as a new messenger, the concept of metastatic niches allows us to remove the contradictions enumerated in the beginning of this section. This concept also opens great prospects for control of metastasis. However, the metastatic niche theory is still not a completed concept. In particular, it remains unclear what specific events trigger the formation of a metastatic niche, i.e. lead to accumulation of fibronectin and changes in the endothelium favorable for BMDC homing. The formation of a metastatic niche is usually described as a process depending on the primary tumor (Peinado et al. 2011). However, it is well known from experimental oncology that intravenous injection of cells of some tumors can induce development of metastases in internal organs, and in some cases with a rather specific (mono-organ) location. This indicates that metastatic niches can be formed due to processes independent of the development of the primary tumor node, but due to some other physiological or pathophysiological reaction. Note that the theory of metastatic niches describes the formation of metastases in general not considering the question why metastases are formed mainly in particular "typical" sites and why the location of metastases and the type of metastatic disease vary in different patients. The known factors synthesized by the tumor and regulating the development of the niche (LOX, MIP-2, VEGFA, TGFβ, TNFα) are not organ-preferential. Nevertheless, Kaplan et al. (2005) and Hiratsuka et al. (2008) have attempted to study this problem experimentally. Mice with grafted Lewis lung carcinomas were injected with conditioned medium from melanoma B16 cells characterized by generalized non-selective metastasis, and as a result metastatic niches and micrometastases developed in various organs and tissues, including such atypical locations as the oviducts (Psaila and Lyden 2009). However, the molecular mechanisms underlying the phenomenon observed by D. Lyden's group are still unclear.

We think that the questions presented above might be answered taking into account the events preceding the formation of a premetastatic niche as it is and to suppose that these events should be based on a physiological or pathophysiological process more or less independent of the development of the primary tumor node. The overall in situ conditions that precede the formation of a premetastatic niche (recruiting BMDCs into the target organ) and determine the localization of a future metastasis is reasonable to term as a "preniche". It is important to note that we think the preniche plays a key role in BMDC homing and also is essential for emigration of tumor cells from the blood flow.

Preniche as the Phenomenon of Physiology and Pathology

Considering the most frequent locations of metastases, it becomes evident that they have one feature in common: they have a large pool of organ-specific macrophages (Kupffer cells in the liver, alveolar macrophages in lungs and microglia in the brain, peritoneal and bone marrow macrophages, lymph node macrophages). This feature is not characteristic of the heart, gastrointestinal tract organs, skeletal muscles, kidneys, or gonads – and in these organs solitary metastases occur relatively seldom. Obviously, the endothelium of these organs has to be adapted to an active immigration into them of macrophage precursors under both normal and inflammation conditions when the need for restitution of the physiological macrophage pool is especially strong.

It seems very likely that regeneration of specialized macrophages should occur not only due to mature monocytes of peripheral blood but mainly due to myeloid progenitor cells that are present in the blood circulation (Ronzoni et al. 2010) and capable of specific differentiation under the influence of specific microenvironmental factors in the correspondent organs. And just these cells can form a premetastatic niche. It seems that under normal conditions they can emigrate from the blood flow at a low frequency, but this emigration can markedly increase on development in an organ of a chronic persistent inflammation. If the mechanism of restitution of specialized macrophages corresponds to that described above, the endothelium of these organs has to constantly express certain adhesive molecules providing for the recruiting of myeloid progenitor cells; and under conditions of inflammation in the sites of "typical" metastasizing the endothelium reaction has to qualitatively and/or quantitatively differ from the reaction of the microcirculatory vessels, e.g. of the heart. In fact, some observations confirm that such organs as the liver and lungs have leukocyte recruiting mechanisms other than those of "usual" tissues (Doerschuk 2001). On the liver and bone marrow endothelium such molecules as vascular cell adhesion molecule 1 (ICAM1), vascular adhesive protein 1 (VAP1), SDF-1, and P- and E-selectins are presented constitutively, and they can be important at certain stages of BMDC homing (Doerschuk 2001). Activation of BMDC homing under conditions of inflammation in organs with "typical" metastasizing seems to be caused by an additional expression of VCAM1 type proteins that together with fibronectin are major molecules interacting with integrin VLA4 on the surface of BMDCs.

Thus, the innate increased ability of the microcirculation vessels for BMDC homing seems to be the first physiological component of a preniche. However, the ability of the endothelium for intensive selective adhesion of BMDCs in organs containing specialized macrophages is insufficient for formation of preniches because BMDCs normally must rapidly differentiate into macrophages and are not accumulated in the tissue. We supposed that for development of premetastatic niches the prerequisite should be the formation of a cluster of immature BMDCs capable of being committed by the tumor cells and of creating, under their influence, a microenvironment that would be adequate for the development of metastasis. We think that this occurs due to another already pathophysiological component of the preniche due to development of a persistent chronic productive inflammation when the formation of a cell cluster is due to their accumulation within the inflammatory infiltrate under the influence of macrophage migration inhibitory factor (MIF)-type factors. Consequently, the niche formation in organs with "typical" metastasizing seems to depend, first, on the adhesion molecules constitutively synthesized by the endothelium and increased ability of the microcirculation vessels for homing and accumulation of BMDCs; and, second, on development of persistent chronic productive inflammation providing conditions necessary for BMDC cluster formation. A full value preniche can arise when these two conditions are combined.

We consider it very important that under conditions of inflammation immature myeloid cells can be recruited as material for commitment into organ-specific macrophages and replenishing the pool of local macrophages only in organs where such pool is rather large. Therefore, under conditions of adequate inflammation, BMDCs can be recruited only in these organs. For metastasis in atypical locations, such as the heart or kidneys, as well as during generalized metastasis, a superphysiological activation of the endothelium in corresponding organs is necessary, which would result in appearance of a receptor phenotype (and in saturation of the interstitium with fibronectin) similar to that in the organs with "typical" metastasis under conditions of adequate inflammation. This can occur either during a long-term local productive inflammation or as a result of systemic cytokine stimulation similar to that, which is observed on development of the syndrome of systemic inflammatory response. In such cases, under conditions favorable for BMDC accumulation and clustering, a full value preniche is also produced, and under conditions of systemic cytokine activation any site of the organism's microcirculation vessels can become a preniche. Obviously, BMDCs must be constantly present in a certain amount in peripheral blood and also be mobilized from the bone marrow under the influence of VEGFA, which is known to be synthesized in inflammation foci.

The extreme importance of a persisting chronic inflammation for the preniche formation is supported by many indirect data. First, organs with typical metastasizing, such as lungs, liver, lymphatic nodes, and serous membranes, often contain so-called cold lymphohistiocytic infiltrates even in patients and SPF-laboratory animals without clinically detectable signs of disease (in the brain where such morphological findings are relatively rare the role of "cold infiltrates" can be due to microglial reactions). These infiltrates can be caused by a persisting infection, by an alteration due to a transient ischemia, etc. Second, histologically metastatic niches are formed in peribronchial zones of the lungs or in periportal zones of the liver (Peinado et al. 2011) where inflammatory infiltrates of these organs are usually developed. Third, it has been known for long that tumor metastasis into atypical regions often coincides with the presence in these organs of a long-standing chronic inflammation ("metastasis into scar"). Fourth, all molecules known to participate in the niche formation are factors involved in development of inflammatory reactions that occur in the absence of any tumorigenesis (e.g. in psoriasis). Fifth, non-steroid anti-inflammatory drugs are shown to be successful in inhibiting tumor metastasis into lungs (e.g. Lewis carcinomas).

It should be emphasized that not all foci of a chronic productive inflammation possess a complete set of features necessary for formation of a preniche. Thus, in some cases the infiltrate can be a result of an effector immune reaction either of the Th1-type or of the Th2-type with a corresponding set of cytokines. Participants of chronic inflammation can be M1- or M2-type macrophages, etc. It is reasonable to expect that such essential differences can influence the ability of the inflammatory infiltrate to function as a preniche. Such speculations are appropriate as the comparison of organs with constitutive macrophages sharply different in probability of development of hematogenous metastases (liver, lungs, bone marrow on one hand; spleen on the other). Later on it will be necessary to more precisely determine the cell composition and intercellular relations within inflammatory foci (including organs with a constitutive pool of macrophages) that limit the formation of preniches and niches.

Preniche and Recruiting Tumor Cells from Microcirculation

The metastatic niche concept suggests that BMDC clustering should precede the formation of a micrometastasis. But this does not mean that tumor cells cannot be recruited into tissues before the formation of premetastatic niches. On one hand, tumor cells are known to enter the circulation long before the formation of a clinically detectable metastasis, but on the other hand there are many adhesive molecules capable of retaining circulating tumor elements in a particular organ. These molecules can be both constitutive and induced first of all due to development of an inflammatory reaction. Both constitutive and activated (by proinflammatory cytokines, metalloproteinases, hypoxic factors) expression of P- and E-selectins, VCAM1, ICAM1, and SDF1 is well known to promote the adhesion and homing of tumor cells (Ronzoni et al. 2010). There are very interesting observations that E-selectin responsible for the initial stage of adhesion and expressed only on the endothelium retains its activation also in foci of chronic inflammation. And due to coincidence of some participating molecules, the tumor elements can be retained just in the sites with a preformed preniche (a chronic persistent inflammation in a site of "typical" or "atypical" metastasis) or in sites with conditions favorable for its arising (constitutively expressed adhesive molecules in sites of "typical" metastasis).

Thus, not only BMDCs but also tumor cells can be accumulated in the sites of the future metastasis due to arising in them of preniches. Certainly, not every locus containing adhesive molecules for tumor cells can also recruit BMDCs. Therefore, we think that the retention of tumor cells in the site of development of an acute inflammation in an organ "atypical" for metastasis or due to constitutive ligands in the absence of inflammation will not lead to metastasis. Nevertheless, such a locus can retain tumor cells ("inefficient metastasis") in the G_0 phase of the cell cycle for an indefinitely long time, until conditions suitable for recruiting BMDCs develop in this place. This phenomenon, at least in some cases, seems to underlie the so-called late metastases observed tens of years after the extirpation of the primary tumor. Note also that the microcirculatory system of such organ as bone marrow, which is frequently affected by metastases, constitutively expresses both E-selectin and SDF-1 (which is important for the cell rolling change-over to stable adhesion to the endothelium), i.e. the whole receptor apparatus required for homing tumor elements. Having in mind that the bone marrow is a source of BMDCs, it can be considered to be a persistently acting as a constitutive premetastatic niche that does not need inflammation for arising.

Preniche and Primary Tumor

For the "preniche" concept under consideration it is important that in some cases conditions determining the development and localization of future metastases do not depend on influences of the primary tumor node. In fact, chronic persistent inflammation underlying the formation of a preniche and then of a premetastatic niche can arise long before the development of the primary tumor. Just the presence of preexisting inflammatory foci can be an explanation of arising of metastatic nodes in the lungs and liver of laboratory animals injected with a suspension of tumor cells without "preconditioning" by the primary tumor. But this does not mean that the preniche cannot be initiated under the proinflammatory influence of the primary tumor. We think that in the above-cited experiment by Lyden (Psaila and Lyden 2009) the injection of the medium from the melanoma B16 cell culture (a tumor characterized by wide and nonselective metastasis) has demonstrated just a possibility of formation of a preniche (and later also of a niche) under the influence of factors secreted by the primary tumor. Some data confirmed that the presence of a systemic inflammatory response determined by the C-reactive protein level in oncological patients was associated with an increased probability of tumor dissemination and bad prognosis. But because proinflammatory cytokines are not organ-preferential, they cannot induce the formation of solitary or multiple metastases in a certain organ but can be responsible only for development of "generalized" preniches and nonselective generalized metastasis.

Thus, the "secretory" phenotype of the primary tumor can determine only a general type of metastatic disease according to the "all or nothing" principle – the disseminated metastasizing of tumors capable of secreting proinflammatory cytokines (it is reasonable to term them "inflammatory tumor", Ti+) or the absence of such metastasizing in tumors unable to systemically activate the endothelium (Ti–). Note that the inflammatory activation of the endothelium and formation of preniches seem not only to arise due to the distant secretion of cytokines but can be also provoked in situ by Ti+ cells occurring at sites of the future metastases owing to adhesion and homing on interaction with the constitutive receptors or simply because of a mechanical "sticking" in the microcirculatory system. The formation of solitary and multiple mono-organ metastases is determined only by local processes of persisting chronic inflammation independently of proinflammatory cytokine influences of tumor cells.

"Niche as It Is" Is a Macrophage "Hybridoma"

We were considering the metastasizing processes taking as axiom that clustering BMDCs should be a key condition for survival of tumor cells and formation of a micrometastasis. However, there is at least one exception when a tumor cell seems to have no need in a classical premetastatic niche and, consequently, also in a preniche. More and more evidences are accumulated that a tumor cell can in vivo form hybrids with macrophages or with immature cells of the macrophage series, and that this process can significantly influence tumor progression (Pawelek and Chakraborty 2008). In fact, it is reasonable to suppose that such tumor–macrophage hybridoma (Tmh) should be able to express surface molecules and soluble factors inherent in macrophages and, respectively, to perform their functions as a regulator of the stromal microenvironment. It seems very likely that hybridization with the tumor cell committed the resulting hybridoma for requirements just of the tumor parenchyma. Thus, first, Tmh as a cell of the macrophage series has a broad ability for homing and can occupy both typical target organs and various inflammation foci in the sites of "atypical" metastasis; second, not needing BMDCs, Tmh can independently form multiple metastases in these organs (but without generalized nonselective metastasis). If such tumor has the Ti+ phenotype, it will inevitably take the pathway of nonselective generalization of tumorigenesis.

Preniche and Types of Metastatic Disease

On describing the essence of the "preniche" concept, we have already mentioned different clinical variants of metastatic disease. To more clearly demonstrate how different variants of metastasis follow from the preniche concept, we shall attempt to present them briefly and generally. Local and multiple metastases arise in several situations:

– first, in all cases of productive (subacute and chronic) inflammation in organs which possess specialized macrophages. Metastases are formed in "typical" sites (liver, lungs, brain) and, depending on localization and spreading of the inflammation and also on the intensity of mobilization of BMDCs into the blood flow, can be solitary or multiple, mono-, two-, three-organ, etc. Although metastases into lymph nodes are formed first of all due to lymphogenous dissemination of tumor cells, the mechanism of preniche formation in them can be similar to that which acts in other organs possessing a specialized pool of macrophages;

- second, metastasis occurs on development of productive (subacute and chronic) inflammation in the sites of atypical localization of hematogenous metastases ("metastasis into scars"), adhesion in them of Ti– cells and a pronounced activation of the endothelium recruiting BMDCs from the blood flow, and also in the case of VEGFA secretion in the tumor or in the inflammatory focus mobilizing BMDCs from the bone marrow. In such cases metastases are limited only by the inflammatory focus location or by the bone marrow where metastases seem to always occur;
- third, local metastases are formed at the primary adhesion of TmhTi– in foci of productive (subacute and chronic) inflammation. In such cases a micrometastasis can be formed in the absence of a preceding cluster of myeloid progenitor cells. The clinical consequence is a formation of metastases in "typical" sites and in every focus of chronic inflammation.

Generalized nonselective metastasis is developed on secretion of proinflammatory cytokines and cytokines recruiting myeloid progenitor cells in the primary tumor Ti+ with an inflammatory activation of the microcirculatory vessels of all organs or at the primary disseminated adhesion of TmhTi+. The absence of metastases may be declared only by convention, because micrometastases always seem to form in the bone marrow and individual vagabond tumor cells can be retained for a long time in other organs. Clinically detectable metastases can appear from these foci first of all depending on angiogenesis. However, the absence of clinical manifestation of extramedullary metastases can also be associated with a "metastatically inefficient" state of the tumor cells inside a target organ until a full-value niche is formed in the site of their localization.

Preniches and Metastatic Niches in Brain

Particular Qualities of Preniches and Niches in Brain

Brain is the organ with a rich macrophage population as liver and lung are. The macrophage elements of CNS were called "microglia". Microglial cells have a lot of specific qualities which could be matter for preniche, metastatic niche and brain metastases formation.

These features are:

1. Unique "trophic" properties of microglial cells and its participation in neuronal differentiation of CNS stem cells;
2. Activation and phenotypic mobility after negligible stimulation, for example, short-time transitory hypoxia;
3. Immune privilege of neural tissue and much more regulatory than antigen presenting function of "resting" microglia and also rarity of chronic productive inflammation development in CNS;
4. Small microglia renewing by the BMDCs in comparison with macrophage populations of liver and lung (microglia are relatively autonomic).

Therefore, firstly, brain has rich population of macrophages with changeable phenotype. This population seems to have ability to niche formation like bone marrow. Secondly, brain is the area of immune privilege and it could be the place of metastases formation without interference of immune reactions that it is very similar to situation in bone marrow constitutive metastatic and stem cell niches. Hence, there are all factors and components for metastatic niche formation in brain. It is possible that metastatic niches (which are like constitutive stem cell niches of bone marrow) can be induced in brain matter and preniche (i.e., inflammatory activation of endothelium for BMDCs recruiting) is of far less importance for brain metastases development in comparison with metastases in lung and liver. Nevertheless, inflammatory lesions in CNS could increase BMDCs and tumor cells recruiting. Undoubtedly, it can promote the metastases development.

Brain Is a "Convenient Niche": Immune Privilege in CNS

Downregulated macrophage phenotype of microglial cells as well as absence of typical lymphatic system and isolation of brain matter from peripherical blood by hematoencephalic barrier is essential for immune privilege status of brain (Lewis et al. 2012). Uncontrolled immune reactions can damage neurons (cells with restricted potency for regeneration) because immune privilege of brain provides suppression and rigorous regulation of immune response. Resident dendritic cells are absent in brain parenchyma under normal condition although microglial cells increase expression of major histocompatibility complexes (MHC) class II molecules, becoming proficient antigen-presenting cells (APCs) after the activation. Activated microglia express MHC class II and costimulatory molecules CD86, but APCs-specific CD80 expression is not induce on surface of microglial cells. Repeated stimulation of T-cells by such "immature" APCs with defective costimulation phenotype can lead to anergy of T-cells and/or differentiation regulatory/Th3-like cells and high levels of anti-inflammatory cytokines (TGFβ and IL-10) secretion in CNS (Kwidzinski et al. 2003). Neurons and other CNS cell populations have no MHC class I molecules. Neuronal expression of MHC class I can be induced under inflammation and, consequently, neurons can become probable targets for cytotoxic $CD8^+$ T cells. However, all CNS cells express Fas-ligand constitutively because interaction between such cells and Fas-expressing $CD8^+$T cells leads to T cells apoptosis. Thus, apoptosis, anergy and suppression of T cells provide immune tolerance if myelin-specific or neuron-specific T cell appear in area of brain damage (Kwidzinski et al. 2003).

Microglia as an Autonomic Niche Cells Population

Microglia are a part of organism macrophage system which also includes bone marrow hematopoietic precursors, blood monocytes, dendritic cells and numerous populations of tissue macrophages. There are two populations of microglial cells. One of them is myeloid/mesenchymal-derived (not only monocytes-derived). The other population is produced from transitory fetal macrophages and these cells are related to amoeboid microglia described in postnatal brain of rodents (Rezaie et al. 2005). Although it is known that most tissue macrophage populations are being renewed by blood monocytes recruiting, microglia are being made up by myeloid cells only under some situations. Migration to nervous tissue and microglial differentiation of BMDCs has very low activity while regeneration of meningeal and perivascular macrophage populations of CNS are carried out by BMDCs much more intensively. It has been demonstrated that only 30% of microglial cells were of a donor origin in irradiated mice with bone marrow graft after a year. Donor-derived microglia were found around vessels and in pia mater, but not in brain parenchyma. At same time, spleen had 90% of donor-derived macrophage cells after a month and a half of alveolar and Kupffer macrophages was changed after a year. Thus, most of the parenchymatous microglial cells are autonomic, BMDCs-independent population (Wu et al. 1995).

Microglia can be "resting" and "active". "Resting" cells have long processes which are growing away round bodies of the cells ("stellate" morphology). It has been supposed that CNS microglia of health humans are functionally passive because it had low expression or absence of microglial cell activation-associated molecules and "stellate" morphology. Now, it has been established that "resting" microglia were a permanent-active compartment. These cells are highly dynamic and they supervise brain microenvironment by continuous cycles of processes extension and constriction. "Resting" microglial cells scan the microenvironment permanently without disruption of neuron system and, if danger factors are revealed, cells will be activated immediately (Nimmerjahn et al. 2005). It is very interesting that phenotype of "resting" microglial cells is more similar to that of BMDCs than phenotype of other resident tissue macrophages. Microglial elements express low levels of CD45 and MHCs on cell surface and they are ineffective

APCs (Conrad and Dittel 2011). The typical markers of BMDCs (CD34 and CD11b/18) have been found on surface of microglial macrophage (Kettenmann et al. 2011). Eventually, all these features of microglia (especially if it is activated) allow it to form metastatic niches for tumor cells in brain.

Inflammatory Activation of Brain Endothelium and Glia: Preniche Forming

Injury of nervous tissue and activation of microglia result in changes of microglial cells morphology and expression markers. Microglia are not passively involved in inflammatory processes. Otherwise, it modulates inflammation actively by pro-inflammatory and anti-inflammatory cytokines production. Morphological and functional changes of microglia can be induced not only by severe disorders of CNS homeostasis related to infection, trauma, and ischemic necrosis, but also by variations of neuronal activity and transitory ischemia (Graeber and Streit 2010). So, it has been elucidated that level of microglia-activating toll-like receptor (TLR) expression increased in brain after hypoxia (Ock et al. 2007) or transitory ischemia (Ziegler et al. 2007). Microglia are supposed to be activated by negligible lesions of neurons and vessels (Trapp et al. 2007). It is probable that the processes of microglia activation are the factors triggering metastatic niche development in brain.

Inflammation is a common reaction in the place of ischemic damage. Adhesive molecules (which provide migration of cells from circulating blood to CNS), such as ICAM, E-selectin, and P-selectin appear on endothelium surface under this condition (Danton and Dietrich 2003). It has been found that SDF-1 (CXCL12) is one of chemokines and is important for leukocyte activation and migration to inflamed areas of CNS. SDF-1 can provide wide spectrum of effects such as neural cells CXCR4 (receptor for SDF-1) expression, microglia chemoattraction, cytokines and glutamate secretion by astrocytes and, at last, migration of BMDCs (Lazarini et al. 2003). It is essential that niche-forming elements, BMDCs (Psaila and Lyden 2009) have been found in damaged regions of CNS (Valli'eres and Sawchenko 2003). Nevertheless, niches developed from BMDCs seem to be supplementation only to activated resident microglia-derived niches. Therefore, brain might be the place of preniche and metastatic niche formation as it has been suggested for other preferential metastases development organs. However, many steps of these processes remain unclear and future deep research of the problem is strongly necessary.

Conclusion

In general, adaptation of tumor elements in the sites of metastasis seems to include the following stages: a preniche that is a totality of cellular and molecular events developed in the site of the future metastasis development previously to entrance into it of myeloid progenitor cells; a premetastatic niche mainly characterized by the presence of a cluster of myeloid progenitor cells without tumor elements; a micrometastatic niche characterized by the presence of a cluster of myeloid progenitor cells with tumor elements; a macrometastatic niche possessing characters of the micrometastatic niche plus the presence of myeloid progenitor cells of the endothelium and initial manifestations of angiogenesis. The introduction of "preniche" stage allowed to complete the concept of metastatic niches: to account for the sites of the most frequent localization of solitary metastases, the formation of metastases in laboratory animals injected intravenously with tumor cells, the cause of different types of metastasis (solitary metastases, multiple metastases in the same organ, or a generalized dissemination throughout the whole organism), and to determine the role of the primary tumor and of the target organ in these processes. If the preniche concept is confirmed, it will lead to creation of schemes of anti-inflammatory therapy preventing development of metastases (at least in patients bearing tumors with a low ability for secretion of proinflammatory cytokines), and this will be important for practice.

Niche formation in brain matter might be eased by particular qualities of microglia such as very high responsibility, unique "trophic" function, similarity to metastatic niche BMDCs and ability to keep immune privilege like bone marrow natural suppressors (CD11b+). Due to these properties of microglia it could provide surviving and proliferation of recruited tumor cells as well as bone marrow stem cell niche. It is possible that frequent localization of metastases on the border of grey and white matter is associated with heterogeneity of brain microglia and that is essential for metastasis formation. Role of preniche in CNS appears to be optional and it could increases recruiting BMDCs additionally to microglial niches and provides tumor cell migration to the site of future metastasis. Thus, we think that relatively high incidence of brain metastases would be to explain with above mentioned features metastatic niche formation in CNS. The problem why there is restricted nosological spectrum of frequent brain metastasizing tumors is the subject for future experimental research.

References

Alix-Panabieres C, Riethdorf S, Pantel K (2008) Circulating tumor cells and bone marrow micrometastasis. Clin Cancer Res 14:5013–5021

Bidard FC, Pierga JY, Vincent-Salomon A, Poupon MF (2008) A "class action" against the microenvironment: do cancer cells cooperate in metastasis? Cancer Metastasis Rev 27:5–10

Conrad AT, Dittel BN (2011) Taming of macrophage and microglial cell activation by microRNA-124. Cell Res 21:213–216

Danton GH, Dietrich WD (2003) Inflammatory mechanisms after ischemia and stroke. J Neuropathol Exp Neurol 62:127–136

Doerschuk CM (2001) Mechanisms of leukocyte sequestration in inflamed lungs. Microcirculation 8:71–88

Duchnowska R, Szczylik C (2005) Central nervous system metastases in breast cancer patients administered trastuzumab. Cancer Treat Rev 31:312–318

Gao D, Joshi N, Choi H, Ryu S, Hahn M, Catena R, Sadik H, Argani P, Wagner P, Vahdat LT, Port JL, Stiles B, Sukumar S, Altorki NK, Rafii S, Mittal V (2012) Myeloid progenitor cells in the premetastatic lung promote metastases by inducing mesenchymal to epithelial transition. Cancer Res 72:1–11

Gavrilovic IT, Posner JB (2005) Brain metastases: epidemiology and pathophysiology. J Neurooncol 75:5–14

Graeber MB, Streit WJ (2010) Microglia: biology and pathology. Acta Neuropathol 119:89–105

Hiratsuka S, Watanabe A, Sakurai Y, Akashi-Takamura S, Ishibashi S, Miyake K, Shibuya M, Akira S, Aburatani H, Maru Y (2008) The S100A8-serum amyloid A3-TLR4 paracrine cascade establishes a pre-metastatic phase. Nat Cell Biol 10:1349–1355

Kaplan RN, Riba RD, Zacharoulis S, Bramley AH, Vincent L, Costa C, MacDonald DD, Jin DK, Shido K, Kerns SA, Zhu Z, Hicklin D, Wu Y, Port JL, Altorki N, Port ER, Ruggero D, Shmelkov SV, Jensen KK, Rafii S, Lyden D (2005) VEGFR1-positive haematopoietic bone marrow progenitors initiate the pre-metastatic niche. Nature 438:820–827

Kettenmann H, Hanisch UK, Noda M, Verkhratsky A (2011) Physiology of microglia. Physiol Rev 91:461–553

Kwidzinski E, Mutlu LK, Kovac AD, Bunse J, Goldmann J, Mahlo J, Aktas O, Zipp F, Kamradt T, Nitsch R, Bechmann I (2003) Self-tolerance in the immune privileged CNS: lessons from the entorhinal cortex lesion model. J Neural Transm Suppl 65:29–49

Lazarini F, Tham TN, Casanova P, Arenzana-Seisdedos F, Dubois-Dalcq M (2003) Role of the alpha-chemokine stromal cell-derived factor (SDF-1) in the developing and mature central nervous system. Glia 42:139–148

Lewis CA, Manning J, Rossi F, Krieger C (2012) The neuroinflammatory response in ALS: the roles of microglia and T cells. Neurol Res Int 2012:803701

Nimmerjahn A, Kirchhoff F, Helmchen F (2005) Resting microglial cells are highly dynamic surveillants of brain parenchyma in vivo. Science 308:1314–1318

Ock J, Jeong J, Choi WS, Lee WH, Kim SH, Kim IK, Suk K (2007) Regulation of Toll-like receptor 4 expression and its signaling by hypoxia in cultured microglia. J Neurosci Res 85:1989–1995

Pawelek JM, Chakraborty AK (2008) Fusion of tumour cells with bone marrow-derived cells: a unifying explanation for metastasis. Nat Rev Cancer 8:377–386

Peinado H, Rafii S, Lyden D (2008) Inflammation joins the "niche". Cancer Cell 14:347–349

Peinado H, Lavotshkin S, Lyden D (2011) The secreted factors responsible for pre-metastatic niche formation: old sayings and new thoughts. Sem Cancer Biol 21:139–146

Perelmuter VM, Manskikh VN (2012) Preniche as missing link of the metastatic niche concept explaining organ-preferential metastasis of malignant tumors and the type of metastatic disease. Biochemistry (Mosc) 77:111–118

Pestalozzi BC, Zahrieh D, Price KN, Holmberg SB, Lindtner J, Collins J, Crivellari D, Fey MF, Murray E, Pagani O, Simoncini E, Castiglione-Gertsch M, Gelber RD, Coates AS, Goldhirsch A, International Breast Cancer Study Group (IBCSG) (2006) Identifying breast cancer patients at risk for central nervous system (CNS) metastases in trials of the International Breast Cancer Study Group (IBCSG). Ann Oncol 17:935–944

Psaila B, Lyden D (2009) The metastatic niche: adapting the foreign soil. Nat Rev Cancer 9:285–293
Rezaie P, Dean A, Male D, Ulfig N (2005) Microglia in the cerebral wall of the human telencephalon at second trimester. Cereb Cortex 15:938–949
Ronzoni M, Manzoni M, Mariucci S, Loupakis F, Brugnatelli S, Bencardino K, Rovati B, Tinelli C, Falcone A, Villa E, Danova M (2010) Circulating endothelial cells and endothelial progenitors as predictive markers of clinical response to bevacizumab-based first-line treatment in advanced colorectal cancer patients. Ann Oncol 21:2382–2389
Scott L, Priestly G, Papayannopoulou T (2003) Deletion of alpha4 integrins from adult hematopoietic cells reveals roles in homeostasis, regeneration, and homing. Mol Cell Biol 23:9349–9360
Trapp BD, Wujek JR, Criste GA, Jalabi W, Yin X, Kidd GJ, Stohlman S, Ransohoff R (2007) Evidence for synaptic stripping by cortical microglia. Glia 55:360–368
Valli'eres L, Sawchenko PE (2003) Bone marrow-derived cells that populate the adult mouse brain preserve their hematopoietic identity. J Neurosci 23:5197–5207
Wels J, Kaplan RN, Rafii S, Lyden D (2008) Migratory neighbors and distant invaders: tumor-associated niche cells. Genes Dev 22:559–574
Wilson A, Trumpp A (2006) Bone-marrow haematopoietic-stem-cell niches. Nat Rev Immunol 6:93–106
Wong J, Johnston B, Lee SS, Bullard DC, Smith CW, Beaudet al, Kubes P (1997) A minimal role for selectins in the recruitment of leukocytes into the inflamed liver microvasculature. J Clin Invest 99:2782–2790
Wu L, Vremec D, Ardavin C, Winkel K, Suss G, Georgiou H, Maraskovsky E, Cook W, Shortman K (1995) Mouse thymus dendritic cells: kinetics of development, changes in surface markers during maturation. Eur J Immunol 25:418–425
Yoo J, Lee SH, Lym KI, Park SY, Yang SH, Yoo CY, Jung JH, Kang SJ, Kang CS (2012) Immunohistochemical expression of DCUN1D1 in non- small cell lung carcinoma: it's relation to brain metastasis. Cancer Res Treat 44:57–62
Ziegler G, Harhausen D, Schepers C, Hoffmann O, Rohr C, Prinz V, Konig J, Lehrach H, Nietfeld W, Trendelenburg G (2007) TLR2 has a detrimental role in mouse transient focal cerebral ischemia. Biochem Biophys Res Commun 359:574–579

Computer Systems for Cell Counting in Histopathologic Slides of Tumours of the Central Nervous System: Advantages and Limitations

12

Tomasz Markiewicz, Wojciech Kozlowski, and Stanislaw Osowski

Contents

Abstract

The chapter presents review of the computer systems for cell counting in histopathologic slides of tumours of the central nervous system, pointing to their advantages and limitations. The cells in tissue specimens subjected to IHC reaction are visualized in different ways depending on their immunoreactivity, for example the light blue (immunonegative) and brown (immunopositive). On the basis of these colors it is possible to recognize the set of immunonegative and immunopositive cells. However, the recognition of densely populated cells in the slides is a difficult challenge for the manual recognition. Therefore the computer supported solutions to this task are already under development.

The chapter will present and discuss the actually existing and newly proposed solutions to this particular problem. We will point the partially existent commercial systems and discuss the progress and development of research over the non-commercial approaches to the automatic analysis of IHC slides.

T. Markiewicz (✉)
Warsaw University of Technology and Military Institute of Medicine, Warsaw, Poland
e-mail: markiewt@iem.pw.edu.pl

W. Kozlowski
Military Institute of Medicine, Warsaw, Poland

S. Osowski
Warsaw University of Technology and Military University of Technology, Warsaw, Poland
e-mail: sto@iem.pw.edu.pl

Introduction

For many last years a lot of research has been done to isolate and measure the factors of diagnostic and prognostic significance for patients suffering from the central nervous system tumours. The important role in this field plays the immunohistochemistry (IHC). IHC refers to

M. Hayat (ed.), *Tumors of the Central Nervous System, Volume 13*,
DOI 10.1007/978-94-007-7602-9_12, © Springer Science+Business Media Dordrecht 2014

the process of detecting antigens in cells of a tissue by using the principle of antibodies binding specifically to antigens in biological tissues. It is widely used in the diagnosis of abnormal cells, for example in cancerous tumors, coloring the cells in different ways depending on their immunoreactivity: the light blue (immunonegative) and brown (immunopositive). On the basis of these colors we can recognize the set of immunonegative and immunopositive cells. The results of this recognition play important role in cancer diagnostics.

However, the recognition of densely populated cells in the slides is a difficult challenge for pathologists. The difficulties follow from their extremely high population, non-uniform coloring, partial overlapping of cells and in many cases close similarity of cells to the background. Therefore the results of manual screening are highly dependent on the human interpretation of colors and also experience of the expert. It is well known that the results of counting may differ even for the same expert, depending on the particular day or his actual mental and physical condition. Therefore the automatization of the image analysis by applying specialized computer programs is very important and may greatly accelerate the research in central nervous system tumors.

The chapter will present and discuss the actually existing and newly proposed solutions to this particular problems. We will point the partially existent commercial systems and discuss the progress and development of research over the non-commercial approaches to the automatic analysis of IHC slides.

Importance of IHC in Diagnostics of Tumors of the Central Nervous System Tumour

Diagnostic Significance

The immunohistochemistry is playing important role in diagnosis of tumours of the central nervous system. In some cases the quantitative analysis based on the specific staining can be used also to support the differential diagnosis. For many types of the tumours of central nervous system, the proliferation evaluated by MIB-1/Ki-67 labelling indices are determined. The most significant remarks from the diagnostic point of view are presented below.

Astrocytic tumours: In pilocytic astrocytoma (code 9421/1) WHO grade I MIB-1 labelling indices range from 0 to 3.9% (mean 1.1%) (Giannini et al. 1999) whereas in diffuse astrocytoma WHO grade II mean level is 2.5% (Watanabe et al. 1996). In anaplastic astrocytoma (code 9401/3) WHO grade III the growth factor is usually in the range 5–10% whereas in glioblastoma (code 9440/3) WHO grade IV it increased to 15–20% (Jaros et al. 1992). However, the observed overlapping values provide little use of this quantification in differential diagnosis.

Choroid plexus tumours: In choroid plexus tumours (code 9390/0-3) WHO grade I–III, mean Ki-67/MIB-1 labelling indices have been reported as 1.9% (range 0.2–6%) for choroid plexus papilloma and 13.8% (range 7.3–60%) for choroid plexus carcinoma and only 0–0.1% for normal choroid plexus (Vajtai et al. 1996). Another study described 4.5% (range 0.2–17.4%), 18.5% (range 4.1–29.7%) and 0% respectively (Carlotti et al. 2002).

Meningeal tumours: In meningiomas (code 9530/0) WHO grade I, II or III, the Ki-67 labeling index is highly correlated with the following types: benign (WHO grade I; mean 3.8%), atypical (WHO grade II; 7.2%) and anaplastic (WHO grade III; 14.7%) (Maier et al. 1997).

Prognostic Factors

The quantitative analysis of the immunohistochemical stained specimens is frequently taken into account as a prognostic factor, restricted to proliferation activity of the tumors. Thus a MIB-1/Ki-67 labelling index is counted and applied as an important factor. Following we depict the confirmed cases pointing to the correlation of the proliferation with a clinical outcome.

Astrocytic tumours: In diffuse astrocytoma (code 9400/3) WHO grade II it was found a gross correlation of proliferation with a clinical outcome. A MIB-1/Ki-67 labelling index of >5% was found to constitute a threshold value for predicting shorter survival (Jaros et al. 1992).

Oligodendroglial tumours: In oligodendroglioma (code 9450/3) WHO grade II higher proliferation rates (>3–5% on MIB-1 labelling index) significantly correlate with worse prognosis (Heegaard et al. 1995; Dehghani et al. 1998). The reported 5-year survival rate was 83% for patients whose oligodendroglioma had a MIB-1 labelling index of <5% whereas only 24% for patients with tomours characterized by >5% positive cells (Dehghani et al. 1998). In oligoastrocytoma (code 9382/3) WHO grade II the lower Ki-67 indices are associated with the longer survival (Shaffrey et al. 2005).

Ependymal tumours: In ependymoma (code 9391/3) WHO grade II Ki-67 labelling index less than 4% has been associated with a significantly longer survival time than labeling index greater than 5% (Kurt et al. 2006). However, its prognostic significance has not yet been firmly established.

Neuronal and mixed neuronal-glial tumours: In central neurocytoma (code 9506/1) WHO grade II Ki-67 labelling index >2% (Mackenzie 1999; Soylemezoglu et al. 1997) or >3% (Rades et al. 2004) have been reported as significantly shorter recurrence-free interval.

Embryonal tumours: In neuroblastoma (code 9500/3) WHO grade IV, in 1999 the International Neuroblastoma Pathology Committee proposed the prognostic classification by adopting the Shimada system (Shimada et al. 1984). Tumors in the favourable histology group include following cases (1) age <1.5 years old, neuroblastoma (Schwannian stroma-poor), poorly differentiated subtype with low (<2%) or intermediate (2–4%) mitosis-karyorrhexis index (MKI) signed in MIB-1/Ki-67 staining process; (2) age between 1.5 and 5 years old, neuroblastoma (Schwannian stroma-poor), differentiating subtype with low MKI; (3) ganglioneuroblastoma, intermixed (Schwannian stroma-rich); (4) ganglioneuroblastoma, maturing and mature subtypes (Schwannian stroma-dominant). In the opposite, tumors in unfavourable group include the following cases (1) any age, neuroblastoma (Schwannian stroma-poor), undifferentiated subtype; (2) age between 1.5 and 5 years old, neuroblastoma (Schwannian stroma-poor), poorly differentiated subtype; (3) any age, neuroblastoma (Schwannian stroma-poor) with high (>4%) MKI; (4) age between 1.5 and 5 years old, neuroblastoma (Schwannian stroma-poor) with intermediate MKI; (5) age ≥5 years old, all neuroblastoma (Schwannian stroma-poor) subtypes; (6) ganglioneuroblastoma, nodular (composite Schwannian stroma-rich/stroma-dominant and stroma-poor) (Schwab et al. 2000). As depicted, the MIB-1/Ki-67 plays the significant role in prognosis of the neuroblastoma.

These numbers are the evidence of importance of the mentioned MIB-1/Ki-67 labeling indices in the process of diagnostics of tumors of the central nervous systems. However, in their estimation we have to recognize and count thousands of densely packed cells present in the IHC slides. Most often this tedious task is still done by the human operator. However, following the development of the computer technology it is possible to develop efficient automatic or semi-automatic solutions to this problem.

Commercial and Non-commercial Image Analysis Systems

Automated image analysis is the direction of biomedical research of increasingly higher importance in medical diagnostics. It plays a special role in the pathology, allowing for quick and objective evaluation of the cells and representing an important factor for quantitative assessment of the preparations. Thanks to such assessment we can quickly and accurately determine the optimal treatment, diagnosis and further research, concerning the pathogenesis of human diseases.

The difficult problem for pathologists is the recognition of cells on the basis of the immunohistochemical (IHC) reactions. The slides representing tissues under investigation contain the cells of two colors: the light blue (immunonegative) and brown (immunopositive), depending on their IHC reaction. The manual screening of it is particularly complex and its results rely on the interpretation of the human expert. This is very difficult task requiring high concentration due to the dense packing of cells, large amount of them, partial overlapping, etc. Because of screening subjectivity the errors can occur and may cause some mistakes in diagnosis. Therefore development of the automatic systems independent from human operators are crucial for further progress in this field.

One of the most important moments in the development of the quantitative methods in pathology was the formation of the International Society on Diagnostic Quantitative Pathology in 1981. The methods used in the quantitative assessment of pathology support the diagnosis of diseases, especially cancer, accelerating and exacerbating the medical methods. Development of the computer approaches to the diagnosis of pathology is increasingly visible at international congresses of pathologists, where more papers concerning examples of applications of automatic image analysis are presented. Almost all companies producing microscopic equipment systems offer automatic scanning image of the entire preparation as well as quantitative computer analysis tools, including a quantitative assessment of some selected reactions and certain types of cancer. The actually existing automated or semi-automatic systems for quantitative analysis of preparations, that can be used in pathological diagnosis can be divided into two basic groups:

- commercial systems offered by manufacturers of microscopes or other companies.
- non-commercial specialized programs presented in scientific publications and also sometimes offered via the Internet.

Some of them will be shortly presented below, indicating the possibility of their applications in the investigations of tumors of the Central Nervous System (CNS).

Commercial Systems

The commercial systems used for an automatic image analysis are usually part of the offer of electronic microscopy manufacturers. The main advantage of such approach is the integrated analysis of digital images, starting from their acquisition. Here we can mention the following solutions: the image analysis offered by Olympus with cooperation of the Soft-Imaging Software GmbH, Carl Zeiss AxioVision, QWin Leica, Aperio's ScanScope cooperating with the Nikon microscope, Dako ACIS III and the company's HistoQuant 3DHistech. The commercial systems offer the additional tools allowing to control the process of scanning the entire preparation. Frequently, they are dedicated to the assessment of the estrogen receptor (ER) or progesterone receptor (PR) in the IHC reactions to specimens of the breast cancers, as well as determination of the HER2 gene at IHC and FISH. The interest of such applications follows from the fact that this type of pathologist evaluation is performed most often in practice. For a quantitative assessment of these preparations the available systems require the definition of the area on the image to be analyzed, so called region of interest (ROI), which means high degree of intervention of human operator. The other rare types of reactions are usually not listed or are designated as "research only".

An example of a commercial company is Dako ACIS III that developed the supporting system used in quantitative evaluation in IHC. Their program is based on the microscopic image of the preparation represented in HSV format. The user defines ranges of the individual components of the HSV-class assignments of pixels, representing either blue or brown cells. Following the classification of all pixels in the image the result of the quantitative evaluation is estimated as the ratio of the number of pixels in brown or blue class to the total number of pixels representing both classes together. This is a very cursory assessment, without delving into details of the image. The additional difficulty is that the user has to mark manually the areas containing only the cells which are subject to counting.

A more advanced system is HistoQuant of 3Dhistech company. It performs the quantitative evaluation of images based on the recognition of the individual nuclei of cells and their classification, according to the threshold of assumed values for colors. The color deconvolution is used to increase the differentiation between positive and negative reactive nuclei. However, the threshold level should be adjusted manually. The system is able to count cell nuclei and to measure the nuclear features (morpho- and densitometric). Also the glandular tissue structures can be morphologically characterized. Detection of glands, measurement of glandular morphometric parameters, intraglandular cell number and cell features are supported. Batch mode processing and manual re-scoring are also possible.

The next system is the Leica QWin. It is described as the powerful software suite, with a comprehensive range of image analysis tools available, and allows to get precise results with mega pixel accuracy for a diverse range of life science and industrial applications, such as morphology, fluorescence imaging and compound materials analysis. It is offered in 5 versions: Leica QWin Lite, Leica QWin Plus, Leica QWin Standard, Leica QWin Runner and Leica QWin Pro, ranging from simple interactive measurements to macro – programming for custom applications with multi parameter measurements. However, these systems are oriented more to the visualization and image "hand-on-control" analysis than direct automatic quantitative analysis of the medical images.

More histologically oriented are the image analysis tools available in Aperio Image Analyze Toolbox. In the IHC image analysis area, they provide FDA-cleared In-Vitro Diagnostic (IVD) algorithms for HER2, ER and PR stained breast specimens. The algorithms can be tuned for different tissue types (e.g. breast, colon, prostate), stains (e.g. HER2, ER, PR, Ki-67, P53, EGFR), reagents (e.g. Dako, Ventana) or to correlate their algorithm results with the other test methods (e.g. FISH, CISH). It seems to be one of the most advanced systems to quantitative image analysis in pathology. The integration with the slide scanner makes this system very interesting.

The other system is the Definiens Developer XD (Definiens AG Bernhard-Wicki-Straße 5 80636 München Germany). It provides a comprehensive range of algorithms for multidimensional image analysis applications. A lot of algorithms tailored to image segmentation, classification, measurement and statistics are offered. However, actually their application in IHC are restricted to HER-2 Scoring and Jejunum Crypt Proliferation Index.

Based on the sale information we can state that most of the commercial systems are specialized in ER, PR nuclear stains and HER2 stain in breast cancer. This is because of the highest demand for this type of image quantification in practice. The other types of stains are often described as "research only". Unfortunately, the tools supporting the cases of the central nervous system tumors are rare in commercial systems and it is difficult to assess their effectiveness for this types of IHC image analyses.

Non-commercial Systems

Every year we observe a series of programs dedicated to certain types of reactions, tumors and tissues, published in scientific journals or presented at scientific conferences. Some are also offered on the Internet, usually free of charge as "freeware". However, most of the presented programs are not easily available, some are available on the Internet only for a limited use and no longer available after a short time. Based on the author's descriptions it is possible to analyze how they work. Sometimes it is possible to download the source code. For these reasons they form not only the possible tool, but are the basis for further modifications and may be used as the starting point in the development of new programs for the analysis of other cancers, or tissue reactions. The examples of such programs are: ImageJ with "plug" plug-in Ki-67 developed in the Institute of Neuropathology University of the Saarland, Homburg, Germany (Kim et al. 2006) or programs presented by the group of Lezoray in France (Lebrun et al. 2007; Lezoray et al. 2000; Lezoray and Cardot 2002; Lezoray and Lecluse 2007).

Most of solutions are the specialized programs, with a narrow scope of application (only for selected types of preparations and reactions of cancer). The ImageJ with the Ki-67 "plug-in" is dedicated to the quantitative assessment of cells with Ki-67 reaction in the image type of brain tumor of meningioma. It is available in the form of a program in Java and can be applied for the analysis of the user delivered images. The result of quantitative evaluation of the analyzed image is generated in the form of annotated cells existing in the image of preparation.

The other "development line" form the programs built by the authors of this chapter. Few of them are dedicated to the central nervous system tumors and will be discussed here. First solution concerning neuroblastoma (Markiewicz et al. 2009) was a Matlab-platform program to quantitative analysis of the neuroblastoma Ki-67 stained specimens. Figure 12.1a presents the general block diagram of image processing leading to the recognition and counting the immuno-positive and negative cells.

The main task of an image processing of the histological slides was automatic recognition and counting of cells, belonging to either immunopositive or immunonegative types. This task was solved by segmenting the image into three parts: the brown cells, the blue cells and the rest, regarded as a background. To construct an efficient segmentation algorithm we have to perform the exact analysis of the processed image and take into account any detailed features characteristic for it, that should be reflected in the methodology of the segmentation procedure. In building these programs we followed the expert observations, which can be summarized in the following points:

- the nuclei are stained generally in a homogenous way,
- in most cases the cytoplasm or stroma area between the nuclei of the cells are significantly brighter,
- the image contains practically only cells subjected to counting,
- the size of the cells is stable (small range of changes),
- the background is generally not stained.

These observations are very significant in developing the efficient algorithms to quantitative image analysis in neuroblastoma Ki-67 staining specimens. The proposed algorithm was based on the mathematical morphology approach (Soille 2003) and application of the Support Vector Machine (Vapnik 1998; Schölkopf and Smola 2002). To locate the nuclei of the dominant blue color, the grey scale version of the image, based on the difference between the blue and green components was used. This image is subject to the step threshold operations defined as (Soille 2003)

$$T_{[t_1,t_2]}\left[f(x)\right]=\begin{cases}1 & if \quad t_1 \le f(x) \le t_2 \\ 0 & else\end{cases} \quad (12.1)$$

where $f(x)$ represents the pixel intensity and t_1, t_2 the selected lower and upper threshold values, respectively. The sequential thresholding procedure was developed based on the size restriction of the individual nucleus area and the following conclusion relation:

$$T_{[t_{max},t_{max}]}(f) \subseteq T_{[t_{max-1},t_{max}]}(f) \subseteq \ldots \subseteq T_{[t_0,t_{max}]}(f) \quad (12.2)$$

The recognition algorithm starts from the minimal value of threshold t_1 (close to the white color) and checks if the selected pixels form the objects fully separated from the others, and at the same time fulfills selected size restriction. If yes, they are added to the cell mask. In this process the threshold is changing from the minimal value to the maximal one, corresponding to the darkest pixel in the image. The sequential increase of threshold is repeated in 100 equal steps. If the object was separated in just one step and contained some darker pixels, then in the next few steps these pixels were preserved, forming the whole object. The additional operation of segmentation in the form of the watershed algorithm (Soille 2003) and supplementary brighter nuclei detection procedure were also applied to reduce the number of glued cells.

In the parallel process the other cells in the image (the brown color immunopositive cells and not well segmented blue cells that remained in

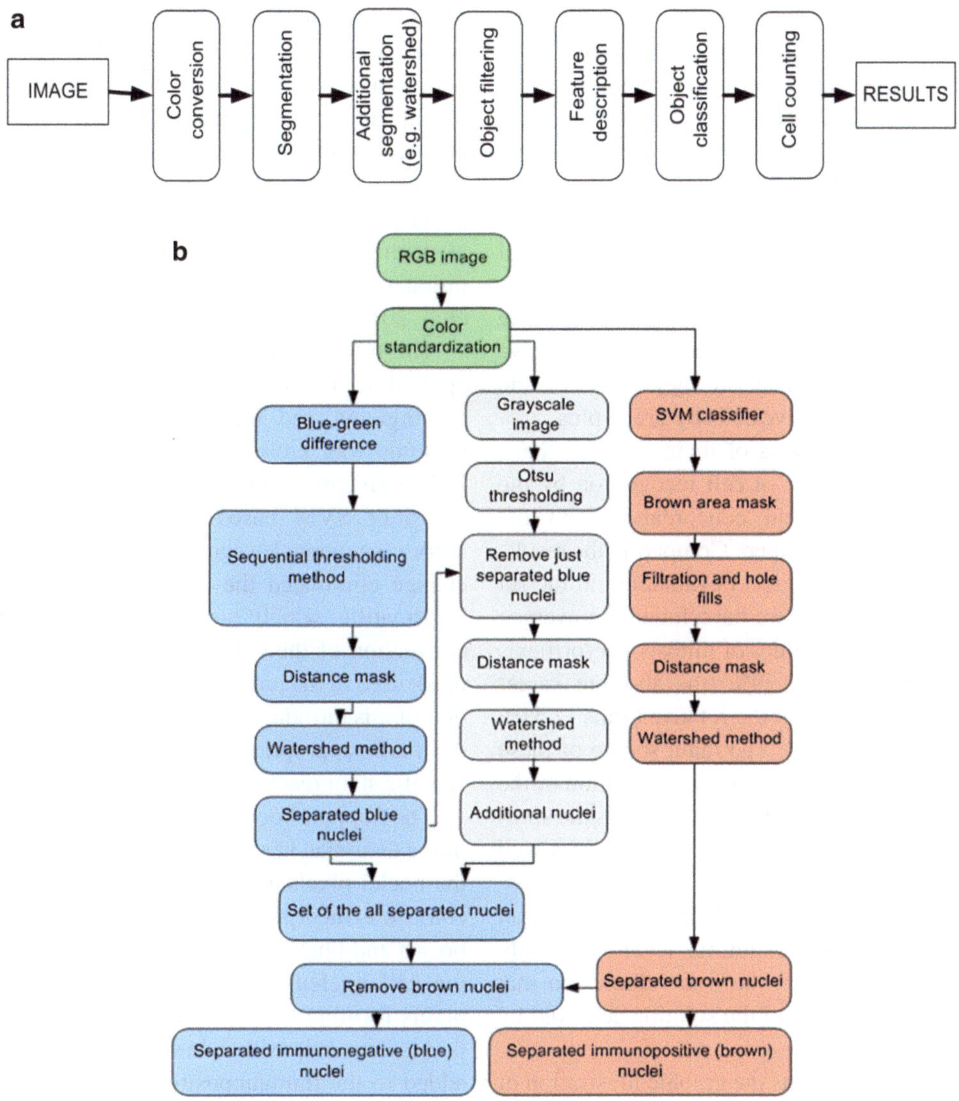

Fig. 12.1 The general block diagram of the applied image processing (**a**) and the detailed scheme of algorithm for counting immunopositive and immunonegative cells (**b**)

the mask) are extracted. We start from the grey scale representation of the RGB image subject to thresholding. The threshold value is determined automatically by using the Otsu method (Otsu 1979). As a result of this step we get the set of cells containing the immunopositive cells and the immunonegative (blue) cells. The blue cells extracted in the first stage of thresholding are removed from this set by subtracting both masks. The differential mask is once again segmented by using the watershed method to separate the glued cells. In this way we get the set of brown cells with some (usually small) amount of blue cells that remained in the image after the segmentation procedure done in the first stage.

The final recognition which cell is immunopositive was done by applying the linear Support Vector Machine, working in the classification mode to recognize the brown pixels from the other pixels. The main idea of this classifier is to create a hyperplane that separates both classes with the maximal margin. The input vector for this network is created by three color components of the pixel, represented in the RBG standard. Figure 12.1b depicts the presented above algorithm, stressing these three parallel processes.

The system delivers the results in the numerical form of MKI indices of the analyzed images as well as graphical form of the annotated images with all recognized cells. The typical graphical output produced by the system, for the images of meningioma and oligodendroglioma are presented in Fig. 12.2a, b. The immunonegative cells depicted in the image are signed by red circle and the immunopositive cells by green plus.

The statistical results of image analysis have shown good accuracy of cell recognition by the developed system and concordance with the results of the human expert. Comparing the results of an automatic and expert evaluation on the set of ten testing images we have found the average discrepancy of the values of mitosis-karyorrhexis index (MKI) equal 0.39% (the absolute value) which means 4.07% in the relative terms (2.51% of standard deviation). The main source of these errors was the limited accuracy recognition of the blue cells. The cause of this problem was the similarity of the blue cells to the background, as well as merging of few cells together in histopathological slides. The minimum (MIN) and maximum (MAX) values of the mitotic level were only slightly different from the true values, and that confirms high quality of the solution achieved by our automatic system.

Figure 12.3 presents the graphic illustration of the minimum and maximum mitotic levels for 13 analyzed images representing meningioma (Fig. 12.3a) and oligodendroglioma (Fig. 12.3b) produced by the system and compared to the human expert results. The system's optimal output is when the lines of automatic system (AS), expert (EX), MIN and MAX are found nearest to each other. Additionally, if the AS line is close to EX line, the mitotic level assessed by the automatic system closely resembles the human expert's results. The difference between the AS and EX values, i.e., the line (AS-EX) denotes the error line. If MIN and MAX lines lie below and above the AS line in equal distances, the errors are well balanced. On the other hand, if the MAX-MIN range is large, the probability of error is greater and the results generated by AS can be not reliable. The presented plots show that in the significant range of the mitotic level, that is between 0 and 15%, the AS results are highly acceptable and compatible with the human expert's results. The detailed description of the algorithm and the numerical results of cell recognition can be found in the paper (Markiewicz et al. 2010).

The other modified version of the algorithm applied to the analysis of two brain tumors: meningioma and oligodendroglioma at Ki-67 staining was presented in (Markiewicz et al. 2010). The important change was introduction of the prior SVM based classification of pixels before sequential segmentation process. This change eliminated the problem of losing some information which were associated with the conversion of the color image to the gray scale mask used in the previous solution. After application of this introductory step we get automatically the mask appropriate to the recognition of the specific stain reaction regions.

The Support Vector Machine (SVM) classifier of the Gaussian kernel, programmed to recognize the nuclei pixels according to their component's color, was incorporated to recognize between the brown and blue cells. If the majority of pixels in the cell nucleus were found to be brown, all pixels forming the extracted nuclei were classified as immunopositive. In such case, the cell was added to the immunopositive cell mask. On the other hand, if the majority of pixels in the cell nucleus were blue, the cell was added to the immunonegative mask. The final step of counting the cells of both groups is a simple task for the Matlab program, since it was transformed to summing up the number of cells gathered in the respective masks, done separately for immunopositive and for immunonegative ones.

One of the most important problems with automatic image evaluation of oligodendroglioma Ki-67 specimens is the incidence of the inflammatory, microglia, lymphocytes and stroma cells in the diagnosed view field. The nuclei of these cells are stained immunonegatively in blue color, similar to the negative neoplastic cells. In the case of

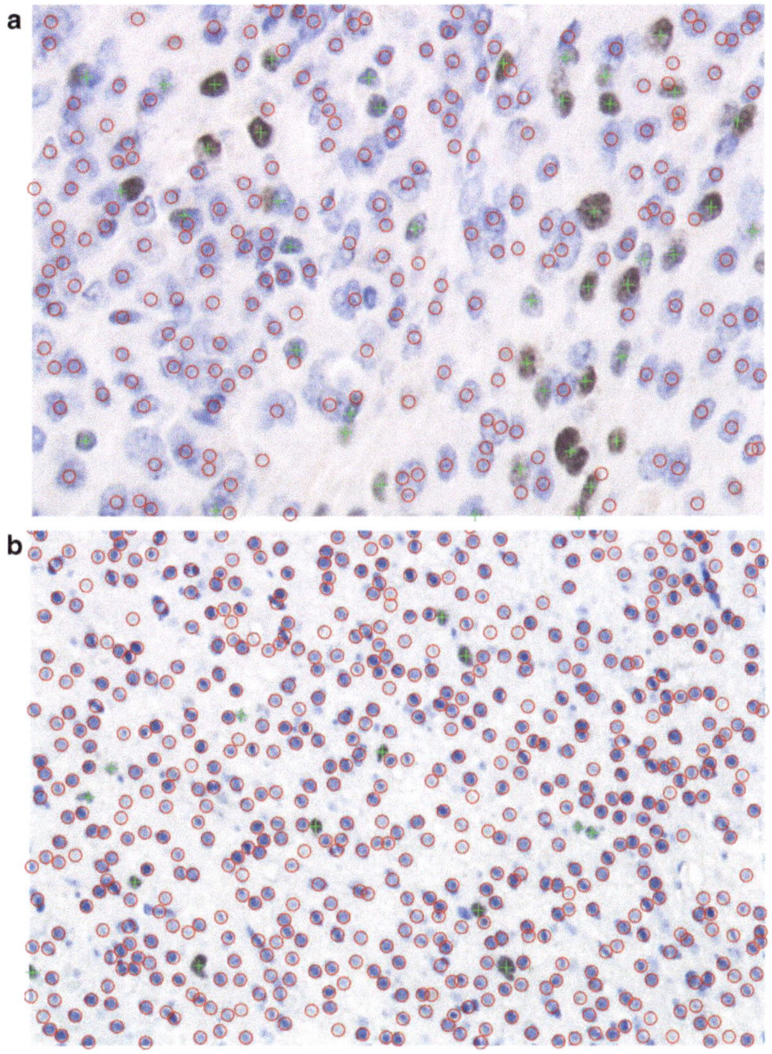

Fig. 12.2 The examples of automatically annotated histological images for the cell counting of (**a**) meningioma and (**b**) oligodendroglioma (Ki-67, ×400)

changing the incidence of these cells, the ratio between immunopositive and immunonegative cells is also changed. For correct diagnosis these cells must be recognized from lesion cells and eliminated. Two different but commonly used methods of texture features generation have been applied to solve the problem. The Markov Random Field model as a source of features (Chellappa and Chatterjee 1985; Wagner 1999) and selected Unser texture features have been exploited. The details of such solution are to be found in the paper (Warowny and Markiewicz 2010).

The actually developed automatic system has enabled to start the statistical quantitative analysis of many patients at H&E and Ki-67 staining. The experiments made on the basis of 24 patients of meningiomas and 26 patients of oligodendrogliomas were directed to the assessment of the tumor cell numbers in the fields of view selected by the pathologist. It was found that all patients suffering from meningioma had the mean of 623 cells in one field of view with the standard deviation of 102 cells. Moreover 95% of investigated fields of view contained between 386 and

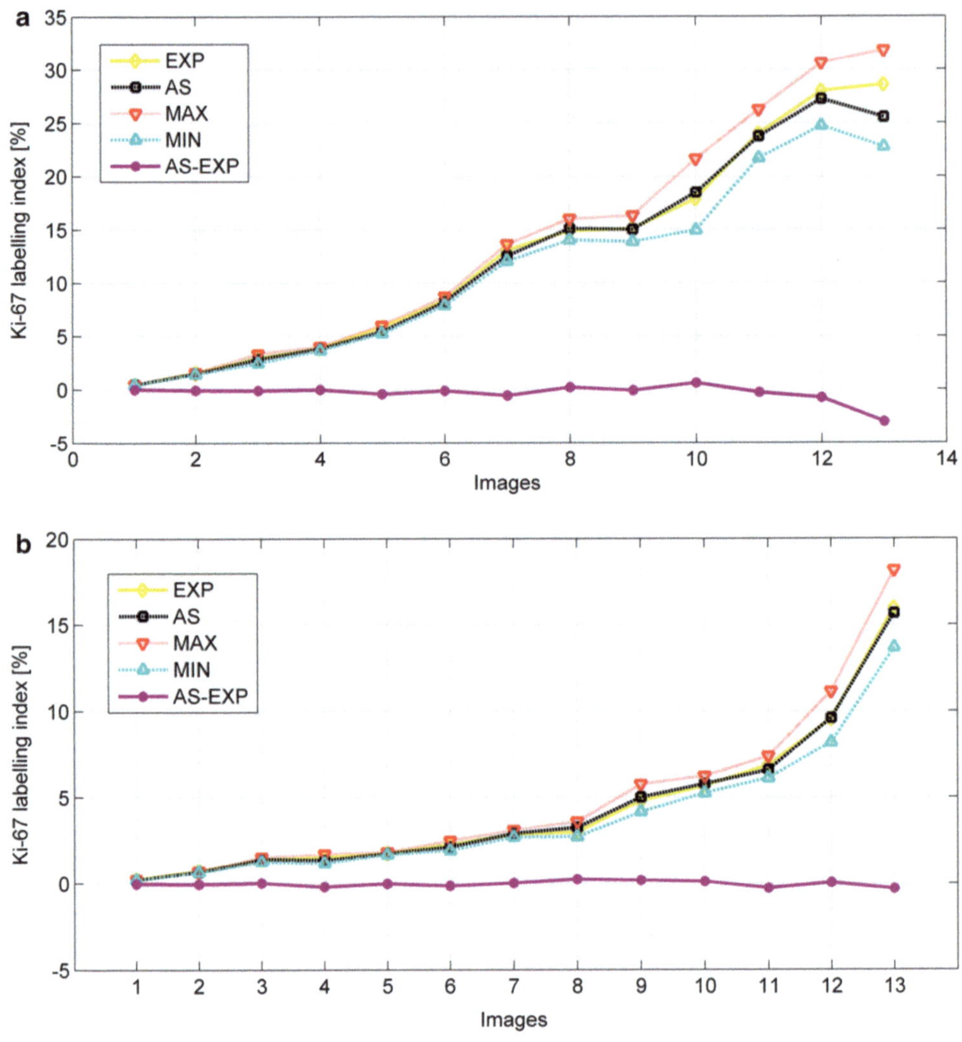

Fig. 12.3 The graphical interpretation of the mitotic level estimation by our system in meningiomas (**a**) and oligodendroglioma (**b**) at Ki-67 staining

781 cells (very wide range). The average size of the nuclear area of the meningioma cells at this staining was equal 184.5 pixels at standard deviation (SD) of 118 pixels.

For patients with oligodendroglioma, the mean was 474 cells with the standard deviation of 152 cells. Among all fields of view 95% contained between 204 and 736 cells, which is a very wide range. In ten different images acquired from the same patient the ratio of the standard deviation to the mean value of cells in meningioma was 10.9%, and in oligodendroglioma this ratio was 18.6%. A large variety of results corresponding to different images acquired from the same patient was shown. Moreover in this case the average size of the nuclear area of the oligodendroglioma cells was equal 183.5 pixels at SD equal 97 pixels.

Similar analysis of the same patients was performed using the developed system at the application of Ki-67 stain and recognition of two groups of cells, depending on their immunoreactivity. The results of cellularity calculated on the basis of one field of view for each patient were as follows. The mean number of cells for meningioma was 652 with SD of 141. The numbers of cells for 95% of patients were in between the 404 and 879 range. The mean number of immunopositive cells was 11.6 per field of view with SD of 8.4 cells (coefficient of variation equal 72.4%).

For oligodendroglioma, the mean cellularity was equal 447 cells with SD of 107 cells (coefficient of variation equal 32.1%). The total number of cells for 95% fields of view was between 268 and 636 cells. The ratio of the standard deviation of the mean value of cells in meningioma was 12.3% when analyzed more than ten different images corresponding to the same patient, and in oligodendroglioma this ratio was as high as 12.4%. Similarly to the case of H&E stained slides, these ratios again depended on highly changing cellularity of tumors and on the localization of the analyzed field of view.

The numerical results have shown a significant variability of density of cells even for a uniform group of tumors, i.e. the uniformity with respect to the cellularity as well as histological type of the malignancy. That variability is the result of the non-stable nature of the biological material. Moreover, the cellular density is sensitive to the way the biological material was obtained, the conditions of its formalin fixation, technology of staining and even the degree of the degenerative changes inside the neoplasm.

Remote Image Analysis and Virtual Slides

In recent years the image analysis systems gradually evolve to the remote applications. Using the system via Internet offers easy update of the programs, makes opportunity to create large image databases and extends analysis by using the available computational servers. One of the first remote system was EAMUS offered by DiagnomX GmbH, Germany (Kayser et al. 2005, 2006). The EAMUS program is available through the Internet when you register as a user. It offers a large selection of product evaluation options (different types of staining, magnification, etc.) and allows the quantitative analysis of IHC preparations, such as DAB and AP, and fluorescence (DAPI, FITC, Texas Red). The program accepts different types of reactions including Comet FISH, nuclear, membrane preparation for analysis. The images subject to analysis are loaded to the server in an on-line mode, and the result are delivered via the Internet. The acceptable magnification of the lens are 20×, 40× or higher for nuclear IHC reaction. An interesting option of the system is the possibility of automatic evaluation and correction of the image quality in terms of such parameters as brightness uniformity and standardization of the histogram.

Another remote Internet platform oriented to image analysis of the meningioma Ki-67 stain specimen was developed by the author of (Markiewicz 2011). It was called the Computerized Analysis of Medical Images (CAMI) software. In presented software implementation the remote image analysis realized using Matlab algorithms was proposed. The algorithms implemented on Matlab platform can be compiled to executable *jar* file with the help of Matlab Builder Java toolbox. The Matlab function must be declared with the set of input data, output structure with numerical results and Matlab web figure. Any function prepared in that manner can be used as a Java function in Java Servlet Pages (JSP). The graphical user interface providing the input data and displaying the results (also in graphical form) should be implemented in JSP. Additionally, the storage of the database can be implemented within algorithms written in Matlab with the help of Matlab Database Toolbox, directly with the image processing. The complete JSP page can be run by Tomcat server. The details of this solution were presented in the paper (Markiewicz 2011).

The other remote platform solution of the non-commercial system, working on the whole-slide images and dedicated to the pathomorphological image analysis was presented in the paper (Kong et al. 2008, 2009; Sertel et al. 2008). It is addressed to the computer-assisted grading of neuroblastic differentiation, based on H&E staining images. In this image analysis system, each tumor image is first segmented into multiple cytologic components. Next, the set of discriminating features (i.e., quantized measures of the physical, histopathologic, and statistical characteristics) are computed. These features are color and texture characterization of the regions in the image. The authors use supervised learning of the classifiers, combining their outputs into one integrated system in the testing mode. They take also into account many technics to improve the computational efficiency, such as gridding, parallel processing and using Graphics Processing Unit (GPU) or PlayStation. The authors applied their system to classify the grade of neuroblastic differentiation and classification of stromal development in neuroblastoma. The 87.9% accuracy of grading of the neuroblastic differentiation was reported, indicating the potential to apply these complex morphometric scheme in the future.

Discussion

The development of the automatic computer systems supporting the diagnostics of the central nervous system is an important trend in this field of research. This type of solution is especially useful for the tasks which require a lot of human effort, like recognizing and counting the cells in histopathologic slides of the central nervous systems.

The important advantage of using automatic systems in these application is the absolute repeatability of scores. This contrasts with the human expert results, which are greatly dependent on the particular expert and his or her actual mental and physical condition. The differences of scores of the human experts are due to their individual interpretations of the tumor cells and may be relatively large. In some extreme cases these differences have grown up to almost 30% in relative terms, which correspond to approximately 3–4% in absolute terms (still acceptable in medical practice). Moreover, the system is very quick in comparison to the human. It needs only seconds for full analysis of the considered image on a PC, compared to over a dozen minutes needed for the human expert analysis of the same image.

Unfortunately the actually existing commercial systems don't offer the solution to this problem in a satisfactory way. They cover only small fraction of the real need. Therefore a lot of effort of scientists is spent now to develop the non-commercial solutions which are able to help the medical staff in easing this tedious and time consuming task.

The actually presented solutions of an automatic or semi-automatic systems for cell recognition have proved their usefulness for the daily analysis of images in the brain tumors research. Their implementation in medical practice may significantly accelerate the process of image analysis, which is usually a very tedious and time consuming daily routine for most pathologists. It provides the evaluation of the proliferation rate in a very fast and easy way delivering the results as accurately as the human expert. Therefore such solutions will be widely used to support the scientific research of tumors in the future.

References

Carlotti CG Jr, Salhia B, Weitzman S, Greenberg M, Dirks PB, Mason W, Becker LE, Rutka JT (2002) Evaluation of proliferative index and cell cycle protein expression in choroid plexus tumors in children. Acta Neuropathol 103:1–10

Chellappa R, Chatterjee S (1985) Classification of texture using Gaussian Markov random field. IEEE Trans Acoust Speech Signal Process ASSP-33:959–963

Dehghani F, Schachenmayr W, Laun A, Korf HW (1998) Prognostic implication of histopathological, immunohistochemical and clinical features of oligodendrogliomas: a study of 89 cases. Acta Neuropathol 95:493–504

Giannini C, Scheithauer BW, Burger PC, Christensen MR, Wollan PC, Sebo TJ, Forsyth PA, Hayostek CJ (1999) Cellular proliferation in pilocytic and diffuse astrocytomas. J Neuropathol Exp Neurol 58:46–53

Heegaard S, Sommer HM, Broholm H, Broendstrup O (1995) Proliferating cell nuclear antigen and Ki-67 immunohistochemistry of oligodendrogliomas with special reference to prognosis. Cancer 76:1809–1813

Jaros E, Perry RH, Adam L, Kelly PJ, Crawford PJ, Kalbag RM, Mendelow AD, Sengupta RP, Pearson AD (1992) Prognostic implications of p53 protein, epidermal growth factor receptor, and Ki-67 labelling in brain tumours. Br J Cancer 66:373–385

Kayser G, Radziszowski D, Bzdyl P, Werner M, Kayser K (2005). Eamus – internet based platform for automated quantitative measurements in immunohistochemistry. In: Conference International Society for Cellular Oncology (ISCO), Belfast, 5–8 Apr 2005

Kayser G, Radziszowski D, Bzdyl P, Sommer R, Kayser K (2006) Theory and implementation of an electronic, automated measurement system for images obtained from immunohistochemically stained slides. Anal Quant Cytol 28:27–38

Kim YJ, Romeike BFM, Uszkoreit J, Feiden W (2006) Automated nuclear segmentation in the determination of the Ki-67 labeling index in meningiomas. Clin Neuropathol 25:67–73

Kong J, Sertel O, Boyer KL, Saltz JH, Gurcan MN, Shimada H (2008) Computer-assisted grading of neuroblastic differentiation. Arch Pathol Lab Med 132:903–904

Kong J, Sertel O, Shimada H, Boyer KL, Saltz JH, Gurcan MN (2009) Computer-aided evaluation of neuroblastoma on whole-slide histology images: classifying grade of neuroblastic differentiation. Pattern Recogn 42:1080–1092

Kurt E, Zheng PP, Hop WC, van der Weiden M, Bol M, van den Bent MJ, Avezaat CJ, Kros JM (2006) Identification of relevant prognostic histopathologic features in 69 intracranial ependymomas, excluding myxopapillary ependymomas and subependymomas. Cancer 106:388–395

Lebrun G, Charrier C, Lezoray O, Meurie C, Cardot H (2007) A fast and efficient segmentation scheme for cell microscopic image. Cell Mol Biol 53:51–61

Lezoray O, Cardot H (2002) Cooperation of color pixel classification schemes and color watershed : a study for microscopical images. IEEE Trans Image Process 11:783–789

Lezoray O, Lecluse M (2007) Automatic segmentation and classification of cells from broncho alveolar lavage. Image Anal Stereol 26:111–119

Lezoray O, Elmoataz A, Cardot H, Gougeon G, Lecluse M, Elie H, Revenu M (2000) Segmentation of color images from serous cytology for automated cell classification. Anal Quant Cytol 22:311–322

Mackenzie IR (1999) Central neurocytoma: histologic atypia, proliferation potential, and clinical outcome. Cancer 85:1606–1610

Maier H, Wanschitz J, Sedivy R, Rössler K, Ofner D, Budka H (1997) Proliferation and DNA fragmentation in meningioma subtypes. Neuropathol Appl Neurobiol 23:997–1000

Markiewicz T (2011) Using MATLAB software with Tomcat server and Java platform for remote image analysis in pathology. Diagn Pathol 6(Supp. 1):18

Markiewicz T, Wisniewski P, Osowski S, Patera J, Kozlowski W, Koktysz R (2009) Comparative analysis of the methods for accurate recognition of cells in the nuclei staining of the Ki-67 in neuroblastoma and ER/PR status staining in breast cancer. Anal Quant Cytol 31:49–62

Markiewicz T, Grala B, Kozlowski W, Osowski S (2010) Computer system for cell counting in selected brain tumors at Ki-67 immunohistochemical staining. Anal Quant Cytol 32:323–332

Otsu N (1979) A threshold selection method from grey-level histograms. IEEE Trans Syst Man Cybern 9:62–66

Rades D, Schild SE, Fehlauer F (2004) Prognostic value of the MIB-1 labelling index for central neurocytomas. Neurology 62:987–989

Schölkopf B, Smola A (2002) Learning with kernels. MIT Press, Cambridge, MA

Schwab M, Shimada H, Joshi V, Brodeur GM (2000) Neuroblastic tumours of adrenal gland and sympathetic nervous system. In: Kleihues P, Cavense WK (eds) Pathology and genetics of tumours of the nervous system. World Health Organization Classification of Tumours, International Agency for Research on Cancer (IARC), Lyon

Sertel O, Kong J, Shimada H, Catalyurek UV, Saltz JH, Gurcan MN (2008) Computer-aided prognosis of neuroblastoma on whole-slide images: classification of stromal development. Pattern Recognit 42:1093–1103

Shaffrey ME, Farace E, Schiff D, Larner JM, Mut M, Lopes MB (2005) The Ki-67 labeling index as a prognostic factor in Grade II oligoastrocytomas. J Neurosurg 102:1033–1039

Shimada H, Chatten J, Newton WA Jr, Sachs N, Hamoudi AB, Chiba T, Marsden HB, Misugi K (1984) Histopathologic prognostic factors in neuroblastic tumors: definition of subtypes of ganglioneuroblastoma and age-linked classification of neuroblastomas. J Natl Cancer Inst 73:405–416

Soille P (2003) Morphological image analysis, principles and application. Springer, Berlin

Soylemezoglu F, Scheithauer BW, Esteve J, Kleihues P (1997) Atypical central neurocytoma. J Neuropathol Exp Neurol 56:551–556

Vajtai I, Varga Z, Aguzzi A (1996) MIB-1 immunoreactivity reveals different labelling in low-grade and malignant epithelial neoplasms of the choroid plexus. Histopathology 29:147–151

Vapnik V (1998) Statistical learning theory. Wiley, New York

Wagner T (1999) Texture analysis. In: Jahne B, Haussecker H, Geisser P (eds) Handbook of computer vision and application. Academic Press, New York, pp 275–309

Warowny M, Markiewicz T (2010) Quick texture generation for multiobject image analysis in brain pathology. Prz Elektrotechn 86:50–52

Watanabe K, Ogata N, von Ammom K, Yonekawa Y, Nagai M, Ohgaki H, Kleihues P (1996) Immunohistochemical assessments of P53 protein accumulation and tumor growth fraction during the progression of astrocytomas. In: Nagai M (ed) Brain tumour research and therapy. Springer, Tokyo, pp 255–262

Part III

Ultrasonography

Intraoperative Ultrasonography in Tumor Surgery

13

Dorothea Miller

Contents

Abstract

Intraoperative ultrasonography (iUS) has been used in neurosurgery since the early 1980s. The intraoperative anatomy and pathology can be displayed in real-time in B-mode imaging whereas color Doppler adds flow information. Recent advances in imaging techniques such as improvements in image quality, navigating the ultrasound (US) probe and the introduction of three-dimensional (3D) US have increased the utility of iUS for neurosurgery. Numerous reports have shown that iUS helps in tumor visualization, location and delineation as well as planning the optimal approach to subcortical or deep-seated lesions. Intraoperative color Doppler can add vascular information, thus facilitating approach planning. Moreover, iUS is a tool that is fast and easy to apply when controlling the extent of tumor resection, as it is highly sensitive in experienced hands to detect tumor remnants. However, the examiner has to be aware, that tumor remnants might be overestimated by iUS during surgery and that a slim hyperechoic rim around the resection cavity might mask residual tumor. Intraoperative US can be an alternative in the guidance of biopsies and has shown to be helpful in spinal tumor cases. iUS is a simple way of intraoperative image update, providing real-time information in different stages of tumor resection.

D. Miller (✉)
Department of Neurosurgery, University Hospital Essen, Hufelandstrasse 55, Essen, Germany
e-mail: dorothea.miller@uk-essen.de

M. Hayat (ed.), *Tumors of the Central Nervous System, Volume 13*,
DOI 10.1007/978-94-007-7602-9_13, © Springer Science+Business Media Dordrecht 2014

Introduction

Ultrasonography (US) is an imaging technique whereby high-frequency sound waves are transmitted into the human body using transducers in contact with the surface of the area of interest. The ultrasound waves produced by the transducer reflect from boundaries between organs and surrounding fluid, and between regions of differing tissue density and are used to visualize inner body structures.

Intraoperative ultrasonography (iUS) has been used in neurosurgical procedures since the early 1980s (Rubin et al. 1980). Due to technical advances, iUS has gained an increasing significance for intraoperative real-time imaging in the past three decades. However, both the image acquisition and its interpretation are subjective and require adequate training and experience. Intraoperative US has therefore never gained the same acceptance as magnetic resonance imaging (MRI) or computed tomography (CT).

Technique

Basics of Ultrasound

Brightness mode (B-mode) is the main ultrasound mode used in neurosurgical procedures to depict the intraoperative anatomy and pathology. In B-mode imaging, an array of transducers simultaneously scans a certain plane through the brain tissue. The amplitude of the reflected echo is coded in a grey scale on a two-dimensional (2D) image.

In cases, where vascular information is needed, either color-coded Doppler sonography (color Doppler) or power Doppler can be applied. Doppler sonography makes use of the Doppler effect in measuring and visualizing blood flow. In color Doppler, the velocity information and the direction of the flow are presented as a color-coded overlay on top of a B-mode image. Power Doppler is an amplitude-coded color Doppler that measures the intensity of the Doppler signal.

Ultrasound Probes

For cranial procedures, mainly convex probes or sector probes with a frequency range of 5–10 MHz are used. Linear array probes with a frequency of 10–15 MHz can be used for superficial cortical lesions. However, the footprint should not exceed 2 cm to allow coupling to the convex cortical surface. For burr hole procedures, different ultrasound producers provide a bayonet-shaped US probe with tip dimensions of 8–12 mm. We use a multi frequency burr hole probe with a frequency range of 3–7.5 MHz and a scan angle of 90° with a specially designed puncture adapter that can be mounted to the probe. In spinal surgery, a high frequency (10–15 MHz) small linear probe that can be placed directly onto the dura allows optimal imaging. Alternatively, a small sector probe will fit into an incision for a hemi-laminectomy and can be used for spinal procedures if the wound is completely filled with saline for acoustic coupling.

Intraoperative Setup

Conventional Two-Dimensional Ultrasound

It is important to consider certain aspects of the intraoperative setup to get good imaging results: The site of the craniotomy should be the highest point of the head, so that the resection cavity can be filled with saline after resection. Sterile saline or sterile gel pads can be used for acoustic coupling. Some manufactures provide probes that can be sterilized. However, most US probes will need sterile draping. Gel should be used between the sterile sleeve cover and the probe to allow acoustic coupling. The focus should be adjusted depending on the depth of the lesion. An overview scan with maximum depths should be performed to gather orientation and to visualize anatomical landmarks. The pathology itself should be scanned in at least two perpendicular planes. A near axial, coronal or sagittal scanning plane (depending on the localization of the lesion and the localization of the craniotomy) is helpful to improve orientation, as these are the planes familiar to the neurosurgeon. It is important to perform a scan before and after resection in all cases, as a proper post-resection exam for tumor remnants needs to be compared to pre-resection images. Areas with a similar echogenicity as the tumor on pre-resection images are suspicious of tumor remnants. The resection cavity should be free of blood or hemostypics and air bubbles for optimal imaging.

In spinal cases, a small linear high-frequency probe that can be applied directly onto the dura will give optimal image quality. Again, sterile saline can be used for acoustic coupling. If the probe does not fit into the bony opening, the wound can be filled with sterile saline and a scan can be performed through the bony opening.

Navigated Ultrasound

In navigated ultrasound (navUS), the ultrasound probe is tracked by a navigation system (Koivukangas et al. 1993). This can be done either by connection the navigation system and the ultrasound platform (see below) or by using an integrated ultrasound-navigation device (Gronningsaeter et al. 2000; Tirakotai et al. 2006).

Three-dimensional (3D) MRI of the patient's head is performed preoperatively for navigation. The patient's head is fixed in a head clamp and a reference star for optical tracking is mounted onto the clamp. Then a patient-to-image registration is performed using anatomical landmarks, skin fiducials or a surface match. The ultrasound system is connected to the navigation system. The ultrasound probe is draped and a tracking device is mounted onto the ultrasound probe. Then, the ultrasound probe is calibrated using a calibration phantom.

Ultrasound images can be displayed side-by-side with the corresponding MRI reconstruction in the same image plane on the navigation screen or as overlay images.

Three-Dimensional Ultrasound

So far, 3D probes (mechanical 3D probes or 3D arrays) have not yet been widely used in neurosurgery as they are often rather bulky or have a convex surface, making coupling to the convex cortical surface more difficult.

Most commonly, multiple tracked 2D-scans of the area of interest are reconstructed into a 3D volume (Trantakis et al. 2002). Axial, coronal, sagittal or any-plane slices are then displayed on the navigation screen (Fig. 13.1). The quality of the 3D image therefore depends on the quality of the 2D scans, tracking accuracy and the algorithm used for 3D reconstruction (Miller et al. 2012).

Contrast-Enhanced Ultrasound

In contrast-enhanced US, a contrast-enhancing agent is applied intravenously during the examination. Contrast-enhancing agents consist of microbubbles filled with gas. Microbubbles have a high echogenicity, thus they have a high ability to reflect echo waves. Moreover, the microbubbles begin to oscillate during insonation in response to the ultrasonic energy field and therefore emit waves with a characteristic (harmonic) spectrum of frequencies. This can easily be distinguished from tissue signal. Contrast-enhanced transcranial ultrasonography has been successfully used for perfusion imaging in cerebral ischemia and in the examination of intracerebral neoplasms. However, there is only limited data on the intraoperative use of contrast-enhanced US. So far, it has not yet reached intraoperative routine.

Applications

Tumor Visualization and Delineation

Conventional Two-Dimensional Ultrasound

Early studies on iUS focused on tumor visualization, characterization and delineation (LeRoux et al. 1992; van Velthoven and Auer 1990). We therefore know today, that most tumors show a higher echogenicity than normal brain tissue, irrespective of whether they show contrast-enhancement on MRI or CT. Surrounding edema can be distinguished from tumor tissue as it is less echogenic, but boundar-

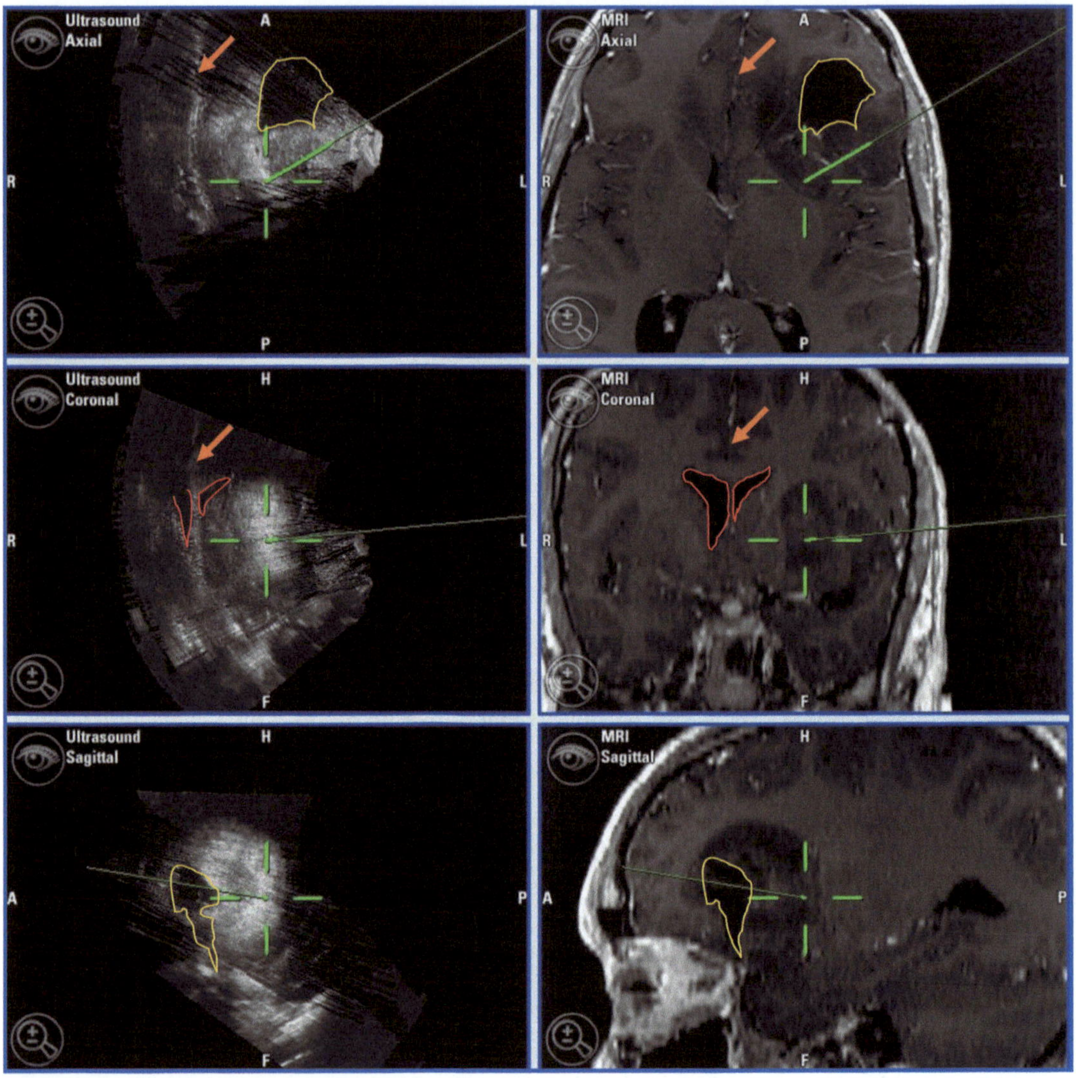

Fig. 13.1 Three-dimensional navUS of a recurrent low-grade glioma is displayed side-by-side with preoperative MRI in axial, coronal and sagittal reconstructions. Ultrasound and MR images correspond exactly to each other. The ventricles are outlined in *red*, the resection cavity of the previous operation is outlined in *yellow* and the falx is marked with an *orange arrow*

ies might be ill demarcated in some cases. Low-grade gliomas are usually homogeneous and have a moderate echogenicity. High-grade gliomas often appear inhomogeneous depending on tumor cysts, necrosis or intratumoral hemorrhage and are usually more echogenic than low-grade gliomas. Hyperechogenic areas within the tumor with an acoustic shadow deep to the lesion are due to calcifications and are more common in oligodendrogliomas. Metastases are highly echogenic and mostly well demarcated. They may appear either homogeneous or inhomogeneous depending on their histopathology. Tumor cysts are usually an- or hypoechoic like the ventricles with a small echoic rim surrounding the cyst (LeRoux et al. 1992). Intratumoral hemorrhage appears highly echogenic in case of a new hemorrhage. Two weeks after hemorrhage, the central parts of the hematoma become iso- to hypoechoic and 6–8 weeks later, the

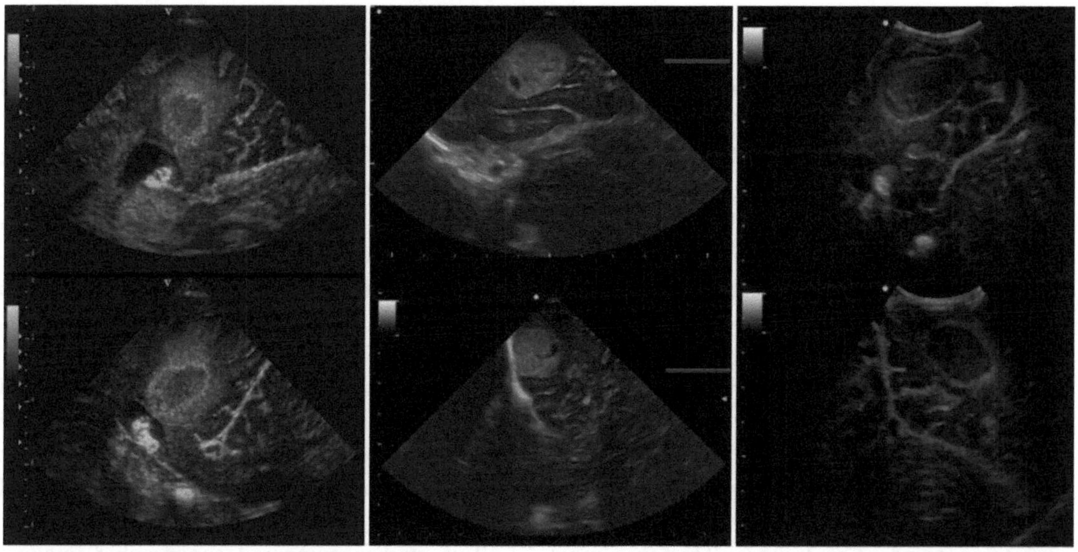

Fig. 13.2 Conventional 2D iUS showing different tumor entities in two perpendicular image planes. *Left*: Left occipital glioblastomas. The tumor is displayed as a hyperechoic rim surrounding a only slightly hyperechoic area that represents the central necrosis. The tumor is surrounded by a hyperechoic edema. Nearby, the dorsal horn of the lateral ventricle with its plexus and the falx are displayed. *Middle*: Right temporal metastasis of a renal cell carcinoma. Please note the rather homogeneous appearance of the rumor with a small hypoechoic cyst. *Right*: Left occipital metastasis of a renal cell carcinoma after intratumoral hemorrhage 3 weeks prior to surgery. The old hematoma is nearly anechogenic by now, surrounded by a hyperechoic rim

hematoma is isoechoic (Enzmann et al. 1981). As an example how different tumor entities may appear on iUS, please refer to Fig. 13.2.

Contrast-Enhanced Ultrasound

Otsuki et al. (2001) presented a coded harmonic angio mode to visualize tumor vessels intraoperatively in a few number of patients. Harmonic imaging with US contrast agents improved the signal-to-noise ratio between brain parenchyma and vasculature in these patients. Kanno et al. (2005) studied 40 patients with an US contrast-enhancing agent during operations. Power Doppler US with contrast enhancement provided better data on the precise real-time position of the tumors and their relationship to adjacent vessels than US obtained before the injection of the contrast agent. Contrast-enhanced power Doppler correlated well with digital subtraction angiography. Engelhardt et al. (2007) performed contrast-enhanced imaging with a low mechanical index during resection of different brain tumors in a small number of patients. Scans were inspected in real-time during tumor resection, and radiofrequency ultrasound data was analyzed off-line. The perfusion pattern of the tumor was semi-quantitatively evaluated and compared with that in normal brain tissue. The distinction of tumor from non-infiltrated parenchyma was facilitated by the influx of contrast agent.

Three-Dimensional Ultrasound

Unsgaard et al. (2005) examined the ability of 3D US to delineate gliomas and metastases in 28 operations by taking 85 biopsies from tumor borders. US findings were in agreement with histopathology in 74% of low-grade astrocytomas, 83% of anaplastic astrocytomas, 77% of glioblastomas and 100% of metastases.

Lesion Localization/Planning the Optimal Approach

Numerous reports on the utility of iUS as a tool for intraoperative navigation exist. Moiydi and Shetty (2011) evaluated the use of conventional

2D iUS in 77 cases of both cranial and spinal lesions. In their hands, iUS was useful in identifying and delineating lesions in 100% of cases, in planning the durotomy in 78% of cases, in planning the corticectomy in 85% of cases and in visualizing adjacent structures in 96% of cases, respectively. They concluded that iUS is a useful tool in planning various stages of tumor resection. In a study comparing conventional iUS and navUS, we could show, that orientation can even be improved by navigating the US probe, thus facilitation anatomical understanding (Miller et al. 2007). Moreover navUS was useful in planning the correct transsulcal approach to subcortical lesions in eloquent areas (Zhou et al. 2009) and in visualizing structures adjacent to tumors. This was most obvious in cystic gliomas (Enchev et al. 2006), where an opening of a cyst during surgery might lead to a local tissue shift. With the aid of navUS, vessels that might be encased or displaced by tumor tissue could be visualized and landmarked as reported by our group (Sure et al. 2005). Rygh et al. (2006) showed that 3D US was helpful in visualizing hidden vessels adjacent to and inside the tumor thus facilitating tailored approaches and increasing the safety of biopsies. Similar to navigated 2D US, navigated 3D US solved the orientation problem experienced with conventional 2D US in neurosurgery as it allowed axial, coronal and sagittal reconstructions and a direct comparison of iUS and preoperative MRI in the same plane of view (Miller et al. 2012; Unsgaard et al. 2002). Three-dimensional US could be used as a navigational tool even without preoperative MRI in cases where no preoperative MRI was available, an intraoperative brain shift occurred or in case of a failure of the navigation system (Miller et al. 2011; Peredo-Harvey et al. 2012).

Resection Control

Two-Dimensional Ultrasound

The use of iUS for resection control during surgery of intracerebral tumors is documented in numerous reports. However, larger randomized controlled trials are lacking.

In an early study by LeRoux et al. (1994), US tumor volume estimates tended to be nonsignificantly larger than T1 gadolinium-enhanced volumes on preoperative MRI. The authors concluded that iUS might therefore aid in detecting areas of tumor extension beyond margins of blood-brain barrier breakdown. Tumor volumes on T2 weighted MRI were larger than US volumes except for low grade gliomas, again this was not statistically significant. US images helped to differentiate edema from solid tumor and normal brain. Overall, the authors concluded that, US information might help to enhance tumor resection. However, there were difficulties in estimating tumor volumes in previously treated lesions where gliosis was present.

Two studies compared sonographic imaging results after resection to histopathology by taking probes from various points at the resection margin. Woydt et al. (1996) acquired a series of 78 biopsies from the resection cavity under continuous sonographic control at the end of surgery in 45 patients with gliomas. 47 out of 53 biopsies (89%) taken from tissue sonographically suspected as tumor revealed solid tumor tissue of which 72% were microscopically assessed as unsuspicious by the surgeon. Six samples contained tumor infiltration zone. Six of twenty-five biopsies taken from sonographically tumor-free borders were diagnosed as tumor tissue. Of 34 cases with gross total removal according to the surgeon's assessment 25 showed sonographic signs of residual tumor tissue, which was confirmed histologically as solid tumor tissue in 22 of these cases. The authors concluded, that iUS following resection of supratentorial gliomas could detect residual tumor tissue with high specificity and thus improve gross total resection. In a second study by Chacko et al. (2003), iUS was useful in defining lesion margins in 71.4% of 35 cases of intracerebral lesions, however in previously radiated cases the margins were ill-defined. 97% of samples from the resection margins were correctly identified as tumor. There were two false negative cases (3%). However, there was a high rate of false positive results (46% of histologically negative probes).

Hammoud et al. (1996) performed a prospective study of 70 patients with a variety of different intraparenchymal brain lesions in 1996 to evaluate the efficacy of iUS in localizing and defining the borders of tumors and in assessing the extent of their resection (as compared to postoperative MRI). In gliomas without previous therapy, tumors were well localized in all patients, margins were well defined in 15/18 cases and the extent of resection was well defined in all patients, respectively. In patients with gliomas that had undergone previous therapy and showed post-radiation changes, the extent of resection was poorly defined. All metastatic lesions were well localized, had well-defined margins and the extent of resection was well defined on iUS.

In a comparative study by Gerganov et al. (2009), iMRI and 2D iUS were compared for resection control in 26 patients with various tumor entities. The authors showed that small tumor remnants (<1 cm) were not detected by the examiner in 2 of 21 patients with tumor remnants identified on 1.5 T iMRI. Suspicious signals on iUS were misinterpreted as tumor remnants in another two cases. A statistical analysis was not performed in this study.

In a recent study by the same group on resection control for low-grade gliomas, tumor borders prior to resection were clear on iUS in 9 of 11 patients as compared to 10 of 11 cases on iMRI (Gerganov et al. 2011). Image quality diminished during surgery, leading to difficulties in interpretation. One tumor remnant was not identified on iUS, and one artifact was falsely interpreted as a tumor remnant. Intraoperative MRI appeared superior for resection control in these cases.

Contrast-Enhanced Ultrasound

There is very little data on the use of contrast-enhanced US for resection control. Preliminary data in a very low number of cases suggested that remaining tumor tissue could be distinguished from surrounding brain tissue more easily as compared to conventional US (He et al. 2008).

Three-Dimensional Ultrasound

Rygh et al. (2008) compared 3D US to histopathological findings at different stages of resection. They could show a very high sensitivity (95%) and specificity (95%) to delineate tumor before resection. During resection, sensitivity dropped slightly to 88%, but specificity was only 42%. After resection sensitivity dropped to 26%, but specificity increased to 88%, while both positive and negative predictive value were 62%.

To date, only one study has compared low-field 0.2-T iMRI and 3D US in seven patients only (Tronnier et al. 2001). The ability to detect tumor was comparable in both imaging modalities in high-grade gliomas and metastases, whereas in the case of a low-grade glioma US was more helpful. The reliability of the ultrasound diminished during tumor resection, so that tumor remnants were not at all or were poorly visible in four patients.

The value of 3D US was evaluated by correlating 3D US and postoperative MRI in a small randomized blinded prospective study by Rohde and Coenen (2011). Three-dimensional US correctly identified tumor remnant in five of seven patients (71%) and correctly confirmed complete tumor removal in three of five patients (60%). Overall the intraoperative situation was correctly visualized in 8 of 12 patients (67%).

Figure 13.3 shows an example of a 3D US of a glioblastoma prior to and after resection.

Visualizing and Compensating Brain-Shift

Deformations of the brain during surgery may be due to loss of cerebro-spinal fluid, swelling or retraction. These deformations may range from 2 to 25 mm depending on the size, the location and the kind of tumor (Lindner et al. 2006).

Several groups worked on compensating brain shift via a co-registration of iUS and preoperative MRI (Coenen et al. 2005; Reinertsen et al. 2007). Coenen et al. (2005) defined ultrasound landmarks in the vicinity of the tumor and the nearby fiber tracts and used a sequential landmark registration to assess fiber tract deformation. They could show on postoperative diffusion-weighted imaging that the actual fiber tract position at the end of surgery corresponded to the sonographically predicted fiber tract position. Reinertsen et al. (2007) validated a vessel-based registration technique for the correction of brain shift. There are multiple reports on other possible co-registration- and deformation algorithms, however

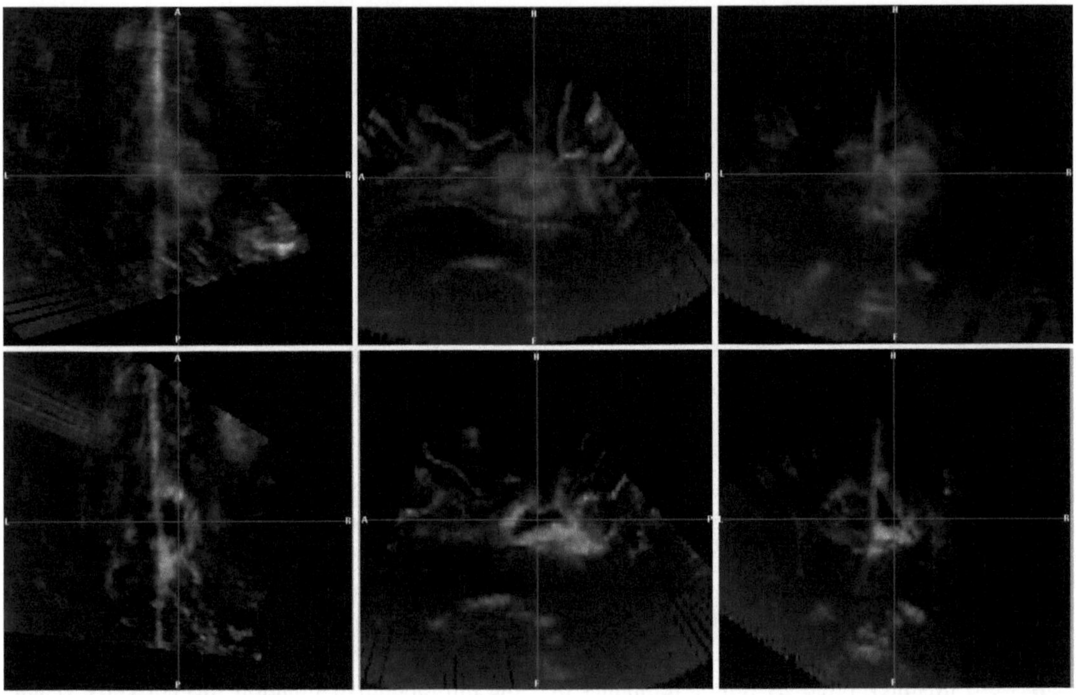

Fig. 13.3 Three-dimensional US of a left parasagittal glioblastomas (*left to right*: axial, sagittal and coronal views). *Upper row*: Glioblastoma before resection. *Lower row*: Resection cavity after resection. Please note, that the resection cavity is partly filled with blood which appears very hyperechoic and should not be mistaken for tumor. The tumor tissue prior to surgery appeared less echogenic than blood

large patient trials are still lacking. Further studies are needed to create robust deformation models to allow fusion of MRI and iUS.

Guidance of Biopsies

Tsutsumi et al. (1989) first achieved real-time imaging in single burr hole procedures in 1989. Histologic diagnosis could be established in 87–95% of cases of US-guided tumor biopsies via a single burr hole trepanation (Benediktsson et al. 1992; Di Lorenzo et al. 1991; Lunardi et al. 1993; Strowitzki et al. 2000).

Ultrasound for Intra- and Extramedullary Spinal Tumors

As shown by our group and others, iUS can be used for a differential diagnosis of different intradural spinal tumors. While intramedullary tumors showed an inhomogeneous pattern, extramedullary tumors were displayed with a homogeneous signal on iUS. Extramedullary tumors were sharply demarcated from the myelon in all cases. Moreover, iUS proved to be useful in displaying the size and exact location of the lesion, thus allowing adjusting the extension of the bony opening (Regelsberger et al. 2005; Zhou et al. 2011).

Discussion

Technical Considerations

iUS allows real-time imaging, thus giving a direct update of the intraoperative anatomy. Only minimal preparation time is needed for scanning. As compared to iMRI, it is less time-consuming, easily repeatable and far less expensive. No special operating room or operating equipment is needed. Ultrasound equipment and disposables are relatively low in cost and basically everywhere available. Moreover, there is no irradiation with the technique. The operating surgeon can perform the examination and can gain an immediate feedback on the intraoperative situation as well as real-time blood flow information. Intraoperative US can even be combined with neuronavigation or endoscopy.

However, there are some major disadvantages using iUS: the limited insonation window of the cranial opening only allows certain planes of view, usually in an oblique orientation. This makes orientation more difficult for the neurosurgeon who is used to axial, coronal and sagittal imaging (Miller et al. 2007). Visualizing anatomical landmarks might as well be limited by the cranial opening. Moreover, image quality and image interpretation depend on the experience and skill of the examiner. Artifact formation due to air bubbles or blood within the surgical field may lead to misinterpretation. Convex probes may reduce the contact area to the brain or dura surface, leading to reduced image resolution. Depth penetration may be limited depending on the probe frequency, lateral resolution is decreased for deep targets. Overall, signal-to-noise ratio is lower than for iMRI. It can be difficult to find the exact view again for monitoring the course of an operation when performing serial images with 2D iUS. Thus in the neurosurgical setting, it can be difficult to compare 2D iUS during ongoing surgery to monitor the extent of a tumor resection. Navigated US addresses some of the drawbacks. We could show that orientation is markedly improved by displaying US images and reconstructed MRI images in the same image plane side by side or as an image overlay (Miller et al. 2007). Comparing MRI and US facilitated the interpretation of the US images. It increased the learning curve and might even aid in resection control. Anatomical landmarks and vessels can be marked (Sure et al. 2005). In 3D-US again orientation is improved, comparison to pre-resection scans is easier. However, a patient-to-image registration as well as a calibration of the US probe is needed for navUS and 3D US. This will increase preparation time before surgery. Signal-to-noise ratio is improved in contrast-enhanced ultrasound as compared to conventional iUS. However, there is only limited data concerning the intraoperative use of contrast-enhanced US in neurosurgery. Further studies are needed to clarify its value in monitoring tumor resection.

Applications

Even though there are numerous reports on different aspects of iUS, most studies only contain small case series. Larger prospective trials are lacking.

Tumor Visualization and Resection Control

A number of imaging methods are used for intraoperative tumor visualization and resection control with high-field iMRI being the gold standard in intraoperative imaging. However, iMRI is a logistically demanding, time-consuming and expensive device. Intraoperative US, on the other hand, is a less expensive and less time-consuming alternative to iMRI for tumor visualization and resection control. Ultrasound provides information on tumor size, distance from the cortical surface and surrounding structures. Most tumors can be visualized well on 2D iUS (LeRoux et al. 1992, 1994) and tumor margins can be defined clearly prior to resection (LeRoux et al. 1994; Moiyadi and Shetty 2011). Tumor margins might be less defined in cases of radiation changes (LeRoux et al. 1992), to our experience however, iUS does not seem to be inferior to preoperative MRI in these difficult cases. In two comparative studies, iUS was only marginally inferior to iMRI in delineating tumor (Gerganov et al. 2009, 2011). Three-dimensional US shows a similarly good ability to visualize tumor as 2D US. The sensitivity to delineate tumor margins was 74–100% prior to resection, depending on histology in a study by Unsgaard et al. (2005). Data from larger trials on contrast-enhanced US for tumor visualization and delineation is still lacking. However, this seems to be a promising technique to improve signal-to-noise ratio (Engelhardt et al. 2007; Kanno et al. 2005; Otsuki et al. 2001). Thus, iUS is a good tool to localize, visualize and delineate intraparenchymal brain tumors.

2D iUS showed to be highly sensitive in detecting tumor remnants (89–97%) when comparing US imaging with histopathology (Chacko et al. 2003; Woydt et al. 1996). However, specificity was only 54–76%, which could lead to an overestimation of tumor remnants. Moreover, a slim hyperechogenic rim is usually seen around the resection cavity and is an unspecific sign. It can mask thin tumor remnants, as pointed out by Woydt et al. (1996). Rygh et al. (2008) showed that 3D US was highly sensitive in the detection of tumor remnants during resection. However, sensitivity dropped by the end of the operation, while the positive predictive value remained acceptable. The drop in sensitivity can be explained by the low number of biopsies classified as tumor at this stage of the operation as compared to normal tissue. Moreover, it is not surprising to find infiltrated tissue in the hyperechogenic rim around the resection cavity. According to Gerganov et al. (2009, 2011) and Tronnier et al. (2001), both 2D and 3D iUS might be inferior in detecting tumor remnants compared to iMRI. However, the studies by Gerganov had several limitations: (1) Comparative analysis was done in a small number of patients only; no statistical analysis was performed. (2) Patients were positioned in a way that was suitable for iMRI, but did not always allow a good post-resection scan with iUS. (3) Intraoperative US is an examination that depends on the experience of the user. The group seemed to be more experienced in iMRI than in iUS. The low number of patients and the fact, that they used a prototype of a now commercially available US-navigation-system, limits Tronnier's study. The interpretation of the US images might have been suboptimal as the examiners were at the beginning of their learning curve with the new technology. Larger studies, that compare 3D US and iMRI are lacking.

We can conclude that, intraoperative 2D US is highly sensitive in detection tumor remnants, randomized prospective trials to compare it to iMRI are lacking. Data on 3D US in a small number of patients also suggest a good sensitivity to detect tumor remnants during resection. However, further studies are needed to confirm these findings. The utility of iUS for resection control depends on the experience of the examiner. Especially post-resection iUS is prone to artifact formation due to blood and air bubbles within the surgical field. Moreover contused adjacent tissue might show up hyperechogenic on iUS and might be mistaken for tumor remnants. Navigated US can possibly improve the learning curve by helping in comparing pre-resection iUS, MRI and post-resection iUS. Moreover, tumor remnants can be landmarked by the navigation system, thus improving orientation.

Up to date, there is no good data on contrast-enhanced US for resection control. Further studies are needed to evaluate post-resection changes of contrast-enhancement for resection control.

Ultrasound as a Tool for Navigation

Numerous reports could show the value of iUS for intraoperative navigation. Moiyadi and Shetty (2011) evaluated iUS as a useful tool in various stages of tumor resection in 78–100% of cases. We could show that navUS was especially useful in planning the correct transsulcal approach to subcortical lesions (Zhou et al. 2009), in improving orientation (Miller et al. 2007), in displaying the exact location of structures surrounding the tumor especially in cases of cystic tumors where an opening of the cyst might lead to local brain shift (Enchev et al. 2006) and in displaying and landmarking vessels that might be displaced or encased by tumor (Sure et al. 2005). Similar to navUS, 3D US facilitates orientation and helps to visualize vessel, thus facilitating tailored approaches (Unsgaard et al. 2002).

Visualizing and Compensating Brain Shift

Deformations of the brain during surgery include small rotations, translations and morphological changes due to swelling, ongoing resection, gravity and loss of cerebro-spinal fluid. Brain shift or brain deformations limit the effectiveness of image guidance systems that rely on preoperative imaging modalities as inaccuracies up to 25 mm may occur (Lindner et al. 2006). Intraoperative navUS or 3D US may enhance the utility of neuronavigation by adding information on brain shift (Lindner et al. 2006). The co-registration of iUS and preoperative MRI is a promising way to compensate brain shift (Coenen et al. 2005; Reinertsen et al. 2007). However, further studies are needed to create robust deformation models to allow fusion of MRI and iUS.

Guidance of Biopsies

Reports on US-guided biopsies have shown a high diagnostic yield. US-guided biopsies are, in experienced hands, a cheap and fast alternative to navigated biopsies or stereotactic biopsies. However, the surgeon has to bear in mind that image quality is reduced as compared to conventional iUS due to the small footprint, a usually slightly lower frequency and possible artifacts by the bone edge of the burr hole. Therefore most authors recommend taking biopsies from supratentorial lesions >15 mm. Stereotactic techniques can be reserved to smaller lesions and more difficult regions (Benediktsson et al. 1992; Di Lorenzo et al. 1991; Lunardi et al. 1993; Strowitzki et al. 2000).

Spinal US

As shown by our group and others, intra- and extramedullary tumors can easily be distinguished on iUS. Moreover, the dura, the spinal cord, the central canal, nerve roots, vertebral bodies, disc spaces and syrinxes can be displayed by iUS. Intraoperative US can help in defining the extension of the tumor and adjust bony opening (Regelsberger et al. 2005; Zhou et al. 2011).

Conclusion

Intraoperative US is a useful tool for tumor visualization, location, delineation and planning of the optimal approach. Moreover, real-time vascular information can be added. In experienced hands, it is an alternative to other techniques to control tumor resection and guide biopsies. However, the examiner has to be aware of the limitations of the technique. Future trials may involve the utility of contrast-enhanced ultrasound in resection control as well as further studies on brain shift compensation by fusion of iUS and preoperative MRI.

References

Benediktsson H, Andersson T, Sjolander U, Hartman M, Lindgren PG (1992) Ultrasound guided needle biopsy of brain tumors using an automatic sampling instrument. Acta Radiol 33:512–517

Chacko AG, Kumar NK, Chacko G, Athyal R, Rajshekhar V (2003) Intraoperative ultrasound in determining the extent of resection of parenchymal brain tumours–a comparative study with computed tomography and histopathology. Acta Neurochir 145:743–748 (Wien)

Coenen VA, Krings T, Weidemann J, Hans FJ, Reinacher P, Gilsbach JM, Rohde V (2005) Sequential visualization of brain and fiber tract deformation during intracranial surgery with three-dimensional ultrasound: an approach to evaluate the effect of brain shift. Neurosurgery 56:133–141

Di Lorenzo N, Esposito V, Lunardi P, Delfini R, Fortuna A, Cantore G (1991) A comparison of computerized tomography-guided stereotactic and ultrasound-guided techniques for brain biopsy. J Neurosurg 75:763–765

Enchev Y, Bozinov O, Miller D, Tirakotai W, Heinze S, Benes L, Bertalanffy H, Sure U (2006) Image-guided ultrasonography for recurrent cystic gliomas. Acta Neurochir 148:1053–1063 (Wien)

Engelhardt M, Hansen C, Eyding J, Wilkening W, Brenke C, Krogias C, Scholz M, Harders A, Ermert H, Schmieder K (2007) Feasibility of contrast-enhanced sonography during resection of cerebral tumours: initial results of a prospective study. Ultrasound Med Biol 33:571–575

Enzmann DR, Britt RH, Lyons BE, Buxton JL, Wilson DA (1981) Natural history of experimental intracerebral hemorrhage: sonography, computed tomography and neuropathology. AJNR Am J Neuroradiol 2:517–526

Gerganov VM, Samii A, Akbarian A, Stieglitz L, Samii M, Fahlbusch R (2009) Reliability of intraoperative high-resolution 2D ultrasound as an alternative to high-field strength MR imaging for tumor resection control: a prospective comparative study. J Neurosurg 111:512–519

Gerganov VM, Samii A, Giordano M, Samii M, Fahlbusch R (2011) Two-dimensional high-end ultrasound imaging compared to intraoperative MRI during resection of low-grade gliomas. J Clin Neurosci 18:669–673

Gronningsaeter A, Kleven A, Ommedal S, Aarseth TE, Lie T, Lindseth F, Lango T, Unsgard G (2000) SonoWand, an ultrasound-based neuronavigation system. Neurosurgery 47:1373–1379

Hammoud MA, Ligon BL, elSouki R, Shi WM, Schomer DF, Sawaya R (1996) Use of intraoperative ultrasound for localizing tumors and determining the extent of resection: a comparative study with magnetic resonance imaging. J Neurosurg 84:737–741

He W, Jiang XQ, Wang S, Zhang MZ, Zhao JZ, Liu HZ, Ma J, Xiang DY, Wang LS (2008) Intraoperative contrast-enhanced ultrasound for brain tumors. Clin Imaging 32:419–424

Kanno H, Ozawa Y, Sakata K, Sato H, Tanabe Y, Shimizu N, Yamamoto I (2005) Intraoperative power Doppler ultrasonography with a contrast-enhancing agent for intracranial tumors. J Neurosurg 102:295–301

Koivukangas J, Louhisalmi Y, Alakuijala J, Oikarinen J (1993) Ultrasound-controlled neuronavigator-guided brain surgery. J Neurosurg 79:36–42

LeRoux PD, Berger MS, Wang K, Mack LA, Ojemann GA (1992) Low grade gliomas: comparison of intraoperative ultrasound characteristics with preoperative imaging studies. J Neurooncol 13:189–198

LeRoux PD, Winter TC, Berger MS, Mack LA, Wang K, Elliott JP (1994) A comparison between preoperative magnetic resonance and intraoperative ultrasound tumor volumes and margins. J Clin Ultrasound 22:29–36

Lindner D, Trantakis C, Renner C, Arnold S, Schmitgen A, Schneider J, Meixensberger J (2006) Application of intraoperative 3D ultrasound during navigated tumor resection. Minim Invasive Neurosurg 49:197–202

Lunardi P, Acqui M, Maleci A, Di Lorenzo N, Fortuna A (1993) Ultrasound-guided brain biopsy: a personal experience with emphasis on its indication. Surg Neurol 39:148–151

Miller D, Heinze S, Tirakotai W, Bozinov O, Surucu O, Benes L, Bertalanffy H, Sure U (2007) Is the image guidance of ultrasonography beneficial for neurosurgical routine? Surg Neurol 67:579–587

Miller D, Benes L, Sure U (2011) Stand-alone 3D-ultrasound navigation after failure of conventional image guidance for deep-seated lesions. Neurosurg Rev 34:381–387; discussion 387–388

Miller D, Lippert C, Vollmer F, Bozinov O, Benes L, Schulte DM, Sure U (2012) Comparison of different reconstruction algorithms for three-dimensional ultrasound imaging in a neurosurgical setting. Int J Med Robot 8(3):348–359

Moiyadi A, Shetty P (2011) Objective assessment of utility of intraoperative ultrasound in resection of central nervous system tumors: a cost-effective tool for intraoperative navigation in neurosurgery. J Neurosci Rural Pract 2:4–11

Otsuki H, Nakatani S, Yamasaki M, Kinoshita A, Iwamoto F, Kagawa N (2001) Intraoperative ultrasound arteriography with the "Coded Harmonic Angio" technique. Report of three cases. J Neurosurg 94:992–995

Peredo-Harvey I, Lilja A, Mathiesen T (2012) Post-craniotomy neuronavigation based purely on intraoperative ultrasound imaging without preoperative neuronavigational planning. Neurosurg Rev 35:263–268; discussion 268

Regelsberger J, Fritzsche E, Langer N, Westphal M (2005) Intraoperative sonography of intra- and extramedullary tumors. Ultrasound Med Biol 31:593–598

Reinertsen I, Lindseth F, Unsgaard G, Collins DL (2007) Clinical validation of vessel-based registration for correction of brain-shift. Med Image Anal 11:673–684

Rohde V, Coenen VA (2011) Intraoperative 3-dimensional ultrasound for resection control during brain tumour removal: preliminary results of a prospective randomized study. Acta Neurochir Suppl 109:187–190

Rubin JM, Mirfakhraee M, Duda EE, Dohrmann GJ, Brown F (1980) Intraoperative ultrasound examination of the brain. Radiology 137:831–832

Rygh OM, Nagelhus Hernes TA, Lindseth F, Selbekk T, Brostrup MT, Unsgaard G (2006) Intraoperative navigated 3-dimensional ultrasound angiography in tumor surgery. Surg Neurol 66:581–592

Rygh OM, Selbekk T, Torp SH, Lydersen S, Hernes TA, Unsgaard G (2008) Comparison of navigated 3D ultrasound findings with histopathology in subsequent phases of glioblastoma resection. Acta Neurochir 150:1033–1041, (Wien) discussion 1042

Strowitzki M, Moringlane JR, Steudel W (2000) Ultrasound-based navigation during intracranial burr hole procedures: experience in a series of 100 cases. Surg Neurol 54:134–144

Sure U, Benes L, Bozinov O, Woydt M, Tirakotai W, Bertalanffy H (2005) Intraoperative landmarking of vascular anatomy by integration of duplex and Doppler ultrasonography in image-guided surgery. Technical note. Surg Neurol 63:133–141

Tirakotai W, Miller D, Heinze S, Benes L, Bertalanffy H, Sure U (2006) A novel platform for image-guided ultrasound. Neurosurgery 58:710–718

Trantakis C, Meixensberger J, Lindner D, Strauss G, Grunst G, Schmidtgen A, Arnold S (2002) Iterative neuronavigation using 3D ultrasound. A feasibility study. Neurol Res 24:666–670

Tronnier V, Bonsanto MM, Staubert A, Knauth M, Kunze S, Wirtz CR (2001) Comparison of intraoperative MR imaging and 3D-navigated ultrasonography in the detection and resection control of lesions. Neurosurg Focus 10:1–5

Tsutsumi Y, Andoh Y, Sakaguchi J (1989) A new ultrasound-guided brain biopsy technique through a burr hole. Technical note. Acta Neurochir 96:72–75 (Wien)

Unsgaard G, Ommedal S, Muller T, Gronningsaeter A, Nagelhus Hernes TA (2002) Neuronavigation by intraoperative three-dimensional ultrasound: initial experience during brain tumor resection. Neurosurgery 50:804–812

Unsgaard G, Selbekk T, Brostrup MT, Ommedal S, Torp SH, Myhr G, Bang J, Nagelhus Hernes TA (2005) Ability of navigated 3D ultrasound to delineate gliomas and metastases–comparison of image interpretations with histopathology. Acta Neurochir 147:1259–1269 (Wien) van Velthoven V, Auer LM (1990) Practical application of intraoperative ultrasound imaging. Acta Neurochir 105:5–13 (Wien)

Woydt M, Krone A, Becker G, Schmidt K, Roggendorf W, Roosen K (1996) Correlation of intra-operative ultrasound with histopathologic findings after tumour resection in supratentorial gliomas. A method to improve gross total tumour resection. Acta Neurochir 138:1391–1398 (Wien)

Zhou H, Miller D, Schulte DM, Benes L, Rosenow F, Bertalanffy H, Sure U (2009) Transsulcal approach supported by navigation-guided neurophysiological monitoring for resection of paracentral cavernomas. Clin Neurol Neurosurg 111:69–78

Zhou H, Miller D, Schulte DM, Benes L, Bozinov O, Sure U, Bertalanffy H (2011) Intraoperative ultrasound assistance in treatment of intradural spinal tumours. Clin Neurol Neurosurg 113:531–537

14 Advantage of Intraoperative Power Doppler Ultrasonography for Intracranial Tumors

Hiroshi Kanno

Contents

Abstract

For positioning intracranial tumors as intraoperative operation supporting system, neuronavigation and ultrasonography are the most common useful systems at present. Power Doppler ultrasonography real-timely shows vasculature in intracranial tumors, while neuronavigation system does not real-timely show intracranial structures due to the base on the preoperative image and is usually affected by brain shift. The information of the preoperative MRI can be compensated by real-time untrasonography. Power Doppler sonography detects vasculature of tumor with 3 dimensional image. Ultrasonographical contrast-enhancing agent enhances vasculature of the tumor, particularly vascular enriched tumors such as hemangioblastoma, glioblastoma, and metastatic tumors, and angioblastic meningioma. In these intracranial tumors, the echo signals obtained using contrast-enhanced power Doppler ultrasonography correlated with digital subtraction angiographic staining. Power Doppler untrasonography with the appropriate contrast agent provided better data on the precise real-time position of the tumors and their relationship to adjacent vessels than ultrasonograms obtained before the injection of the contrast agent. In addition, combination with neuronavigation system and power Doppler ultrasonography with 3 dimensional image contributes to intracranial tumor surgery.

H. Kanno (✉)
Department of Neurosurgery, Yokohama City University School of Medicine, Yoohama, Japan
e-mail: kanno@med.yokohama-cu.ac.jp

M. Hayat (ed.), *Tumors of the Central Nervous System, Volume 13*,
DOI 10.1007/978-94-007-7602-9_14, © Springer Science+Business Media Dordrecht 2014

Introduction

Ultrasonography (US) has been applied to medicine since the 1950s, and recently, intraoperative US has been commonly used to evaluate intracranial tumors (Chandler et al. 1982). Use of intraoperative ultrasonography (US) during surgery of intracranial tumor has been started since in the 1970s. Reid (1978) first described the use of sonography for neurosurgical guidance. At first, resolution of US remained in low level, and only existence of tumor was detected. Sosna J and Barth MM, et al. described in the review as follows: intraoperative US has been used over many years and is an efficient imaging adjunct to neurosurgery. Image resolution as well as the size and engineering of probes have improved considerably since 1978, reported by Reid (1978). Three-dimensional (3D) US with navigation software solves the orientation problem experienced previously with 2-dimensional (2D) US in operation (Unsgaard et al. 2002). Intraoperative US of the brain is most commonly used for localization and characterization of intracranial tumors, and the tumors such as meningiomas, gliomas, lymphomas, and metastases are typically hyperechoic relative to normal brain parenchyma and surrounding vasogenic edema. Anechoic cysts are present in patients with cystic astrocytomas and other cystic tumors, and cystic degeneration of benign or malignant solid lesions is common. Localization and characterization of solid tumors become more difficult in the presence of chronic peritumoral edema, which causes increased echogenicity and could obscure the margins between the mass lesion and adjacent normal brain. Although acute edema is typically less echogenic than tumors, the appearance of chronic edema or previously radiated tissue is less predictable and is occasionally isoechoic or hyperechoic. Infiltrative and aggressive tumors are different from their echogenicity and composition. Low-grade astrocytomas show often heterogeneous, with indistinct margins (Hatfield et al. 1989). However, even a distinct tumor margin on US is occasionally deceptive and a tumor margin is not correlated to be seen on CT or MRI. Primary central nervous system lymphoma appears either hypoechoic or isoechoic to surrounding parenchyma. Despite their infiltrative nature, these lesions tend to displace blood vessels.

Doppler which has been discovered in the 1990s could be visible vasculature of tumor, and at surgery of liver angioma, usefulness of power Doppler has been shown (Young et al. 1998). Real-timely detection of tumor vessels provides important information about tumor-surrounding vessels. High resolution of US was able to detect small vessels as well as large vessels in both arterial and venous systems, and power Doppler US showed vasculature of tumors. Early application to intracranial lesion was directed to vascular lesion. Contrast-enhanced power Doppler US is useful to detect intracranial vessels as well as extracranial carotid or vertebral arteries. Transcranial contrast-enhanced power Doppler US was able to detect intracranial major vessels consisting of Circle of Willis. However, directions of a US probe are limited. Therefore, except for flow of middle cerebral artery and parts of internal cervical artery and anterior cerebral artery, detections of vessel structures are not shown distinct. Compared with transcranial power Doppler US, its intraoperative application is not limited in direction or depth due to use of transdural or cortical probe. Advancement of resolution of equipment provides clear image similar to digital subtraction angiography (DSA). Contrast-enhancing agent (Levovist, Bayel, Berlin, Germany) is useful to detect more clear image for intracranial vascular-enriched tumors, particularly hemangioblastoma, glioblastoma, metastatic tumors, and angioblastic meningioma.

As to the power Doppler US with contrast enhancing agent, He et al. (2008), to study the value of intraoperative contrast-enhanced US in the resection for brain tumors, the feasibility of contrast-enhanced US was investigated to identify the border and residual of the brain tumors in the operation, and the usefulness of contrast enhanced power Doppler US was demonstrated.

More recently, combination with intraoperative power Doppler US, neuronavigation system with diffusion tensor tractography, and neurophysiological monitoring has been often reported

(Rygh et al. 2006; Nossek et al. 2011). Rygh et al. (2006) described that avoiding beeding by damage to blood vessels and is one of numerous concerns for neurosurgeons during intracranial tumor surgery. Neuronavigation system is useful for neurosurgeons to locate the position of vital structures such as vessels. However, not only brain shift but also registration errors limits the overall accuracy of a neuronavigation system (Roberts et al. 1998). To compensate a brain shift, intraoperative imaging has been introduced in neuronavigation (Gronningsaeter et al. 2000). The equipment also has capability of power Doppler US imaging for visualization of vessels. Thus, power Doppler US image data is able to be acquired intraoperatively for neuronavigation.

Intraoperative Application of Power Doppler Ultrasonography to Intracranial Tumors

Intraoperative application of power Doppler US to intracranial tumor surgeries has been reported by several authors. Yasuda et al. (2003) reported intraoperative 3D reconstruction of power Doppler US in 2003. They have developed a simple method for reconstructing a 3D image in the operating room from sequentially scanned intraoperative 2D power Doppler images using a personal computer and commercially available software. Intracranial vessel images were digitally transferred to a personal computer by freehand scanning over the dura mater or the brain surface with a 7.5-MHz linear probe. A series of 2D images were converted to a smaller file, and 3D image was reconstructed with volume-rendering software. In the intracranial tumor case, the anatomical relation between the vascular structures and the tumor was clear. They concluded that this simple 3D reconstruction method provides spatial information about intracranial vascular structures that is useful in intraoperative surgical planning.

Kanno H and Ozawa Y et al. reported the usefulness of power Doppler US for intracranial tumors. They described that particularly B-mode imaging has improved intraoperative orientation as well as the potential for more tumor resection, and three-dimensional intraoperative US has become available for navigation during brain tumor resection (Kanno et al. 2005). Intraoperative US for intracranial tumors is not only more handy than CT or MR imaging, but also affords true real-time imaging, which is helpful to evaluate the precise tumor location and the anatomical relationship between the tumor and surrounding tissues such as vessels and ventricles. In addition, technological advancements in US equipment have produced US resolution that is equal to that of CT or MR imaging. Preoperative transcranial color Doppler US, which can demonstrate vessel flow, is helpful to evaluate not only vascular lesions but also highly vascularized intracranial neoplasms (Bogdahn et al. 1994). A galactose-based transpulmonary US contrast-enhancing agent, Levovist (SH U 508A; Bayel, Berlin, Germany), was shown to provide a clearer assessment of intracranial vascular lesions and vascular structures around intracranial tumors during brain operations. They evaluated intraoperative power Doppler US with and without a contrast agent for the examination of 40 intracranial tumors, Before injection of the US contrast-enhancing agent, intra- and peritumoral power Doppler flow signals were detected in 32 of the intracranial tumors. After the injection, the signals were enhanced in blood vessels around the tumors and in the tumor parenchyma in 36 tumors. The duration of contrast enhancement continued for 70–365 s after the injection. Among the tumors, hemangioblastomas displayed particularly strong contrast enhancement. In these intracranial tumors, the echo signals obtained using contrast-enhanced power Doppler US correlated with DS angiographic staining. Moreover, they described that the information obtained from contrast-enhanced power Doppler US is helpful for surgical navigation because the appropriate route to access the tumor becomes quite apparent, and for an evaluation of peritumoral vessels, particularly for showing the vasculature of highly vascularized tumors such as hemangioblastomas. Practical use of power Doppler US for hemangioblastoma is shown in Fig. 14.1. The mural nodule, tumor substance, is rapidly enhanced after intravenous injection of contrast-enhancing agent, Levovist.

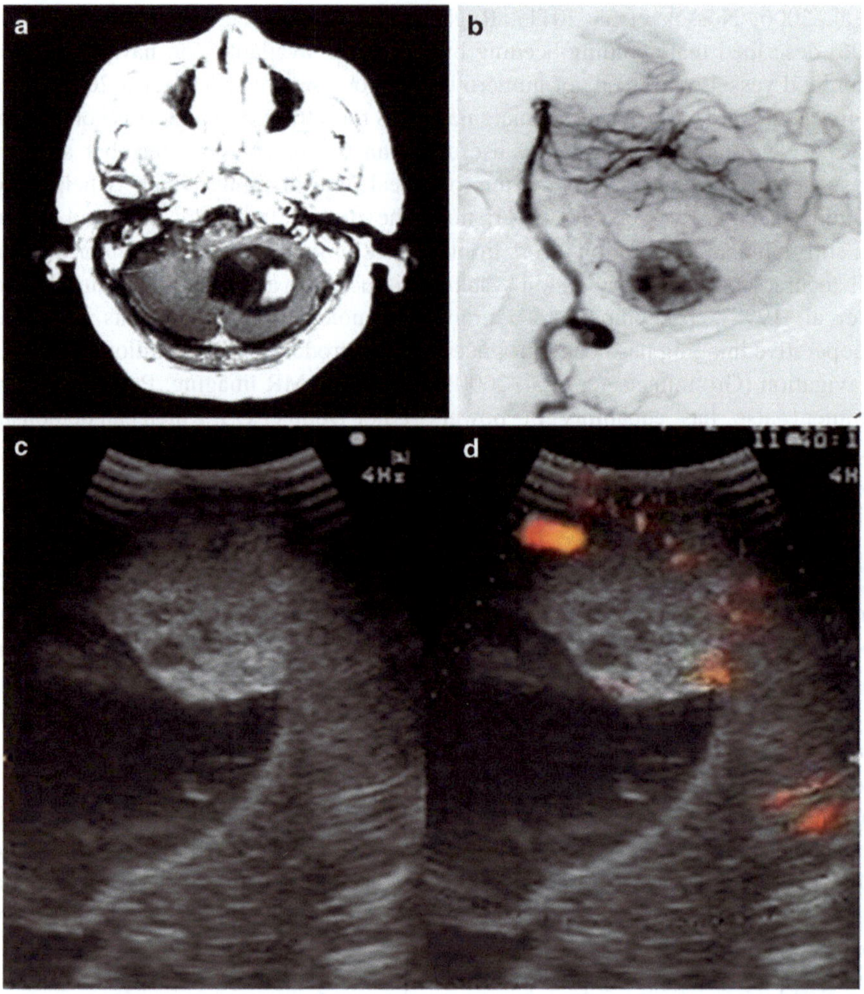

Fig. 14.1 Imaging studies obtained in a patient with a cerebellar hemangioblastoma. (**a**) Axial contrast-enhanced MR image revealing a well enhanced tumor with a cyst in the left cerebellar hemisphere. (**b**) Digital subtraction angiogram (*lateral view*) depicting distinct angiographic staining at the late arterial phase. (**c**) Gray scale B-mode ultrasonography depicting a large mural nodule with a cyst. (**d**) Power Doppler ultrasonogram without contrast agent depicting flow signals in the tumor

Early phase after injection of contrast-enhanced agent shows feeding artery to a tumor, and the late phase shows intratumoral vasculature including intratumoral veins. In addition, Glasker S and Shah M, et al. reported power Doppler US-guided resection of central nervous system hemangioblastomas (Gläsker et al. 2011). They used Sonowand intraoperative navigation system in a consecutive series of 64 hemangioblastoma of patients with von Hippel-Lindau disease as well as sporadic. They described all tumors were visible on power Doppler US, but, in 40 cases, only the pathological vessels and not the solid tumor itself on power Doppler US. They stated that hemangioblastomas appear to be ideally suited for detection with power Doppler US, since the tumors frequently have a cystic component and a small solid mural nodule, which is distinctly detected because of being highly vascularized. To evaluate the usefulness of intraoperative power Doppler US for surgery of hemangioblastomas, they gathered the intraoperative data including the visibility of the tumors before dura opening in power Doppler flow mode. The power Doppler gain were adjusted for optimal vascularization of the tumors and the surrounding neural tissue. After opening the dura mater, they used US again to identify the location of the tumors and to guide surgical resection. All tumors were visible in power Doppler flow mode. In 38% of tumors, the solid tumor nodule were directly visible due to enhancement of tumor tissues. Their tumors appeared demarcated areas of a mixed color pattern consistent with the marked vascularity of these tumors. The remaining 40 tumors were indirectly visible due to enhancement of feeding arteries and draining veins. Power Doppler US was found to be useful for localizing tumors, which were not visible after dura opening in several cases. After complete surgical removal, there was no more vascular enhancement detectable by power Doppler US. They concluded power Doppler US is a sensitive intraoperative equipment to guide the surgical approach and provides reliable resection control in surgery of CNS hemangioblastoma.

As to the power Doppler US with contrast-enhancing agent, He et al. (2008) described that after intravenously administrating ultrasound contrast, the enhancement of tumors started from 9 to 20 s and the peak of the enhancement ranged from 20 to 120 s. The enhancement passed the peak and decreased from hyper-enhancement to hypo-enhancement from 80 to 120 s. The enhancement continued for 3–10 min allowed observation. The time, intensity, and distribution of the contrast enhancement on contrast-enhanced power Doppler US varied in the intracranial tumors with various pathological components. There was no enhancement in the area with cystic degeneration or necrosis of the tumor. The normal brain tissue was hypo-enhanced. The margin of the tumors was clearly identified on contrast-enhanced US images, which was ill-defined on conventional ultrasound. The edema border around the tumor was easily visualized as hypo-enhancement. The signal of power Doppler of the tumor was increased after intravenously administrating contrast-enhanced agent in all cases. The assessment of the extent of resection including the location, size, and margin of the tumors by intraoperative contrast-enhanced power Doppler US showed significant correlation with MRI. There were nine patients with suspected remaining tumor tissue who underwent second contrast-enhanced power Doppler US after the initial resection. Residual tumor tissue was not clearly identified on conventional ultrasound, which presented as an enhanced rim along the resection cavity suggested remained tumor tissue. Intraoperative biopsy was performed on suspected remained tumor tissue in the area that was minimally vascular on conventional color flow image and present as hypervascularity on contrast-enhanced US.

As to the combination with intraoperative power Doppler US and neuronavigation system, Rygh et al. (2006) used an intraoperative US-based neuronavigation system with a 4- to 8-MHz flat phased array probe with optimal focusing properties at 3–6 cm and with the capability of acquiring power Doppler US images. On the power Doppler US images, vessels were displayed as an overlay in shades of red over the tissue image according to the power of the Doppler signal. The ultrasound platform in SonoWand also has triplex imaging, but this cannot be imported and used for neuronavigation with 3D US. For tracking of the ultrasound probe, a tracking frame was attached. In addition, as to integration with the power Doppler US and neurophysiological monitoring, Nossek et al. (2011) reported the integration with motor evoked potential (MEP) monitoring and power Doppler US. They analyzed 55 patients undergoing resection of tumors located within or in proximity to the corticospinal tracts. Corticospinal tract tractography based on diffusion tensor imaging was co-registered to surgical navigation-derived images in 42 patients. Direct cortical-stimulated motor evoked potentials (dcMEPs) and subcortical-stimulated MEPs (scrtMEPs) were recorded intraoperatively to assess function and estimate the distance from the corticospinal tracts. Intraoperative US updated the navigation imaging and estimated resection proximity to the corticospinal tracts. Preoperative clinical motor function was compared with postoperative outcome at several time points and correlated with incidences of intraoperative dcMEP alarm and low scrtMEP values. The threshold level to elicit scrtMEPs was plotted against the distance to the corticospinal tracts based on diffusion tensor imaging tractography after brain shift compensation with intraoperative power Doppler US, generating a trend line that demonstrated a linear order between these variables, and a relationship for every 1 mm of brain tissue distance from the corticospinal tracts. Clinically, 71% of 55 patients had no postoperative deficits, and 9 of the remaining 16 improved to baseline function within 1 month. Seven patients had various degrees of permanent motor deficits. Subcortical stimulation was applied in 45 of the procedures. The status of 32 patients did not deteriorate postoperatively: 84% of them displayed minimum scrtMEP thresholds >7 mA. Six patients who experienced postoperative deterioration quickly recovered and displayed minimum scrtMEP thresholds >6.8 mA. Five of the seven patients who had late or no recovery had minimal scrtMEP thresholds <3 mA. An scrtMEP threshold of 3 mA was found to be the cutoff point below which irreversible disruption of corticospinal tract integrity may be anticipated. Moreover, as to the QOL changes after glioma surgery using power Doppler US, Jakola, A and Unsgård G et al. explored the relationship between QOL and traditional outcome parameters, and examined possible predictors of change in QOL (Jakola et al. 2011). They concluded that the surgical procedures may not significantly alter QOL in the average patient with glioma, but have a major undesirable effect on QOL. The active use of intraoperative US may be associated with a preservation of QOL. The EQ-5D seems like a good outcome measure with a strong correlation to traditional variables while offering a more detailed description of outcome.

As to the experimental study using an animal model, Monome Y and Furuhata H, et al. reported malignant glioma RT2 cells implanted stereotactically into the right caudate nucleus (Manome et al. 2009). They injected microbubble contrast-enhancing agent Levovist, and detected implanted glioma RT cells contrast-enhanced by Levovist.

Discussion

With skilful and technical advancement, 3D directions of feeding arteries can be detected. After removal of the tumor, no residual tumor is easily confirmed with power Doppler image. Only weak point is required for skillfulness and rapid fading of arterial image. It had better re-observe record of power Doppler US with contrast-enhancing agent in every one frame. Although subtraction method of power Doppler is also useful, B-mode image also provide background echo signals, and then power Doppler US without subtraction is enough to evaluate the tumor vasculature. The information obtained from contrast-enhanced power Doppler US is useful for surgical navigation since the appropriate approach route to access the tumor becomes quite apparent, and for an evaluation of peritumoral vessels, particularly for showing the vasculature of highly vascularized tumors such as hemangioblastomas.

As for CNS hemangioblastomas, Glasker S and Shah M, et al. indicate that power Doppler flow sonography is a reliable and useful tool to localize hemangioblastomas intraoperatively (Gläsker et al. 2011). On grayscale imaging B-mode, many hemangioblastomas were only visible if cystic. However, with the use of power Doppler US, all lesions were able to be localized intraoperatively. The distinct sonographic visualization of hemangioblastomas compared with other tumors in the posterior fossa is due to their vascular and cystic nature. To localize cerebellar hemangioblastomas, power Doppler US surpasses MRI navigation since there is no brain shift. Furthermore, US- guided puncture of large posterior fossa cysts was useful for rapid release of pressure in cases where cerebellar tissue herniated after dural opening. In surgery for intramedullary hemangioblastomas, power Doppler US is the only possibility of navigation. This is specifically important in patients with VHL disease who frequently undergo multiple surgeries. Such minimal approaches do not allow extensive searching of a tumor that is not directly visible after dural opening. Despite the fact that Avila and coauthors (Avila et al. 1993) used an older power Doppler technique with inferior resolution compared with the current technique, the authors feel that power Doppler US is helpful for surgery of hemangioblastomas and facilitates the localization of the tumors. The use of power Doppler has advantages and disadvantages compared with other methods of neuronavigation. The advantages include the possibility of real-time imaging, fast availability, non brain shift, and the possibility to visualize vascular structures. It is the only well-established navigation method in the surgery of intra-axial tumors. The disadvantages include limitations in its resolution. More than all other navigation methods, the usefulness of USs depends on the experience of the surgeon. The image planes may be unfamiliar for inexperienced surgeons. The imaging artifacts are different from other imaging methods. In addition, based on their findings, Kanno H and co-authors suggested that data on vascularity obtained using DSA correlated with those obtained using power Doppler US, but that the US contrast- enhancing effect obtained using Levovist were different from that revealed on MR or CT images (Kanno et al. 2005). In vitro studies, animal experiments, and clinical studies have revealed that a US contrast agent Levovist, did not have toxicity or side effects. Furthermore, Levovist improves the signal intensity and the signal- to-nose ratio for several minutes. Nevertheless, there have been few studies on intraoperative power Doppler US for intracranial lesions, in which both non-enhanced and contrast-enhanced images have been examined. In addition, it is not clear what kinds of intracranial tumors might display the enhancing effect. Otsuki H and co-authors demonstrated that intraoperative harmonic subtraction images of intracranial lesions obtained using a US contrast agent were almost equal to MR angiograms and DSAs (Otsuki et al. 2001). Previously, we examined the US contrast-enhancing effect on the tumor parenchyma and assessed the relationship between vascularization seen on DSAs and the contrast enhancement of intracranial tumors seen on power Doppler USs. Our results revealed that most meningiomas, metastatic tumors, and, particularly, hemangioblastomas reflected strong signals on contrast-enhanced power Doppler US images, while some gliomas and malignant lymphomas displayed weak signals despite their contrast enhancement on CT and MR images. Bogdahn U and co-authors found that for all Grade IV gliomas contrast-enhanced transcranial color-coded US revealed not only color Doppler flow signals within the tumor parenchyma, but also atypical arterial and venous spectra (Bogdahn et al. 1994). The mechanism of US contrast enhancement has not yet been fully elucidated, but depends on the microbubble concentration in the blood pool in the tumor parenchyma, which is considered to be related to tumor vascularity and not to the destruction of the blood–brain barrier. On the other hand, contrast enhancement on CT or MR images is considered to be related to destruction of the blood–brain barrier rather than to tumor vascularity. Various abdominal tumors, such as liver cell carcinoma, liver hemangioma, metastatic liver carcinoma, gallbladder carcinoma, pancreatic carcinoma, and, in particular, liver cell carcinoma and metastatic liver carcinoma, display a high level of US contrast enhancement. In addition, these findings indicate that the US contrast-enhancing effect in abdominal tumors depends on vascular enrichment. It is demonstrated that hemangioblastomas, meningiomas, metastatic brain tumors, and some high-grade gliomas also show high-grade contrast enhancement during power Doppler US when a contrast agent is used, and that the angiographic grade was correlated with this level of contrast enhancement. In particular, hemangioblastomas, even if small, demonstrate the US contrast-enhancement effect, which is useful to determine the orientation of a tumor during an operation. Thus, the use of power Doppler US with a contrast-enhancing agent has several advantages to facilitate tumor removal. The correct approach route to access the deep-seated tumor becomes immediately apparent because feeding arteries, draining veins, and tumor vasculature are clearly shown in color images. In addition, the extent of resection in parenchymal brain tumors is easily evaluated, particularly in tumors that have abundant vessels, and finally total removal of the tumor is confirmed with the disappearance of power Doppler signals in the tumor vasculature. Although we used the US contrast agent as a single bolus injection, other enhancement protocols such as repeated bolus injections or continuous infusion should be tried. In addition, other imaging modes such as pulse inversion and 3D US may be practical when performing power Doppler US with the contrast agent. Although a neuronavigation system based on data obtained from MR or CT images makes errors because of brain shift, US provides true real-time imaging without any brain shift. A neuronavigation system that could be revised by data obtained using power Doppler US with contrast enhancement would provide more superior information during surgery.

As to the power Doppler US with contrast enhanced agent, He W and Jiang X et al. demonstrated that intraoperative contrast-enhanced power Doppler US was helpful to evaluate the tumor location, defining the border and observing the vascularity of the intracranial tumors (He et al. 2008). The enhancement indicates an actively viable tumor proved by pathology, which needs to be completely removed during the operation. Non-enhancement indicates necrosis or cystic degeneration. While the edema surrounding the brain tumor showed hypo-enhancement, which can differentiate the edema from the brain tumor tissue (Smith et al. 1985; Bogdahn et al. 1994; Kanno et al. 2005). Appropriate neurosurgical intervention often provides the patient with a brain tumor an improved neurological status, improved quality of life, and possibly prolonged survival. Optimal resection can be achieved when the tumor is specifically localized, the borders are clearly elucidated, and any residual tumor is readily identified. Intraoperative contrast-enhanced power Doppler US showed the obvious advantage of determining these factors in real time. Two major avoidances in the resection for brain tumors are incomplete resection and excessive resection. An incomplete resection could result in recurrence of the tumor. On the other hand, over-resection may lead to damage healthy brain that may shorten the time of patient survival and diminish quality of life. Furthermore, it is very difficult to recognize recurrent brain tumors with conventional ultrasound at previous surgical site where the local thrombosis and scar tissue are all hyperechoic that mimic tumor tissue. It is often impossible to distinguish the border and remained tissue of the tumor under naked eyes. MRI and CT are not portably applicable in operating room in all medical institutes. Intraoperative contrast-enhanced power Doppler US, a widely available, efficient, nonconstraining, and relatively inexpensive imaging technique, allows not only identifying the border of tumors but also distinguish remaining or recurrent tumor from scar tissue or local thrombosis, which is helpful in designing surgical procedure pre resection as well as planning an extent surgery to remove remained tumor tissue after the initial resection. Suspicious remaining tumor tissue was visualized as the enhancement in the local area on contrast-enhanced power Doppler US that guides a biopsy for pathological confirmation. Moreover, ultrasound images helped differentiate edema, as seen on T2 images from solid tumor and normal brain (Leroux et al. 1994.). The intraoperative contrast-enhanced power Doppler US much depends on the performer. It is important to master the skills of scanning and the surgical anatomy for improving feasibility and accuracy of intraoperative contrast-enhanced power Doppler US. Cranial MRI and CT are superior to ultrasound in the diagnosis of the brain tumors pre operation. For brain tumor resection control, intraoperative contrast-enhanced power Doppler US provides a useful imaging method to locate the lesion, define the border between the tumor and adjacent healthy brain, and detect residual tumor tissue after the initial resection. Nevertheless, intraoperative contrast-enhanced power Doppler US has dramatic potential for future applications in neurosurgery.

As to the integrated combination with power Doppler US and neuronavigation system, Rygh O and co-authors (Rygh et al. 2006) described as follows: intraoperative power Doppler US imaging may be necessary to maintain the accuracy in neuronavigation. Both intraoperative MRI and US are modalities for acquiring intraoperative angiographic image data for use in neuronavigation. Intraoperative US integrated with neuronavigation is a convenient alternative, since modern power Doppler technology can depict intracranial vasculature with sufficient image quality (Gronningsaeter et al. 2000; Tekula et al. 2001; Hernes et al. 2003). Sure U and co-authors has reported making the position of vessels using color flow Doppler and described that marking the position of vessels adjacent to a tumor facilitated image-guided tumor resection (Sure et al. 2000). The power Doppler US image enables navigation directly without need for landmarking on a preoperative MRI data set and also gives angiographic images of sufficient quality for navigation during the different stages of the resection of intracranial tumors. In addition, in cases with tumors in the medial part of the temporal lobe, it was shown that visualization of vessels in the basal cistern was helpful for safer resection and biopsy sampling. Power Doppler US was not occasionally found helpful, such as in superficial gliomas not close to larger vessels. Nevertheless, even in such cases, unexpected vessels may give the surgeon unpleasant surprises, and confirmation that important vessels are not close to the tumor is reassured by the surgeon. In skull base surgery, since the vessels are usually attached to the skull base, brain shift mostly has no problem. On the other hand, in MRI-based neuronavigation, inaccuracy in image registration may still lead to inaccurate targeting of skull base vessels. However, no image registration is needed in neuronavigation based on power Doppler US (Unsgaard et al. 2006). One limitation of the 3D ultrasound–based neuronavigation is that covering the whole area of interest during 3D ultrasound image data acquisition is possibly difficult. Furthermore, in skull base surgery, the skull base itself may hinder ultrasound imaging of the entire tumor. Still, careful planning of the surgical approach and keeping in mind optimal positioning of the ultrasound probe reduces such problems to a minimum. Moreover, simultaneous display of MRI and 3D ultrasound in neuronavigation is useful and also gives overview and anatomical orientation in cases with large tumors that are difficult to cover entirely with ultrasound. In contrast to color flow Doppler US imaging, power Doppler US does not have flow velocity or direction information but is less angle-dependent, and there is no aliasing and the modality is more flow sensitive (Riccabona et al. 2002). Due to less angle dependence, vessel continuity is also better with power Doppler than with color flow Doppler. In an application with free-hand 3D ultrasound image acquisition, the ultrasound beam will almost always hit a vessel with several different angles; therefore, there is minimal risk of missing a vessel when acquiring power Doppler US image data. In contrast to MR angiography, power Doppler US shows both arteries and veins at the same time. However, it is not possible to discriminate between arteries and veins using the power Doppler imaging technique alone. Using triplex display may also be helpful to discriminate between arteries and veins by evaluating the Doppler spectrum. Using the US probe as a pointer will enable targeting of a vessel with neuronavigation while using triplex display, and in this way, it is possible to evaluate the targeted vessel with triplex. With larger vessels, it was often possible to decide whether the targeted vessel was an artery or a vein by using anatomical knowledge and comparing with arterial MR angiography. In power Doppler US imaging, the ultrasound data are used for simultaneous imaging of both tissue and angiographic imaging. The high sensitivity of power Doppler may lead to the visualization of numerous vessels, as small and surgically less important vessels are depicted. Occasionally, this may result in confusing images that are difficult to interpret. Reducing the gain setting on the ultrasound scanner overcomes this problem by filtering out the smaller vessels. For better orientation, simultaneous display of preoperative MRI and MR angiography is also useful. Blooming may also be a problem with power Doppler, as the power Doppler signal tends to expand beyond the vessel walls. In addition, the 2D ultrasound image plane itself has a certain thickness that varies with the depth; thus, the resolution in the elevation direction of the ultrasound beam affects the image quality in 3D ultrasound. Therefore, the vessels in reconstructed power Doppler US images tend to appear bulkier when viewed on slices orthogonal to the scan plane. In addition, power Doppler US is relatively sensitive to flash artifacts, which may occur when the probe is in a cavity filled with saline and the motion in the saline is detected by the power Doppler. This is another consequence of the high motion sensitivity of power Doppler, which can occur relatively often during free-hand 3D ultrasound angiography acquisition. Flash artifacts can be reduced with gentle movement of the probe during image acquisition and by adjustments on the ultrasound scanner. The overall clinical accuracy of the SonoWand system may be as low as 2 mm in a clinical setting when using intraoperative imaging to compensate for brain shift (Lindseth et al. 2002). This technical evaluation study of navigated power Doppler US demonstrated the straightforward application of neuronavigation with this imaging modality in many cases.

In future prospects of power Doppler US, Rygh et al. (2006) described as follows: power Doppler US still has potential for significant improvements, and developments along several other equipments are to be expected: New signal processing methods such as BFI (Løvstakken et al. 2006) may better visualize flow inside vessels and hopefully reduce the problem of blooming. Ultrasound contrast-enhancing agents have been reported to be helpful in tumor neurosurgery for assessing the vasculature close to and inside the tumor (Kanno et al. 2005; Otsuki et al. 2001) and may improve the quality of ultrasound angiographic imaging. New probes will have a more optimal beam shape with better resolution in the elevation direction. This will further improve the image quality of ultrasound angiography because elevation resolution is a limiting factor for image quality with the present technology. Real-time 3D probes will be able to acquire 3D ultrasound data sets directly without free hand movement, and this will minimize flash artifacts. New multimodal visualization techniques where preoperative MR and intraoperative ultrasound are integrated in the same 3D scene are already available (Lindseth et al. 2003). This may probably further enhance the surgeon's perception of anatomic and spatial relationships between tumor and adjacent vessels. Diffusion tensor imaging of tracts and functional MRI of eloquent cortex are also visualized. Furthermore, robust volume-to-volume registration techniques for registration of preoperative MR angiography data to intraoperative ultrasound angiography data can make it possible to adjust preoperative MR image data in cases of brain shift.

References

Avila NA, Shawker TH, Choyke PL, Oldfield EH (1993) Cerebellar and spinal hemangioblastomas: evaluation with intraoperative gray-scale color Doppler flow US. Radiology 188:143–147

Bogdahn U, Frohlich T, Becker G, Kronea A, Schlief R, Schurman J, Jachmczak P, Hofmann E, Roggendorf W, Roosen K (1994) Vascularization of primary central nervous system tumors: detection with contrast-enhanced transcranial color-coded real-time sonography. Radiology 192:141–148

Chandler WF, Knake JE, McGillicuddy JE, Lillehei KO, Silver TM (1982) Intraoperative use of real-time ultrasonography in neurosurgery. J Neurosurg 57: 157–163

Gläsker S, Shah MJ, Hippchen B, Neumann HP, van Velthoven V (2011) Doppler-sonographically guided resection of central nervous system hemangioblastomas. Neurosurgery 68(suppl 2):267–275

Gronningsaeter A, Kleven A, Ommedal S, Aarseth TE, Lie T, Lindseth F, Langø T, Unsgård G (2000) SonoWand, an ultrasound-based neuronavigation system. Neurosurgery 47:1373–1379

Hatfield MK, Rubin JM, Gebarski SS, Silbergleit R (1989) Intraoperative sonography in low-grade gliomas. J Ultrasound Med 8:131–134

He W, Jiang XQ, Wang S, Zhang MZ, Zhao JZ, Liu HZ, Ma J, Xiang DY, Wang LS (2008) Intraoperative contrast-enhanced ultrasound for brain tumors. Clin Image 32:419–424

Hernes TA, Ommedal S, Lie T, Lindseth F, Langø T, Unsgaard G (2003) Stereoscopic navigation-controlled display of preoperative MRI and intraoperative 3D ultrasound in planning and guidance of neurosurgery: new technology for minimally invasive image-guided surgery approaches. Minimal Invasive Neurosurg 46:129–137

Jakola A, Unsgård G, Solheim O (2011) Quality of life in patients with intracranial gliomas: the impact of modern image-guided surgery. J Neurosurg 114:1622–1630

Kanno H, Ozawa Y, Sakata K, Soto H, Tanabe Y, Shimizu N, Yamamoto I (2005) Intraoperative power Doppler ultrasonography with a contrast-enhancing agent for intracranial tumor. J Neurosurg 102:295–301

Leroux PD, Winter TC, Berger MS, Mack LA, Wang K, Elliott JP (1994) A comparison between preoperative magnetic resonance and intraoperative ultrasound tumor volumes and margins. J Clin Ultrasound 22:29–36

Lindseth F, Langø T, Bang J, Nagelhus Hernes TA (2002) Accuracy evaluation of a 3D ultrasound-based neuronavigation system. Comput Aided Surg 7:197–222

Lindseth F, Kaspersen JH, Ommedal S, Langø T, Bang J, Hokland J, Unsgaard G, Hernes TA (2003) Multimodal image fusion in ultrasound-based neuronavigation: improving overview and interpretation by integrating preoperative MRI with intraoperative 3D ultrasound. Comput Aided Surg 8:49–69

Løvstakken L, Bjaerum S, Martens D, Torp H (2006) Blood flow imaging – a new real-time, 2-D flow imaging technique. IEEE Trans Ultrason Ferroelectr Freq Control 53:289–299

Manome Y, Furuhata H, Hashimoto A, Funamizu N, Suzuki R, Ishizawa S, Akiyama N, Kobayashi T, Watanabe M (2009) Application of therapeutic insonation to malignant glioma cells and facilitation by echo-contrast microbubbles of levovist. Anticancer Res 29:235–242

Nossek E, Korn A, Shahar T, Kanner AA, Yaffe H, Marcovici D, Ben-Harosh C, Ben Ami H, Weinstein M, Shapira-Lichter I, Constantini S, Hendler T, Ram Z (2011) Intraoperative mapping and monitoring of the corticospinal tracts with neurophysiological assessment and 3-dimensional ultrasonography-based navigation. J Neurosurg 114:738–746

Otsuki H, Nakatani S, Yamasaki M, Kinoshita A, Iwamoto F, Kagawa N (2001) Intraoperative ultrasound arteriography with the "Coded Harmonic Angio" technique. Report of three cases. J Neurosurg 94:992–995

Reid MH (1978) Ultrasonic visualization of a cervical cord cystic astrocytoma. AJR Am J Roentgenol 131: 907–908

Riccabona M, Resch B, Eder HG, Ebner F (2002) Clinical value of amplitude-coded colour Doppler sonography in paediatric neurosonography. Child Nerv Syst 18: 663–669

Roberts DW, Hartov A, Kennedy FE, Miga MI, Paulsen KD (1998) Intraoperative brain shift and deformation: a quantitative analysis of cortical displacement in 28 cases. Neurosurgery 43:749–758

Rygh OM, Nagelhus Hernes TA, Lindseth F, Selbekk T, Brostrup Müller T, Unsgaard G (2006) Intraoperative navigated 3-dimensional ultrasound angiography in tumor surgery. Surg Neurol 66:581–592

Smith SJ, Vogelzang RL, Marzano MI, Cerullo LJ, Gore RM, Neiman HL (1985) Brain edema: ultrasound examination. Radiology 155:379–382

Sure U, Alberti O, Petermeyer M, Becker R, Bertalanffy H (2000) Advanced image-guided skull base surgery. Surg Neurol 53:563–572

Tekula F, Pritz MB, Kopecky K, Willing SJ (2001) Usefulness of color Doppler ultrasound in the management of a spinal arteriovenous fistula. Surg Neurol 56:304–307

Unsgaard G, Ommedal S, Muller T, Gronningsaeter A, Hernes TA (2002) Neuronavigation by intraoperative three-dimensional ultrasound: initial experience during brain tumor resection. Neurosurgery 50:804–812

Unsgaard G, Rygh OM, Selbekk T, Müller TB, Kolstad F, Lindseth F, Hernes TA (2006) Intra-operative 3D ultrasound in neurosurgery. Acta Neurochir (Wien) 148:235–253

Yasuda K, Nakajima S, Wakayama A, Oshino S, Kubo S, Yoshimine T (2003) Intraoperative three-dimensioned reconstruction of power Doppler vascular image. Minimal Invasive Neurosurg 46:323–326

Young LK, Yang WT, Chan KW, Metreweli C (1998) Hepatic hemangioma: quantitative color power US angiography – facts and fallacies. Radiology 207:51–57

Intraoperative Ultrasound in Neurosurgical Oncology – Scope and Utility

15

Aliasgar Moiyadi

Contents

Abstract

Safe maximal resection is the guiding principle in the surgical management of brain tumors. Given the highly eloquent nature of the brain, anatomical and functional guidance during surgery are crucial. Various technological adjuncts are currently at the disposal of the neurosurgeon. Intraoperative ultrasound (IOUS) has historically been widely used in the neurosurgical operating room. Limitations of conventional 2D ultrasound have resulted in restricted applicability of this tool. Technological advances have resulted in navigable ultrasound becoming available. This combines the benefits of easy and convenient, yet powerful intraoperative imaging, along with navigational capabilities and hence overcomes most of the perceived drawbacks of standalone ultrasound machines. Understanding the various technical nuances and applications of the technique can facilitate optimal utilization of its capabilities for brain tumor resections.

Introduction

Intra-axial tumors of the brain and spinal cord, especially malignant gliomas, continue to pose a serious challenge to neurooncologists in general and neurosurgeons in particular. Advances in surgical techniques and adjuncts notwithstanding, prognosis presently remains dismal. The enthusiasm and optimism promised by current multimodality therapy is dampened by the stark reality of the

A. Moiyadi (✉)
Neurosurgery Services, Department of Surgical Oncology, Tata Memorial Hospital, Room 48 Main Building, E Borges Road, Parel, Mumbai 400 012, India
e-mail: aliasgar.moiyadi@gmail.com

M. Hayat (ed.), *Tumors of the Central Nervous System, Volume 13*,
DOI 10.1007/978-94-007-7602-9_15,

inevitability of recurrence especially in glioblastomas. As with most solid tumors, surgery remains the mainstay of management of intra-axial brain tumors. Surgical management of CNS tumors differs in many respects from general oncosurgery principles. *En bloc* resections are the norm in oncosurgical practice. Such radical resections are, however, rarely, if ever, possible in the majority of CNS tumors especially malignant gliomas. The brain and spinal cord are highly eloquent areas, where the risk of neurological deficits at the slightest insult is high. Most intra-axial tumors do not have a discrete plane of demarcation from the surrounding normal parenchyma. Even where a plane does exist there may be admixed normal tissue, precluding a radical excision. The surgical access through the rigid bony skull is limited, often necessitating working through narrow corridors. Deep seated lesions further require traversing normal tissue to reach the site of the tumor. The normal parenchyma's low threshold to withstand mechanical pressure necessitates minimization of retraction. The safest route to a given lesion needs to be individualized based on a thorough preoperative assessment of imaging combined with sound knowledge of microsurgical anatomy. Safe maximal resection remains the underlying tenet.

Extent of Resection – The Controversy

Given the almost impossible chance of resecting a glioma radically (without microscopic residual disease), the issue of how radical a neurosurgeon should be has been strongly debated (Sanai and Berger 2008). In malignant gliomas it was a strongly held belief that unless a very radical (>98%) volumetric resection was achieved, survival benefit was unlikely (Lacroix et al. 2001). There are two problems with this contention. Firstly, it is very difficult to be sure upfront how much of a radical resection would be possible. To prove the point, in the recent 5-aminolevulinic acid (5-ALA) study (Stummer et al. 2006) which included only so-called "resectable" malignant gliomas, the control arm achieved gross-total resection (<0.175 cc of tumor) in only 36%, and this improved to 65% when using 5-ALA. Secondly, with this "all or none" attitude, neurosurgeons may be pushed towards pessimism and in the bargain forget many other benefits of debulking surgery (which include relief of neurological symptoms, optimization of patient for adjuvant therapy, and probably facilitating adjuvant therapy, opportunity for delivering local therapies, and last but not the least providing valuable source of tissue for research). In this context a more recent study (Sanai et al. 2011) demonstrated that even debulking of ~78% has a survival benefit, and this benefit improves incrementally with increasing extent of resection above 78%. Taking together all the evidence a radical debulking (even short of a gross total resection) without compromising neurological outcomes is favoured. In low-grade gliomas too, there is enough evidence to suggest that when possible a radical resection should be attempted (Soffietti et al. 2010). In any case, the goal of this approach is to obtain maximum resection without compromising safety. To this end every form of intraoperative guidance (both anatomical image-based and functional) is crucial and should be used whenever available. In fact, it may be justified to suggest that management of tumors be referred to centres with access to these adjuncts so as to offer the best initial surgical therapy to these patients.

History of Intraoperative Ultrasound in Neurosurgery

Though A-mode ultrasound was used earlier, the defining period for intraoperative ultrasound in neurosurgery was probably the introduction of real-time B mode US and the initial pioneering work by Rubin and Dohrmann (2001) in the 1970s. Subsequent work by others confirmed the promise and the proof of the principle was established, paving the way for further application of 2D US in the neurosurgical operating theatre. In fact, the use of US in neurosurgery was one of the earliest applications of US in the intraoperative setup.

Since its initial use, numerous studies have demonstrated the efficacy of the US in intraoperative imaging for brain and spinal tumors (Gooding et al. 1984; Montalvo and Quencer 1986; Quencer and Montalvo1986; van Velthoven and Auer 1990). Not only is the efficacy of the US in localizing well circumscribed lesions (like metastases and high grade tumors) good (Machi et al. 1984; Kumar et al. 1993; Sun and Zhao 2007) but contrary to expectations even for low-grade diffuse gliomas, the US is better able to demarcate the hyperechoic tumor which may not be discernable on CT or MR and difficult to localize with the naked eye at surgery (Hatfield et al. 1989; LeRoux et al. 1992). There remain concerns, however, regarding the ability of the US to resolve differences between peritumoral edema, infiltrative margin, and normal parenchyma (van Velthoven 2003). US is also less reliable in post-treatment cases where diffuse changes related to the treatment effect cannot be differentiated from recurrence of tumor (Hammoud et al. 1996). US, however, is an excellent tool to differentiate solid and cystic lesions. It can also be used for real-time guidance to target lesions either for biopsy, drainage or for catheter placements. Attempts have even been made to perform volumetric studies using the IOUS. However, its efficacy vis a vis MRI remains to be proven (Hammoud et al. 1996; LeRoux et al. 1989; Renner et al. 2005). Our own experience with 2D US has been very encouraging (Moiyadi and Shetty 2011). We found that it was very useful in more ways than one, at every step during resection of brain and spinal cord tumors. It is this multipurpose attribute which makes IOUS very attractive and appealing.

Evolution from Real-Time 2-D to Navigable 3-D Ultrasound

One of the main limitations of 2D US is its inability to help in planning the craniotomy. This is not surprising as presently transcranial (through an intact bony skull) insonation is not practical. With accumulating experience using 2D US and with expanding indications for use, it was soon realized that beyond its many advantages there exist obvious limitations. This has been well illustrated in an article by Unsgaard et al. (2002a).

One of the major problems encountered when using the IOUS is the difficulty in anatomical orientation. This is primarily because the field of view in the US image is limited (unlike the full head views that neurosurgeons are more familiar with on MRI). This is particularly acute in the initial phase of the learning curve. Identification of known anatomical landmarks (such as the falx, ventricles, etc.) can help orientation and should always be attempted on the initial scan performed. Subsequently, with experience it becomes a routine step and facilitates the entire process improving confidence levels.

Another drawback is that it is very difficult and impractical to perform surgery with the US probe in the field. Though the IOUS provides quick and easy updates during surgery, once the probe is removed the information is lost and further image guidance is not possible. It is still not truly *online*. Moreover, lack of the third dimension means that the surgical guidance outside the plane of insonation is not possible. Consequently, for performing targeted procedures (such as biopsy or draining a cyst or placing a catheter), the probe needs to be held on the exposed surface and the instrument (biopsy forceps, ventricular cannula) has to be passed parallel to the probe. When this is done, the instrument trajectory is not always parallel to the plane of the US image and hence the tip of the instrument may not be correctly identified. On the contrary, it may misrepresent the tip (the point along the length of the instrument where the US plane cuts it may be erroneously interpreted as the tip) and lead to problems. Increasing the size of the craniotomy or adding a second craniotomy has been suggested to overcome this problem. This solution, however, seems too radical. Probe-mounted biopsy guides do help perform such targeted procedures, but they are still not absolutely accurate and practical and hence not widely used.

Advances in brain imaging technology (CT and MR), image-processing and computational capabilities coupled with refined stereotactic principles gave birth to navigation which transformed the way neurosurgical procedures were performed (Wadley et al. 1999). Undoubtedly the image-guidance provided was a significant step forward. Much of the benefit was related to planning customized, "tailored" craniotomies and orientating the surgeon during the surgery. Soon the problem of brain shift emerged, obviating the usefulness of the preoperatively acquired images (Dorward et al. 1998). Many approaches were adopted to overcome this; but none solved the problem. Moreover, inaccuracies related to image and patient registration added to the problem. Thus, purely navigation-based systems were not truly able to provide "real-time intraoperative image guidance". Responding to this need, the MR and CT, routinely used in neurosurgery preoperatively, were introduced into the operating theatre. There is unequivocal evidence to support the benefit of Intraoperative magnetic resonance imaging (IOMR) in improving resections in gliomas as well as in improving overall outcomes (Mehdorn et al. 2011; Kubben et al. 2011; Senft et al. 2011). Unfortunately, it is a very costly technology and presently still remains inaccessible to the majority of neurosurgical centres, and thus its usefulness cannot be widely applied. At almost the same time as the MR was being adapted to the operating theatre, the US was already being used along with navigation as a means to correct brain-shift (Jödicke et al. 1998; Ohue et al. 2010). This approach essentially involved obtaining real-time intraoperative 2D-US images and superimposing them on the corresponding preoperatively obtained MR images. The disparity between them was then corrected by mathematical algorithms and the MR data recalibrated to provide a more updated image. This, however, was an indirect use of the US with the MR images still serving as the primary source for navigation.

Navigable 3D US

Different solutions for navigable US have been developed by different groups independently (Nikas et al. 2003; Bozinov et al. 2011; Unsgaard et al. 2002a.) We have been using the SonoWand system [M/s SONOWAND, Trondheim, Norway] described by Unsgaard et al. which is a navigable 3D US system. Navigable 3D US essentially combines navigation technology with a high-end dedicated cranial insonation probe capable of generating 2D as well as 3D images. The cranial probe is precalibrated and registered to the navigation system and is automatically recognized as such once connected. It can rapidly (30–40 s) acquire a series (~200–300) of 2D images which are computed automatically into a 3D volume set. Both these features combine to provide a preregistered 3D volume of high resolution (Figs. 15.1 and 15.2) which can then be displayed on the navigation system in either the traditional ACS (axial, coronal, sagittal) planes, or a more user-friendly and intuitive "dual-anyplane" mode. The US data can also be superimposed on preoperative MR (when available). Using this US dataset, the neurosurgeon is able to navigate. This dataset can then be repeatedly updated (as and when necessary) during the course of the surgery. The system can be used as either a stand-alone navigation system using preoperative CT and MR images; or as a stand-alone ultrasound machine providing real-time intraoperative 2D images as well as a navigable 3D ultrasound (which allows navigation based solely on the IOUS without requiring preoperative MR images); or in a combined mode using both preoperative images and intraoperative US. The use of the same has been extensively described elsewhere (Unsgaard et al. 2002b). Problems with orientation especially in the initial phase of the learning curve, may be overcome by using the combined mode (along with preoperative MR images). In this mode, superimposition of the US images on the corresponding MR images is very useful. This feature is also helpful to demonstrate and appreciate brainshift when it is present (Fig. 15.1). One big advantage of using US images directly is that it obviates the need for acquiring preoperative MR images [especially if they have already been done elsewhere or if the patient is not cooperative enough for an MRI]. In such cases the "3D Direct" mode is very useful. Besides, using the direct mode eliminates the inaccuracies due to image and patient registration.

It must be reiterated that this 3D US is not truly real-time. This means that the 3D US is first acquired and the images are reformatted and then subsequently used to navigate. Operating while a 3D scan is in progress is not possible. Solutions

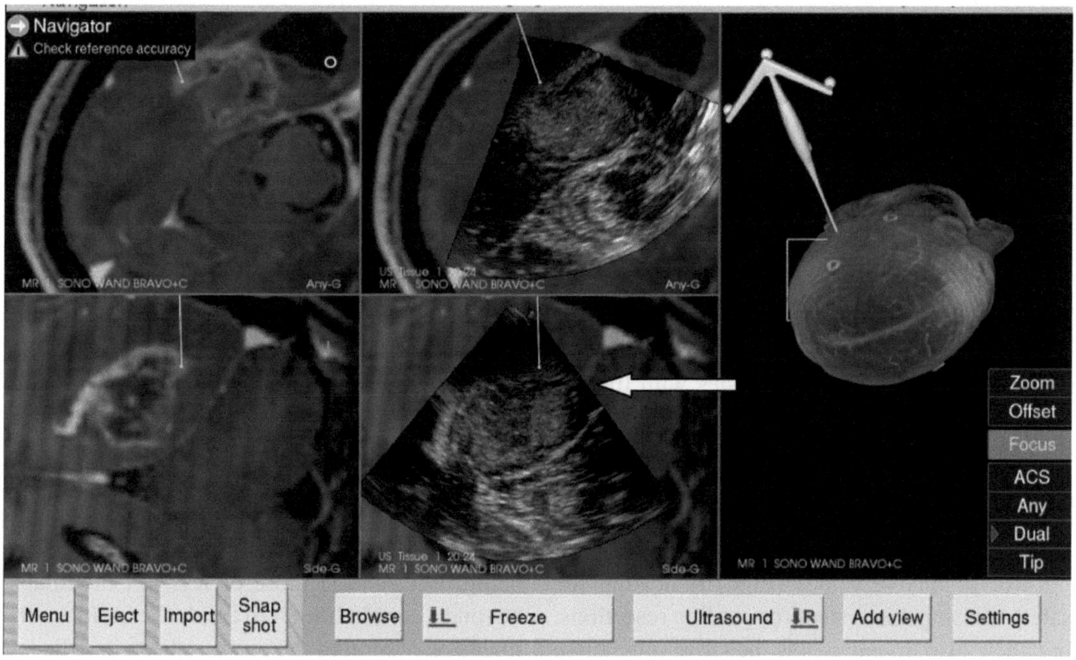

Fig. 15.1 Screenshot of a right temporal high-grade glioma during combined sononavigation. The preoperative contrast enhanced MRI images are seen in the *left panel* (dual anyplane view). The *central panel* shows the US images superimposed on the corresponding MR images. The tumor is seen to be hyperechoic. The brainstem and ipsilateral temporal horn can be appreciated on both US and MR images to facilitate orientation. Also note the brainshift depicted on the US images (*white arrow*) as evidenced by the change in position of the tentorial edge between the US and the MR images

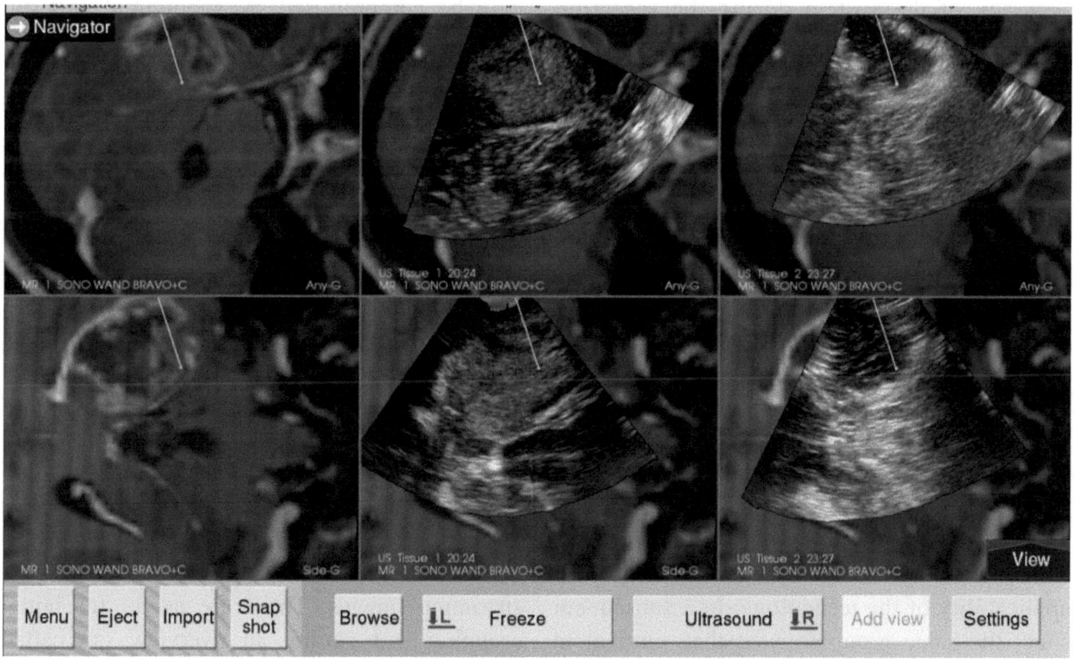

Fig. 15.2 Screenshot of the same patient as in Fig. 15.1 depicting various stages of the resection. *Left panel* shows the preoperative MR. *Central panel* shows the pre-resection US overlaid on the corresponding MR. The *right panel* shows the post-resection US overlaid on the MR. Note the resection cavity in the last scan with a hyperechoic rim (artefacts due to blood in the wall of the cavity). Any suspicious residue can be identified and reached using the navigator

for real-time 3D US are available and used in gynecology; however its role in neurosurgical guidance remains to be established (Bozinov et al. 2011).

Accuracy and Impact of 3D US

The Trondheim group showed (using meticulous histological correlation of biopsies with the US as well as MR images) that this system using navigable 3D US was as good and reliable as navigated MR for delineating high- and low-grade gliomas as well as metastases (Unsgaard et al. 2005). They reported high specificity and positive predictive values (PPV) indicating the safety of using this system for guiding resections, but they also found a low negative predictive value implying that when the IOUS was negative there was a possibility of tumor left behind. Improvements in image resolution capabilities in the future are expected to resolve these issues. Interestingly, the same study also found a higher PPV for low-grade gliomas. In a follow-up study the same group from Trondheim evaluated the accuracy of the system during the resection (in the subsequent phase of the surgery) (Rygh et al. 2008). They found that due to imaging artifacts imparted by blood and other changes in the adjacent tissue due to handling, the specificity and PPV dropped. Careful attention during hemostasis and tissue handling are essential to ensure optimal image quality.

With respect to the clinical impact of the 3D US system, it was effectively used in an unselected consecutive cohort of high-grade gliomas (Solheim et al. 2010). In this cohort the authors were able to achieve acceptable results (37% gross total resections [GTR] with 13% morbidity). They also noted that the system was routinely used in a majority of their surgeries and by a wide range of surgeons (including residents) attesting to the ease-of-use and wide applicability of the technology. In a subsequent study the same group showed that survival in GBMs improved in the years after the routine introduction of the SonoWand system (Saether et al. 2012).

Assessment of Utility of Intraoperative Ultrasound

The goal of using IOUS or any other intraoperative imaging tool, is to obtain accurate, comprehensive, and easily interpretable, and usable images, as frequently as needed. This is the endpoint or "gold standard" with which to measure the utility of the tool. There are many aspects that need to be assessed when determining the overall utility.

Efficacy: Efficacy implies the performance of the technology under *ideal* circumstances. For the ultrasound related imaging tools, this heavily depends on the resolution of the probe being used, and is more or less not modifiable beyond a certain limit. Given ideal conditions the image resolution of the probe reflects its efficacy. It must be borne in mind though, that "ideal" conditions, may not always be present especially in real-life situations where more often than not this is the case. Conditions may be suitably modified to improve the efficacy and this is an important learning step when initiating oneself to the technology. For example, ultrasound images may be compromised by suboptimal acoustic coupling and artifacts. Ensuring a vertical cavity (by intelligent patient positioning), using acoustic coupling (saline or gel) and eliminating air bubbles, and paying heed to hemostasis (avoiding extraneous hemostatic agents) can improve the image quality and thereby the efficacy of the tool. The efficacy of the probe will also determine the accuracy of the images. The measure of accuracy is reflected in the sensitivity, specificity, and predictive values of the imaging tool. This requires histological correlation. Fortunately, dedicated studies have addressed this issue with respect to the navigable 3D US system (Unsgaard et al. 2005) Modes/Functions: Different imaging modes can be used either with the same probe or using a combination of probes. 2D-scans are the commonest mode of use and provide real-time streaming images. These are useful for initial scanning and getting an overview (so called "bird's eye-view") of the field of interest. It helps orient oneself to the field. In the navigable US system this can be used superimposed on the

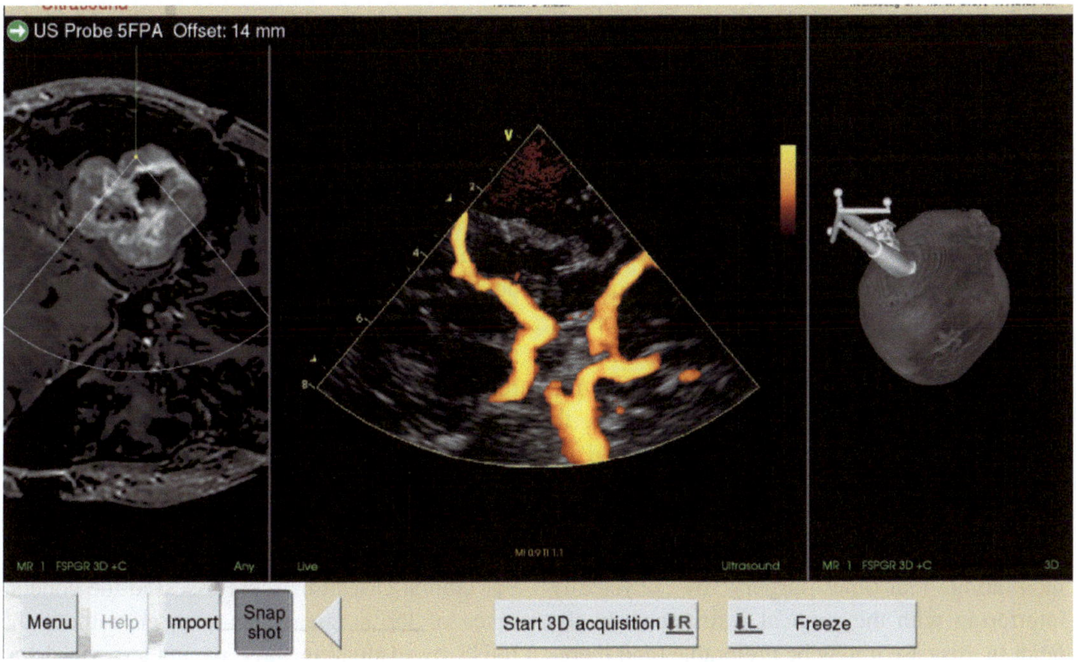

Fig. 15.3 Intraoperative Power Doppler angiography. Screenshot of a case of right temporal chondroscarcoma (preoperative MR images seen in *left panel*). The *centre panel* depicts the intraoperative angiogram obtained using the PD function. *Right panel* shows the 3D surface view. Note the excellent delineation of the circle of Willis. With kind permission from Springer Science+Business Media: Acta Neurochirurgica:Usefulness of three-dimensional navigable intraoperative ultrasound in resection of brain tumors with a special emphasis on malignant gliomas. Ahead of print DOI:10.1007/s00701-013-1881-z. Moiyadi AV et al, Figure 2

corresponding MR images and provide increased confidence and comfort in orienting the surgeon (especially in the initial phase of the learning curve). As described above, there are certain significant limitations of 2D scans in the context of intraoperative image-guided surgery. The 3D US mode seems to overcome some (though not all) of these limitations. For one, it permits a volume imaging rather than just a planar imaging. Combination with navigation technology (navigable ultrasound) then allows representation of the 3D volume in various planes as desired (multiplanar imaging), a very important and crucial step in planning and guiding surgical procedures. Besides the ultrasound imaging special functions which are particularly useful are the Power Doppler (PD), colour flow imaging. These allow visualization of important vascular structures, knowledge of which can be crucial during surgery. The PD function in particular permits high resolution angiography intraoperatively and coupled with navigation is a very handy tool (Fig. 15.3). Scope of use: The scope of application of the US reflects the range of procedures it could be used for. This, in turn, is a function of the ability to insonate the pathology/region of interest. It could range from (though not limited to) tumors, inflammatory processes, and cysts to vascular lesions (hematomas and malformations), both of the brain and spinal cord. It can be used for image-directed procedures such as biopsies (frameless biopsy), catheter, shunt placements, and delivery of local therapies. It can also be used for identifying deep-seated lesions (localization) and controlling the resection of many intra-axial tumors (resection control). Effectiveness: This measures the usefulness of the technique in daily real-life situations and in that sense truly reflects the "practical utility" of the tool. Though difficult to establish objectively, it is a combination of the efficacy, modes of function, and scope of application which governs the overall utility. It may so be that a technique or tool has excellent efficacy (in ideal situations); but if the ideal conditions are very unlikely most of the times, the tool would have limited utility and hence is not very effective. Even if it may be possible to have ideal circumstances, the modes of function may be limited and hence scope of use restricted to a small percentage of routine procedures. Again, though efficacy would be high, applicability would be low and hence the overall utility would be limited. For example, if there is very high-resolution probe but the footprint is too large, it may not be possible to use it in smaller craniotomies. This could be circumvented by having a range of probes with different footprints. Again, the resolution of a 2D scan could be excellent but it may be limited in facilitating real-time surgery by physically interfering with the surgical instruments in the field. In this case it may be preferable (and therefore more useful) to acquire a 3D volume which can be registered and then navigated, permitting removal of the probe and introduction of the surgical instruments with freedom to manoeuvre them as desired. So, a combination of accessories and functionalities, each individually efficacious in certain situations may improve the overall effectiveness of the tool. For a technology to be acceptable, it should have easy accessibility, widespread applicability, and reasonable efficacy. This would ensure a better overall efficiency (effectiveness or output in relation to costs or input). It is always preferable to have an efficient tool, rather than a very efficacious one with limited efficiency. Surgeon Comfort: This is an very subjective parameter. The surgeon's comfort with the tool is related to his/her experience with the particular tool. It is especially so with IOUS where there is a steep learning curve. This is primarily because neurosurgeons in general are more familiar with MR images than US images of the brain. As mentioned earlier, navigable 3D US overcomes most of the perceived drawbacks of 2D US. Repeated use and experience with the application (as with any surgical tool or technique) is imperative to reach a "comfort-zone".

Objective assessment of utility of surgical adjuncts is very difficult and scarcely reported. Unlike regulations for new medicinal drug which mandate phase 1 and 2 studies to establish efficacy and safety of the drug prior to phase 3 clinical studies, no such stipulations dictate the introduction of new surgical techniques. Most clinical trials and studies dealing with surgical adjuncts concentrate on other outcome measures (extent of resection, immediate perioperative outcomes, or survivals). Few studies have been reported dealing specifically with efficacy and utility.

Machi et al. (1984) had objectively assessed the role of IOUS in brain and spinal surgery and reported that it was useful for localization of the lesion, for delineation of tissue features as well as assessment of spatial relationships. This was a valuable though concise and incomplete assessment. Kumar et al. (1993) have reported a three point scoring system to assess the utility of IOUS. For cranial cases this score assessed the concurrence of the surgical plan with or without the IOUS by evaluating three parameters viz. location, depth, and planned trajectory to the lesion. The more the discordance in the plan, the higher the score, and better the utility of the IOUS. This was primarily an assessment of the IOUS for the purpose of biopsy of deep seated lesions. The spinal scoring system (also a 3-point score) assessed different parameters (adequacy of laminectomy, adjacent neural elements, and characteristics of the tumor). The authors concluded that IOUS was useful in the cases they studied, although no validation of this score has been reported.

Our group has devised a "utility score" for IOUS applications for both cranial and spinal cases (Moiyadi et al. 2011). We identified various parameters which reflect different aspects of the perceived utility of the 2D -IOUS. These were assessed and documented in the surgical notes. A scoring system was devised based on these parameters. Each parameter was given a score of 0 [no utility] or 1 [useful] (Table 15.1). These individual scores were totaled to obtain an overall "utility score" (maximum 7, minimum 0). We found that the IOUS was useful in multiple ways at various stages during the surgical procedure with >90% cases demonstrating utility in five of the seven parameters assessed. The major limitation of 2D US is the inability to plan craniotomy as has been discussed earlier. This can be easily overcome by having navigation

Table 15.1 Scoring system for assessing the utility of IOUS (Moiyadi and Shetty 2011)

	Parameter	Interpretation	Score
1	Lesion identification	Lesion discernable	1
		Not discernable on IOUS	0
2	Lesion delineation	Well defined margins	1
		Poorly defined margins but discernably distinct from normal brain	1
		Imperceptibly diffuse. no use of IOUS	0
3	Utility in craniotomy/laminectomy modification	IOUS prompted a modification/extension in the craniotomy/laminectomy	1
		No modification of bone removal needed	0
4	Use in durotomy planning	Useful/helped optimize the durotomy site and extent	1
		Not needed/useful	0
5	Use in corticectomy/myelotomy planning	Useful to plan the entry site	1
		Not needed/useful	0
6	Use for assessment of extent of resection	Used for assessing extent of resection/residue	1
		Not needed/useful	0
7	Visualization of adjacent structures	Useful and needed	1
		Not needed/useful	0

capabilities in addition to the US. Our scoring system is more comprehensive and applicable to both cranial and spinal cases. As compared to the other scoring systems, it incorporates more aspects to assess the utility of the IOUS at successive stages in the operation. Utilization of this scoring system as a regular exercise during surgeries performed with IOUS would sensitize young neurosurgeons regarding the potential applications of this adjunct.

In conclusion, intraoperative imaging is crucial in neurosurgical oncology practice. Localization of the lesion and online resection control facilitates safe maximal resection. The intraoperative ultrasound is a very useful tool. Newer navigable 3D US technology overcomes most of the limitations of 2D US as well as conventional navigation systems with a wide range of applications. Mastering the technique of performance as well as interpretation of the US images is essential to ensure maximal utility of the technique.

References

Bozinov O, Burkhardt JK, Fischer CM, Kockro RA, Bernays RL, Bertalanffy H (2011) Advantages and limitations of intraoperative 3D ultrasound in neurosurgery. Technical note. Acta Neurochir Suppl 109:191–196; In Pamir MN (eds) Intraoperative Imaging

Dohrmann GJ, Rubin JM (2001) History of intraoperative ultrasound in neurosurgery. Neurosurg Clin N Am 12:155–166

Dorward NL, Alberti O, Velani B, Gerritsen FA, Harkness WF, Kitchen ND, Thomas DG (1998) Postimaging brain distortion: magnitude, correlates, and impact on neuronavigation. J Neurosurg 88(4):656–662

Gooding GA, Boggan JE, Weinstein PR (1984) Characterization of intracranial neoplasms by CT and intraoperative sonography. AJNR Am J Neuroradiol 5(5):517–520

Hammoud MA, Ligon BL, elSouki R, Shi WM, Schomer DF, Sawaya R (1996) Use of intraoperative ultrasound for localizing tumors and determining the extent of resection: a comparative study with magnetic resonance imaging. J Neurosurg 84(5):737–741

Hatfield MK, Rubin JM, Gebarski SS, Silbergleit R (1989) Intraoperative sonography in low-grade gliomas. J Ultrasound Med 8(3):131–134

Jödicke A, Deinsberger W, Erbe H, Kriete A, Böker DK (1998) Intraoperative three-dimensional ultrasonography: an approach to register brain shift using multidimensional image processing. Minim Invasive Neurosurg 41(1):13–19

Kubben PL, ter Meulen KJ, Schijns OEMG, ter Laak-Poort MP, van Overbeeke JJ, van Santbrink H (2011) Intraoperative MRI-guided resection of glioblastoma multiforme: a systematic review. Lancet Oncol 12:1062–1070

Kumar P, Sukthankar R, Damany BJ, Mishraa J, Jha AN (1993) Evaluation of intraoperative ultrasound in neurosurgery. Ann Acad Med Singapore 22(3):422–427

Lacroix M, Abi-Said D, Fourney DR, Gokaslan ZL, Shi W, DeMonte F, Lang FF, McCutcheon IE, Hassenbusch SJ, Holland E, Hess K, Michael C, Miller D, Sawaya R (2001) A multivariate analysis of 416 patients with

glioblastoma multiforme: prognosis, extent of resection, and survival. J Neurosurg 95(2):190–198
LeRoux PD, Berger MS, Ojemann GA, Wang K, Mack LA (1989) Correlation of intraoperative ultrasound tumor volumes and margins with preoperative computerized tomography scans. An intraoperative method to enhance tumor resection. J Neurosurg 71(5):691–698
LeRoux PD, Berger MS, Wang K, Mack LA, Ojemann GA (1992) Low grade gliomas: comparison of intraoperative ultrasound characteristics with preoperative imaging studies. J Neurooncol 13:189–198
Machi J, Sigel B, Jafar JJ, Menoni R, Beitler JC, Bernstein RA, Crowell RM, Ramos JR, Spigos DG (1984) Criteria for using imaging ultrasound during brain and spinal cord surgery. J Ultrasound Med 3(4):155–161
Mehdorn HM, Schwartz F, Dawirs S, Hedderich J, Dörner L, Nabavi A (2011) High-field iMRI in glioblastoma surgery: improvement of resection radicality and survival for the patient? Acta Neurochir Suppl 109:103–106
Moiyadi AV, Shetty P (2011) Objective assessment of utility of intraoperative ultrasound in resection of CNS tumors – a cost-effective tool for intraoperative navigation in neurosurgery. J Neurosci Rural Pract 2(1):4–11
Montalvo BM, Quencer RM (1986) Intraoperative sonography in spinal surgery: current state of the art. Neuroradiology 28:551–590
Nikas DC, Hartov A, Lunn K, Rick K, Paulsen K, Roberts DW (2003) Coregistered intraoperative ultrasonography in resection of malignant glioma. Neurosurg Focus 14(2):E6
Ohue S, Kumon Y, Nagato S, Kohno S, Harada H, Nakagawa K, Kikuchi K, Miki H, Ohnishi T (2010) Evaluation of intraoperative brain shift using an ultrasound-linked navigation system for brain tumor surgery. Neurol Med Chir (Tokyo) 50(4):291–300
Quencer RM, Montalvo BM (1986) Intraoperative cranial sonography. Neuroradiology 28(5–6):528–550
Renner C, Lindner D, Schneider JP, Meixensberger J (2005) Evaluation of intra-operative ultrasound imaging in brain tumor resection: a prospective study. Neurol Res 27(4):351–357
Rygh OM, Selbekk T, Torp SH, Lydersen S, Hernes TA, Unsgård G (2008) Comparison of navigated 3D ultrasound findings with histopathology in subsequent phases of glioblastoma resection. Acta Neurochir (Wien) 150(10):1033–1041
Saether CA, Torsteinsen M, Trorp SH, Sundstrom S, Unsgård G, Solheim O (2012) Did survival improve after the implementation of intraoperative neuronavigation and 3D ultrasound in glioblastoma surgery? A retrospective analysis of 192 primary operations. J Neurol Surg A Cent Eur Neurosurg 73(02):73–78
Sanai N, Berger MS (2008) Glioma extent of resection and its impact on patient outcome. Neurosurgery 62(4):753–764; discussion 264–266
Sanai N, Polley MY, McDermott M, Parsa AT, Berger MS (2011) An extent of resection threshold for newly diagnosed glioblastomas. J Neurosurg 115:3–8
Senft C, Bink A, Franz K, Vatter H, Gasser T, Seifert V (2011) Intraoperative MRI guidance and extent of resection in glioma surgery: a randomised, controlled trial. Lancet Oncol 12:997–1003
Soffietti R, Baumert BG, Bello L, von Deimling A, Duffau H, Frénay M, Grisold W, Grant R, Graus F, Hoang-Xuan K, Klein M, Melin B, Rees J, Siegal T, Smits A, Stupp R, Wick W, European Federation of Neurological Societies (2010) Guidelines on management of low-grade gliomas: report of an EFNS-EANO Task Force. Eur J Neurol 17(9):1124–1133
Solheim O, Selbekk T, Jakola AS, Unsgård G (2010) Ultrasound-guided operations in unselected high-grade gliomas – overall results, impact of image quality and patient selection. Acta Neurochir 152:1873–1886
Stummer W, Pichlmeier U, Meinel T, Wiestler OD, Zanella F, Reulen HJ, ALA-Glioma Study Group (2006) Fluorescence-guided surgery with 5-aminolevulinic acid for resection of malignant glioma: a randomised controlled multicentre phase III trial. Lancet Oncol 7(5):392–401
Sun H, Zhao JZ (2007) Application of intraoperative ultrasound in neurological surgery. Minim Invasive Neurosurg 50(3):155–159
Unsgård G, Gronningsaeter A, Ommedal S, Nagelhus HTA (2002a) Brain operations guided by real-time two-dimensional ultrasound: new possibilities as a result of improved image quality. Neurosurgery 51(2):402–411
Unsgård G, Ommedal S, Muller T, Gronningsaeter A, Hernes TAN (2002b) Neuronavigation by intraoperative 3D ultrasound, initial experiences during brain tumor resections. Neurosurgery 50(4):804–812
Unsgård G, Selbekk T, Brostrup MT, Ommedal S, Torp SH, Myrh G, Bang J, Nagelhus HTA (2005) Ability of navigated 3D ultrasound to delineate gliomas and metastases – comparison of image interpretation with histopathology. Acta Neurochir (Wien) 147:1259–1269
van Velthoven V (2003) Intraoperative ultrasound imaging: comparison of pathomorphological findings in US versus CT, MRI and intraoperative findings. Acta Neurochir Suppl 85:95–99
van Velthoven V, Auer LM (1990) Practical application of intraoperative ultrasound imaging. Acta Neurochir (Wien) 105(1–2):5–13
Wadley J, Dorward N, Kitchen N, Thomas D (1999) Preoperative planning and intraoperative guidance in modern neurosurgery: a review of 300 cases. Ann R Coll Surg Engl 81:217–225

Part IV

Surgery

Resection of Brain Tumors: Intraoperative Confocal Microscopy Technology

16

Nader Sanai and Robert F. Spetzler

Contents

Abstract

The ability to diagnose brain tumors in vivo and reliably identify tumor margins in the course of resection are two innovations that could impact the neurosurgical oncologist's ability to maximize resection and minimize morbidity. Recent advances in optical imaging and miniaturization have enabled the production of a hand-held intraoperative confocal microscope. We present a first-look feasibility analysis of the intraoperative confocal microscope as an adjunct for brain tumor resection. Intraoperative confocal microscopy is an emerging and practicable technology for the resection of human brain tumors. Our preliminary assessment indicates the reliability of this technique for a variety of lesions in identifying tumor cell populations, as well as the tumor-brain interface, in situ. Further refinement of this technology may depend upon the approval of tumor-specific fluorescent contrast agents for human use.

Introduction

Maximizing the extent of resection is the primary objective of most surgery for brain tumors. Yet, normal tissue can only be distinguished from abnormal tissue definitively on pathological analysis conducted after surgery. For some extraaxial lesions, the gross appearance of the tumor is sufficient to establish the tumor-brain tissue planes with microdissection techniques. Other lesions, however, are less easily distinguished from

N. Sanai (✉) • R.F. Spetzler
Department of Neurosurgical Oncology, Barrow Brain Tumor Research Center, Barrow Neurological Institute, St. Joseph's Hospital and Medical Center, Phoenix, AZ 85013, USA
e-mail: nader.sanai@bnaneuro.net

M. Hayat (ed.), *Tumors of the Central Nervous System, Volume 13*,
DOI 10.1007/978-94-007-7602-9_16, © Springer Science+Business Media Dordrecht 2014

normal brain tissue, particularly in patients who have undergone prior treatment or in those with cerebral edema or microscopic infiltration. This problem is particularly true with gliomas and higher-grade meningiomas, where defining extent of resection on the basis of gross tissue characteristics is insufficient and neuronavigation can be unreliable because of brain shift. To overcome these persistent challenges in the resection of complex intra- and extraaxial brain tumors, recent work has been directed toward adapting routine postoperative neuropathology methods into a real-time intraoperative technique.

Beyond identifying tumor margins, the opportunity to define tumor grade and histologic subtype intraoperatively is critical, particularly for intracranial gliomas, where tumor grade cannot be predicted reliably with either preoperative magnetic resonance (MR) imaging (Kondziolka et al. 1993) or stereotactic biopsy (Muragaki et al. 2008). Likewise, intraoperative frozen-section analysis can be misleading or nondiagnostic, especially when tissue is disrupted mechanically from the resection process (Tilgner et al. 2005; Uematsu et al. 2007). Such diagnostic unpredictability is further complicated by the inherent heterogeneity of gliomas, which can contain high-grade cellular populations nested within a low-grade stroma (Dowling et al. 2001). Collectively, these complexities represent a significant challenge for neurosurgical oncologists. However, they could be overcome by multiple in vivo optical biopsies during the course of tumor exposure and resection.

Intraoperative Confocal Technology

Intraoperative confocal microscopy is an emerging technology that enables the cytoarchitecture of live tissue to be visualized with spatial resolution on a cellular level (Becker et al. 1996; Khoshyomn et al. 1998; Tadrous 2000). Until recently, the size of requisite apparatus limited the use of confocal microscopy to examination of excised tissue samples or isolated cells in a laboratory. The latest incarnation of this technology, however, features fiber-optic and microscopic miniaturization, substantially expanding its portability and in vivo applicability in a clinical setting (Delaney et al. 1994; Flusberg et al. 2005, 2008; Helmchen 2002). A single optical fiber can be used as both the illumination point source and detection pinhole to acquire high-resolution images and to combine them with miniaturized scanning and optical systems (Hoffman et al. 2006).

Consequently, this device has recently been integrated into the distal tip of conventional video endoscopes and combined with intravenous fluorophores to screen the gastrointestinal mucosa for cancer (Kiesslich et al. 2005; Polglase et al. 2005). The bladder mucosa, skin, and eye have also been studied with in vivo confocal microscopy (Brezinski et al. 1998; Bussau et al. 1998; Koenig et al. 2001; Papworth et al. 1998). We have recently used a glioma model to study the capacity of handheld confocal imaging to discern microvasculature, the grey-white junction, and tumor margins in the rodent cortex (Eschbacher et al. 2012). This initial feasibility analysis of intraoperative confocal microscopy establishes its potential value during the microsurgical resection of both intra- and extra-axial tumors.

Histological Resolution

Examination of tissue integrity during confocal visualization demonstrates excellent preservation of cellular and subcellular structures. In normal and tumor parenchyma, large and small vasculature is evident and intact, both on the tissue surface and as deep at 500 μm to the surface. Within this microvasculature, the systolic and diastolic flow of erythrocytic transit can be appreciated. Notably, areas of neovascularization from tumor-induced angiogenesis distinguish tumor from adjacent normal tissue.

Glioma Visualization

Among World Health Organization (WHO) grade III and IV gliomas, vascular neoproliferation seen on confocal imaging enables both surgeon and neuropathologist to identify abnormal tissue reliably. In our experience, these areas of suspicion correlate well with both neuronavigational imaging and histological evidence of tumor. High-grade human glioma specimens are associated with specific confocal features, including neovascularization, dense cellularity, and irregular cellular phenotypes. Waves of varied density, likely corresponding to regions of necrosis, are also visible. Because, these features are evident to both surgeon and neuropathologist, the confocal microscopic findings can quickly be integrated into the operative plan.

Low-grade gliomas are also distinctive with intraoperative confocal microscopy. Although neovascular proliferation is less evident, distinctions in cellular density and morphology correspond with T2-weighted signal abnormalities on MR imaging. All specimens analyzed from regions where a confocal optical biopsy was obtained have demonstrated diagnostic patterns of low-grade histology on permanent sections. The pattern associated with pure oligodendrogliomas on confocal microscopy is distinct from other astrocytomas of comparable grade. However, mixed oligoastrocytomas do not appear to have a unique signature on intraoperative imaging and could not be reliably distinguished from low-grade astrocytomas.

Other Intra-axial Lesions

Other intraaxial lesions also have unique confocal signatures associated with their histologic phenotype. For central neurocytomas, in vivo imaging demonstrates uniform round cells organized in a honeycomb conformation, embedded against a background of arborized capillaries. Hemangioblastomas are similarly distinctive on confocal imaging and are identified by large stromal cells mixed with a dense capillary network. Within these tumors, both cystic changes and areas of microhemorrhage can be appreciated.

Meningioma Visualization

Meningiomas are common extraaxial lesions amenable to characterization with confocal microscopy. Fluorescein-enhanced confocal images of these tumors highlight fine histopathologic detail. The configuration of tumor cells in classic meningothelial meningiomas is largely uniform. The cells are organized as dense sheets with no evidence of whorls or psammoma bodies. In contrast, fibrous meningiomas contain cells with spindle-shaped morphology, which is easily distinguished from adjacent normal parenchyma. These tumors also contain fibrous bundles of matrix interlacing with adjacent tumor cells and dura.

Intraoperative Confocal Microscope: Emerging Applications

For a variety of tumor histologies, including gliomas, meningiomas, hemangioblastomas, and central neurocytomas, this handheld device generates a real-time, fluorescein-enhanced pathological image that is of sufficient resolution for a neuropathologist to establish a preliminary diagnosis. Mounting evidence supports the value of maximizing the extent of resection of many intracranial tumors, including gliomas (Sanai and Berger 2008). Consequently, several intraoperative techniques have been developed to help achieve this goal: intraoperative ultrasonography (LeRoux et al. 1989), intraoperative MR imaging (Alexander et al. 1997), optical spectroscopy (Toms et al. 2005), and fluorescent dyes such as 5-aminolevulinic acid (Stummer et al. 2006). However, none of these technologies can detect infiltrating tumor margins at a microscopic level. In contrast, our preliminary experience with intraoperative confocal microscopy not only suggests a correlation between imaging and tumor grade but has demonstrated a capacity to distinguish tumor margins from adjacent parenchyma.

Tumor heterogeneity and biopsy sampling error remain considerable sources of inaccuracy and primary causes of undergrading gliomas. The ease with which a handheld confocal microscope can generate numerous optical biopsies during the course of tumor exposure and resection may effectively neutralize such diagnostic imprecision. Furthermore, the ability of the device to scan as deeply as 500 μm through the visible tissue surface permits analysis of an even broader spectrum of tissue. This unique capability also raises the possibility of examining subependymal regions through an intraventricular corridor, allowing intraoperative confocal microscopy to detect pockets of subependymal tumor migration that are radiographically undetectable and remote from the primary tumor site. It has been postulated that subependymal spread is both a negative prognostic sign and a primary route of glioma migration (Lim et al. 2007). The ability to detect its occurrence at an early stage may permit anticipatory and focused interventions as well as lead to insights into basic mechanisms of gliomagenesis (Sanai et al. 2005).

Routine neuropathological diagnostic analysis depends on excision and processing of tumor tissue, which can be altered both mechanically and chemically during the analysis itself. Use of intraoperative confocal microscopy circumvents these limitations with real-time in situ visualization of tumor cytoarchitecture and microvasculature. The opportunity to observe active blood flow through tumor capillaries also provides a unique visual dimension to tissue assessment. The probe can quickly survey the visible extent of a tumor with the potential to detect disconnected, cell-dense islands that may portend worse histologic grade and heterogeneity. Future pilot studies will not only enhance our familiarization with this technique, but they may demonstrate a correlation with frozen section findings that could allow confocal microscopy to replace conventional intraoperative diagnostic methods. This modern approach to intraoperative diagnosis would also lend itself to remote, internet-based analyses, expanding the accessibility of diagnostic expertise beyond centers with dedicated tumor neuropathologists.

Future versions of the device may benefit from an auto-irrigation mechanism, similar to the endoscope. Furthermore, while fluorescein contrast allows detailed assessment of overall tissue cellularity, the ability to discern cellular cytoplasm remains limited, as does its specificity for nuclear morphology. Since tumor cell nucleus-to-cytoplasm ratios are critical to histopathological diagnosis, other contrast agents should be investigated. One candidate agent is 5-aminolevulinic acid (5-ALA), which allows highly specific localization of tumor cells. Currently, however, the technique is only available for clinical use in Europe.

Intraoperative Confocal Microscopy and 5-Aminolevulinic Acid

The orally administered prodrug, 5-ALA, is metabolized intracellularly to form the fluorescent molecule protoporphyrin IX (Duffner et al. 2005; Stummer et al. 1998). This heme synthesis pathway substrate accumulates preferentially in tumor cells and epithelial tissues and emits a red-violet light ($\lambda = 635–704$ nm) when excited with blue light ($\lambda = 400–410$ nm) (Ishihara et al. 2007; Stummer et al. 2003). Successful neurosurgical integration of 5-ALA-induced fluorescence for high-grade gliomas was demonstrated by a European randomized, controlled trial conducted by Stummer et al. (2006). This Phase IIIa clinical trial was halted after an interim analysis of 270 patients indicated a 65% vs. 36% rate of gross-total resection and a 41% vs. 21.1% rate of 6-month progression-free survival benefiting high-grade gliomas patients that received 5-ALA. The study lacked sufficient power to assess overall survival, nor were surgeons permitted to use intraoperative neuronavigation during tumor resection. Notwithstanding, intraoperative 5-ALA fluorescence has since emerged as a valuable surgical adjunct for high-grade gliomas (Liao et al. 2012; Nabavi et al. 2009; Tonn and Stummer 2008) in the international community. In the United States, its merits will be the subject of a planned prospective, multicenter trial (RTOG 1105) (Roberts et al. 2011; van Meir et al. 2010).

For low-grade gliomas, however, standard intraoperative 5-ALA fluorescence remains ineffective, because it does not produce visible fluorescence for most tumors. In a few cases, heterogeneous tumor fluorescence has been noted in focal areas of anaplastic transformation. However, the vast majority of low-grade gliomas are invisible with 5-ALA (Floeth et al. 2011; Ishihara et al. 2007; Stockhammer et al. 2009; Widhalm et al. 2010). Interestingly, protoporphyrin

IX fluorescence can be measured ex vivo in low-grade gliomas tissue after 5-ALA administration (Floeth et al. 2011; Ishihara et al. 2007). In these analyses, the resultant fluorescent intensity of the tumor tissue is significantly higher than similarly treated normal tissue (Ishihara et al. 2007) and increases proportionally with both tumor grade and MIB-1 proliferative index (Floeth et al. 2011; Stummer et al. 1998). Together, these reports suggest that 5-ALA tumor fluorescence can be localized microscopically to low-grade gliomas even when it is not evident macroscopically.

Recently, we reported the first combined effort using intraoperative confocal microscopy to visualize 5-ALA tumor fluorescence in low-grade gliomas during the course of microsurgical resection (Sanai et al. 2011). Our initial experience demonstrated several important findings: (1) integration of intraoperative confocal microscopy and 5-ALA fluorescence into a practical and efficient operating room workflow, (2) detection of 5-ALA fluorescence at a cellular level within WHO grade I and II low-grade gliomas, (3) cellular identification of low-grade glioma margins, and (4) equivalence between in vivo and ex vivo use of the confocal microscope when evaluating tumor fluorescence.

This initial report on intraoperative confocal microscopy and 5-ALA tumor fluorescence suggested that a combined approach may expand the utility of 5-ALA beyond glioblastomas. Specifically, we described ten consecutive patients with WHO grade I and II gliomas, including a WHO grade II oligodendroglioma with histological evidence of anaplastic transformation. In all ten patients, the tumors were invisible using conventional methods for 5-ALA tumor fluorescence detection. In all patients, however, intraoperative confocal microscopy identified tumor fluorescence at a cellular level, a finding that corresponded to tumor infiltration on matched histological analysis. This combined strategy was as robust in vivo as it was ex vivo. Together, these findings represented the first successful effort at real-time, intraoperative detection of WHO grade I and II gliomas using 5-ALA.

However, the potential impact of 5-ALA on glioma surgery when combined with neuronavigation, intraoperative confocal microscopy, or both remains unclear. To address this issue, we are accruing patients for the Barrow ALA Intraoperative Confocal Evaluation (BALANCE) study. This Phase IIIa randomized, placebo-controlled trial consists of two study arms (1) to assess the efficacy of 5-ALA for HGGs when used in conjunction with intraoperative neuronavigation and (2) to evaluate the impact of 5-ALA on low-grade glioma resection when combined with intraoperative confocal microscopy and intraoperative neuronavigation. The BALANCE trial is thus intended to determine the clinical value of 5-ALA tumor fluorescence for high-grade glioma surgery using modern era practice principles and for low-grade glioma surgery using next-generation confocal technology.

In conclusion, the initial reported experience with intraoperative confocal microscopy in humans demonstrates its feasibility in examining a variety of intra- and extraaxial brain tumors. A preliminary analysis of the technology demonstrates its capacity not only to identify tumor cell populations in situ but potentially to replace the practice of frozen intraoperative specimen collection. The technique may complement conventional neuropathological diagnostic techniques, while reducing operative time and increasing sample size, particularly at the tumor margins. Intraoperative confocal microscopy also offers the unique ability to directly examine human brain tumor biology in vivo, without relying on xenograft animal models or surrogate imaging.

References

Alexander E III, Moriarty TM, Kikinis R, Black P, Jolesz FM (1997) The present and future role of intraoperative MRI in neurosurgical procedures. Stereotact Funct Neurosurg 68:10–17

Becker DE, Ancin H, Szarowski DH, Turner JN, Roysam B (1996) Automated 3-D montage synthesis from laser-scanning confocal images: application to quantitative tissue-level cytological analysis. Cytometry 25:235–245

Brezinski ME, Tearney GJ, Bouma B, Boppart SA, Pitris C, Southern JF et al (1998) Optical biopsy with optical coherence tomography. Ann N Y Acad Sci 838:68–74

Bussau LJ, Vo LT, Delaney PM, Papworth GD, Barkla DH, King RG (1998) Fibre optic confocal imaging (FOCI) of keratinocytes, blood vessels and nerves in hairless mouse skin in vivo. J Anat 192(Pt 2):187–194

Delaney PM, King RG, Lambert JR, Harris MR (1994) Fibre optic confocal imaging (FOCI) for subsurface microscopy of the colon in vivo. J Anat 184(Pt 1): 157–160

Dowling C, Bollen AW, Noworolski SM, McDermott MW, Barbaro NM, Day MR et al (2001) Preoperative proton MR spectroscopic imaging of brain tumors: correlation with histopathologic analysis of resection specimens. AJNR Am J Neuroradiol 22:604–612

Duffner F, Ritz R, Freudenstein D, Weller M, Dietz K, Wessels J (2005) Specific intensity imaging for glioblastoma and neural cell cultures with 5-aminolevulinic acid-derived protoporphyrin IX. J Neurooncol 71:107–111

Eschbacher J, Martirosyan NL, Nakaji P, Sanai N, Preul MC, Smith KA et al (2012) In vivo intraoperative confocal microscopy for real-time histopathological imaging of brain tumors. J Neurosurg 116:854–860

Floeth FW, Sabel M, Ewelt C, Stummer W, Felsberg J, Reifenberger G et al (2011) Comparison of (18)F-FET PET and 5-ALA fluorescence in cerebral gliomas. Eur J Nucl Med Mol Imaging 38:731–741

Flusberg BA, Cocker ED, Piyawattanametha W, Jung JC, Cheung EL, Schnitzer MJ (2005) Fiber-optic fluorescence imaging. Nat Methods 2:941–950

Flusberg BA, Nimmerjahn A, Cocker ED, Mukamel EA, Barretto RP, Ko TH et al (2008) High-speed, miniaturized fluorescence microscopy in freely moving mice. Nat Methods 5:935–938

Helmchen F (2002) Miniaturization of fluorescence microscopes using fibre optics. Exp Physiol 87:737–745

Hoffman A, Goetz M, Vieth M, Galle PR, Neurath MF, Kiesslich R (2006) Confocal laser endomicroscopy: technical status and current indications. Endoscopy 38:1275–1283

Ishihara R, Katayama Y, Watanabe T, Yoshino A, Fukushima T, Sakatani K (2007) Quantitative spectroscopic analysis of 5-aminolevulinic acid-induced protoporphyrin IX fluorescence intensity in diffusely infiltrating astrocytomas. Neurol Med Chir (Tokyo) 47:53–57

Khoshyomn S, Penar PL, McBride WJ, Taatjes DJ (1998) Four-dimensional analysis of human brain tumor spheroid invasion into fetal rat brain aggregates using confocal scanning laser microscopy. J Neurooncol 38:1–10

Kiesslich R, Goetz M, Vieth M, Galle PR, Neurath MF (2005) Confocal laser endomicroscopy. Gastrointest Endosc Clin N Am 15:715–731

Koenig F, Knittel J, Stepp H (2001) Diagnosing cancer in vivo. Science 292:1401–1403

Kondziolka D, Lunsford LD, Martinez AJ (1993) Unreliability of contemporary neurodiagnostic imaging in evaluating suspected adult supratentorial (low-grade) astrocytoma. J Neurosurg 79:533–536

LeRoux PD, Berger MS, Ojemann GA, Wang K, Mack LA (1989) Correlation of intraoperative ultrasound tumor volumes and margins with preoperative computerized tomography scans. An intraoperative method to enhance tumor resection. J Neurosurg 71:691–698

Liao H, Noguchi M, Maruyama T, Muragaki Y, Kobayashi E, Iseki H et al (2012) An integrated diagnosis and therapeutic system using intra-operative 5-aminolevulinic-acid-induced fluorescence guided robotic laser ablation for precision neurosurgery. Med Image Anal 16:754–766

Lim DA, Cha S, Mayo MC, Chen MH, Keles E, VandenBerg S et al (2007) Relationship of glioblastoma multiforme to neural stem cell regions predicts invasive and multifocal tumor phenotype. Neuro Oncol 9:424–429

Muragaki Y, Chernov M, Maruyama T, Ochiai T, Taira T, Kubo O et al (2008) Low-grade glioma on stereotactic biopsy: how often is the diagnosis accurate? Minim Invasive Neurosurg 51:275–279

Nabavi A, Thurm H, Zountsas B, Pietsch T, Lanfermann H, Pichlmeier U et al (2009) Five-aminolevulinic acid for fluorescence-guided resection of recurrent malignant gliomas: a phase ii study. Neurosurgery 65:1070–1076

Papworth GD, Delaney PM, Bussau LJ, Vo LT, King RG (1998) In vivo fibre optic confocal imaging of microvasculature and nerves in the rat vas deferens and colon. J Anat 192(Pt 4):489–495

Polglase AL, McLaren WJ, Skinner SA, Kiesslich R, Neurath MF, Delaney PM (2005) A fluorescence confocal endomicroscope for in vivo microscopy of the upper- and the lower-GI tract. Gastrointest Endosc 62:686–695

Roberts DW, Valdes PA, Harris BT, Fontaine KM, Hartov A, Fan X et al (2011) Coregistered fluorescence-enhanced tumor resection of malignant glioma: relationships between delta-aminolevulinic acid-induced protoporphyrin IX fluorescence, magnetic resonance imaging enhancement, and neuropathological parameters. Clinical article. J Neurosurg 114:595–603

Sanai N, Berger MS (2008) Glioma extent of resection and its impact on patient outcome. Neurosurgery 62:753–764

Sanai N, Varez-Buylla A, Berger MS (2005) Neural stem cells and the origin of gliomas. N Engl J Med 353:811–822

Sanai N, Snyder LA, Honea NJ, Coons SW, Eschbacher JM, Smith KA et al (2011) Intraoperative confocal microscopy in the visualization of 5-aminolevulinic acid fluorescence in low-grade gliomas. J Neurosurg 115:740–748

Stockhammer F, Misch M, Horn P, Koch A, Fonyuy N, Plotkin M (2009) Association of F18-fluoro-ethyl-tyrosin uptake and 5-aminolevulinic acid-induced fluorescence in gliomas. Acta Neurochir (Wien) 151:1377–1383

Stummer W, Stepp H, Moller G, Ehrhardt A, Leonhard M, Reulen HJ (1998) Technical principles for protoporphyrin-IX-fluorescence guided microsurgical resection of malignant glioma tissue. Acta Neurochir (Wien) 140:995–1000

Stummer W, Reulen HJ, Novotny A, Stepp H, Tonn JC (2003) Fluorescence-guided resections of malignant gliomas – an overview. Acta Neurochir Suppl 88:9–12

Stummer W, Pichlmeier U, Meinel T, Wiestler OD, Zanella F, Reulen HJ (2006) Fluorescence-guided surgery with 5-aminolevulinic acid for resection of malignant glioma: a randomised controlled multicentre phase III trial. Lancet Oncol 7:392–401

Tadrous PJ (2000) Methods for imaging the structure and function of living tissues and cells: 3. Confocal microscopy and micro-radiology. J Pathol 191:345–354

Tilgner J, Herr M, Ostertag C, Volk B (2005) Validation of intraoperative diagnoses using smear preparations from stereotactic brain biopsies: intraoperative versus final diagnosis—influence on clinical factors. Neurosurgery 56:257–265

Toms SA, Lin WC, Weil RJ, Johnson MD, Jansen ED, Mahadevan-Jansen A (2005) Intraoperative optical spectroscopy identifies infiltrating glioma margins with high sensitivity. Neurosurgery 57:382–391

Tonn JC, Stummer W (2008) Fluorescence-guided resection of malignant gliomas using 5-aminolevulinic acid: practical use, risks, and pitfalls. Clin Neurosurg 55:20–26

Uematsu Y, Owai Y, Okita R, Tanaka Y, Itakura T (2007) The usefulness and problem of intraoperative rapid diagnosis in surgical neuropathology. Brain Tumor Pathol 24:47–52

Van Meir EG, Hadjipanayis CG, Norden AD, Shu HK, Wen PY, Olson JJ (2010) Exciting new advances in neuro-oncology: the avenue to a cure for malignant glioma. CA Cancer J Clin 60:166–193

Widhalm G, Wolfsberger S, Minchev G, Woehrer A, Krssak M, Czech T et al (2010) 5-Aminolevulinic acid is a promising marker for detection of anaplastic foci in diffusely infiltrating gliomas with nonsignificant contrast enhancement. Cancer 116:1545–1552

Brainstem Cavernomas, Accessible Lesions: Surgery

17

Adib A. Abla and Robert F. Spetzler

Contents

Abstract

Brainstem cavernous malformations are vascular neoplasms located in highly eloquent brain matter. These lesions often hemorrhage and cause a relapsing and remitting course of symptoms followed by gradual improvement after each hemorrhage. Surgery is greatly protective in preventing future bleeds and further neurologic deterioration. The five essential approaches to the brainstem include the orbitozygomatic, retrosigmoid, far-lateral, suboccipital, and supracerebellar infratentorial approach. None of these approaches involve going through brain to enter the brainstem. Entry into the middle cerebellar peduncle or brachium pontis, which can be considered a part of the brainstem, allows access to the pons. At our institution these approaches have been used in the world's largest surgical experience with brainstem cavernous malformations in both adults and children with excellent outcomes. Nuances of the experience and approaches are described.

Introduction

Cavernous malformations, or cavernomas, are low-flow vascular lesions found throughout the central nervous system. They can be considered vascular neoplasms, although they are benign rather than malignant entities. They can occur in the brainstem where they are more likely to be symptomatic and less likely to cause seizures. It

A.A. Abla • R.F. Spetzler (✉)
Department of Neurosurgical Oncology, Barrow Brain Tumor Research Center, Barrow Neurological Institute, St. Joseph's Hospital and Medical Center, Phoenix, AZ, USA
e-mail: neuropub@dignityhealth.org

M. Hayat (ed.), *Tumors of the Central Nervous System, Volume 13*,
DOI 10.1007/978-94-007-7602-9_17, © Springer Science+Business Media Dordrecht 2014

is unclear whether cavernous malformations are present since birth in all cases because de novo cases have been demonstrated. Therefore, it is difficult to determine the exact natural history regarding retrospective calculations of annual hemorrhage risk when left untreated. However, once they hemorrhage, these lesions tend to hemorrhage in clusters and are more likely to bleed again. Our patients are often from an international referral base and tend to have aggressive lesions. Based on our experience, it is difficult to ascertain patients' actual natural history given this referral bias.

The experience at the Barrow Neurological Institute represents the world's largest experience treating these lesions with more than 300 patients treated in total, including more than 40 pediatric patients (Abla et al. 2010a, 2011). Given the delicate nature of crossing motor, sensory, and cranial nerve fibers in the brainstem as well as the presence of cranial nerve nuclei, operating in this region is associated with considerable risk. Such surgery requires the utmost attention to detail and application of technological advances, including neuronavigation and instrumentation. This chapter outlines our approach to the surgical treatment of these lesions.

Patient Selection

When a brainstem cavernous malformation demonstrates the classic "popcorn" appearance on T2-weighted magnetic resonance imaging (MRI) and symptoms are referable to the lesion, the patient is considered to have a cavernous malformation that has bled. Blood products from a hemorrhage can be demonstrated on T1- and T2-weighted MRI sequences and on CT. Signs and symptoms of the hemorrhage commonly include cranial nerve deficits, sensory complaints, headaches, hemiparesis, diplopia, ataxia, and vertigo in decreasing order of likelihood (Abla et al. 2010b). The symptoms of patients who have clinical or radiographic hemorrhages often improve with time; however, these patients are not protected from future clinical deterioration associated with future hemorrhages.

In our opinion, the cycle of neurologic deterioration and gradual improvement can be stopped only by removing the offending lesion; we do not recommend radiosurgery to patients with a cavernous malformation. Cavernous malformations deep to the surface of the brainstem from all angles (and that cause mild symptoms) can be observed until they enlarge enough to reach the brainstem surface. Our indications for surgery include the following factors: acute hemorrhages outside the cavernoma, exophytic or pial abutting lesions, repeat hemorrhages with progressive symptoms, and clinical deterioration (Abla et al. 2010b).

All lesions treated surgically are removed through one of the following five approaches or a combination thereof (e.g., hybrid approach): orbitozygomatic (OZ) pterional craniotomy, far-lateral craniotomy, retrosigmoid craniotomy, suboccipital craniotomy, or lateral supracerebellar infratentorial craniotomy. None of these approaches to the brainstem require cortical entry; that is, the brainstem can be accessed without violating any brain matter with all five approaches.

The craniotomy can often be selected by using the two-point method (Brown et al. 1996). One point is made in the center or limit of the cavernoma requiring surgical access and the other point at the surface of the brainstem where the lesion reaches the pial surface. Connecting these two lines and extending them to the surface of the cranium approximates the required craniotomy (Fig. 17.1).

Orbitozygomatic Approach

The OZ approach is the least utilized of the five major approaches, but it is still an essential tool in the armamentarium for approaching these lesions. This approach is useful for cavernous malformation in the anterior half of the midbrain. It involves a pterional craniotomy and some variation of the OZ osteotomy (Lemole et al. 2003). The dissection involves wide arachnoid dissection of the opticocarotid cistern and a trajectory through the oculomotor-carotid window

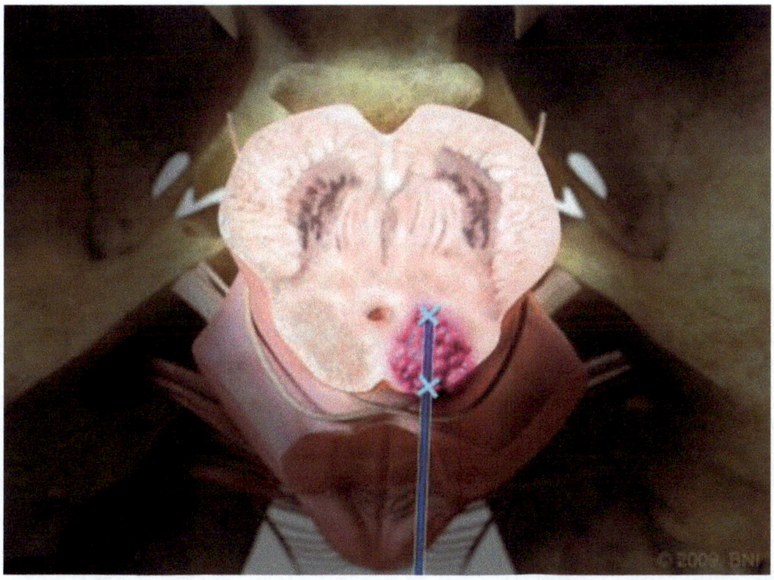

Fig. 17.1 The two-point method involves creating a point at the proposed limit of resection and one where the cavernous malformation abuts the pial surface; connecting the two points creates a line that can be extended to the cranial surface and that helps choose the appropriate craniotomy (Used with permission from Barrow Neurological Institute)

to Lillequist's membrane. Once Lilliquist's membrane has been traversed, the basilar artery and midbrain are exposed. Great care must be taken to avoid basilar perforators when the cavernous malformation is entered and to avoid the motor fibers of the cerebral peduncle. As in all cases, neuronavigation is useful in locating the brainstem 'cortical' opening point when no telling hemosiderin stain is present. The OZ approach is useful for lesions that can be accessed in the midline in the interpeduncular cistern or for lesions that can be accessed with this approach that sit lateral to the peduncle and can be removed by creating an opening in the midbrain lateral to the pyramidal tracts.

Retrosigmoid Approaches

The retrosigmoid approach is the second most commonly used approach at our institution (Abla et al. 2011). The largest proportion of cavernomas in the brainstem lies in the pons, and this approach is extremely useful in accessing these lesions. Cavernous malformations of the brainstem often expand the pons but do not reach a pial surface. The middle cerebellar peduncle, a relatively noneloquent structure, has been used to access such formidable lesions.

The middle cerebellar peduncle is often accessed behind the brainstem entry zone of the fifth cranial nerve. Staying superior is critical; an inferior entry into the middle cerebellar peduncle likely involves the seventh cranial nerve as it crosses around the sixth cranial nerve nucleus on its way to exit the brainstem in the cerebellopontine angle.

In approaching the pons in this fashion, we have abandoned more anteriorly situated craniotomies that were previously used to access the pons, including the transpetrous approaches and subtemporal approaches involving a medial petrosectomy. In our experience the retrosigmoid approach is very well tolerated and provides robust access to the pons without jeopardizing some of the more anteriorly situated motor fibers or cranial nerve nuclei. A previously reported case of an 11-year-old girl demonstrates the retrosigmoid approach (Fig. 17.2) (Abla et al. 2010a).

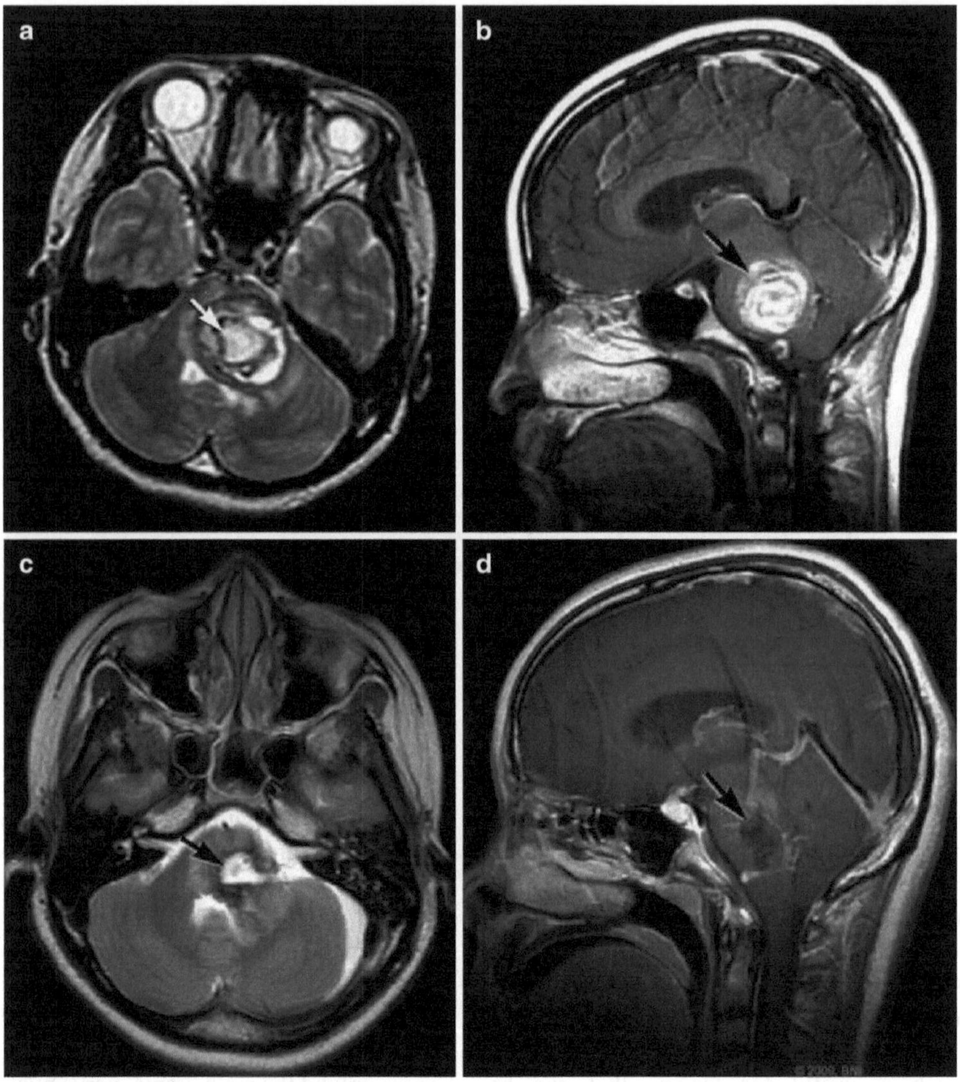

Fig. 17.2 Preoperative (**a**) axial T2-weighted and (**b**) sagittal contrast-enhanced MRI of an 11-year-old girl with a giant pontine cavernoma (*arrows*). Resection was performed via a left retrosigmoid craniotomy and entry through the middle cerebellar peduncle. Postoperative (**c**) axial T2-weighted and (**d**) sagittal contrast-enhanced MRIs confirm gross total resection (*arrows*) (Used with permission from Barrow Neurological Institute)

Lateral Supracerebellar Craniotomy

The lateral supracerebellar infratentorial craniotomy is useful for approaching cavernomas of the posterior and/or lateral midbrain. The use of the supracerebellar infratentorial approach to cavernous malformations of the brainstem in 45 patients treated at our institution has been reported elsewhere (de Oliveira et al. 2010). In that report, median, paramedian, and extreme lateral supracerebellar approaches were discussed (Fig. 17.3). Now, however, the supracerebellar approach most often used at our institution is the lateral supracerebellar approach.

This incision is placed similar to that for a retrosigmoid craniotomy, but exposure of the sigmoid sinus is not critical. The craniotomy is placed more superior to allow exposure of the transverse sinus. The craniotomy can extend superiorly

Fig. 17.3 Illustration showing three variations of the supracerebellar infratentorial approach: a median, paramedian, and extreme lateral variation (Used with permission from Barrow Neurological Institute)

above the transverse sinus. The craniotomy allows the cerebellum to be retracted inferiorly away from the transverse sinus and provides a trajectory from lateral (superficially) to more medially at the depth of the operating field usually aimed at the collicular plate.

The approach is useful for reaching lateral and posterior situated cavernomas of the midbrain. It has also been used for cavernous malformations that extend into the thalamus rostrally or into the pons inferiorly. The trochlear nerve is a useful and dependable landmark crossing around the posterior midbrain toward the ambient cistern. The park-bench position is often used (de Oliveira et al. 2010).

Far-Lateral Approach

The far-lateral approach allows access to the anterior portion of the medulla. The patient is positioned in a prone, or more preferably, a park-bench position (de Oliveira et al. 2010). A U-shaped incision was once used for this approach, but we now prefer a laterally situated vertical incision, which minimizes obstruction of the surgical corridor by the surgical flap itself.

The approach involves a suboccipital craniotomy situated lateral to a typical suboccipital opening. This approach also can involve some degree of drilling of the occipital condyle. Typically, as much as one-third of the posterior condyle can be drilled safely to increase the degree of dural retraction and a lateral-to-medial trajectory. A C1 laminectomy is also performed. The dura is opened in a curvilinear fashion and flapped laterally. Great care must be taken to avoid injury to the vertebral artery as it crosses the atlas; however, we do not routinely mobilize or skeletonize the vertebral artery in this approach to brainstem cavernomas. Dissection of the arachnoid proceeds lateral to the medulla and around the vertebral artery, twelfth cranial nerve rootlets, and the exit of the posterior inferior cerebellar artery. The brainstem is usually entered at the lateral to anterolateral aspect of the medulla.

Suboccipital

The suboccipital approach is the most commonly used approach and includes the telovelar variant. Of utmost importance with this approach is respect for the floor of the fourth ventricle near the facial colliculus. Rarely, if ever, are lesions approached through the facial colliculus if they do not reach the surface. The only exception would be patients with permanent unilateral sixth and seventh cranial nerve palsy. Even so, the contralateral side could still be injured, which would be devastating in these patients. Lesions deep to the facial colliculus that are not exophytic are among the lesions most commonly observed until they hemorrhage again and reach a pial surface.

The suboccipital approach is useful for lesions in the posterior medulla and for those that extend into the lower aspect of the fourth ventricle. Lesions in the middle cerebellar peduncle that are more medial and inferior in the middle cerebellar peduncle can be accessed with the telovelar variant of the suboccipital approach, which involves opening the cerebellomedullary fissure via dissection of the inferior medullary velum and tela choroidea of the fourth ventricle on one side. Both sides can be opened to increase visualization and mobilization of the cerebellar tonsils, but doing so is seldom required.

Various Techniques, Concepts, and Instrumentation

New techniques to brainstem cavernous malformation include endoscopic approaches (Kimball et al. 2012). However, this approach is not used at our institution. The concern for cerebrospinal fluid leakage is still a valid detracting factor for this approach. Subtemporal approaches to brainstem cavernous malformation have been used at our institution and are also used by other authors to access the pons (Gross and Day 2010). However, we have moved away from these approaches, which involve petrosal drilling, including those requiring either medial petrosectomy, transcochlear, or translabyrinthine approaches. Although they provide a remarkable and unhindered view of the brainstem, this degree of bony exposure is unnecessary for exposing brainstem cavernomas.

More recently, as an adjunctive measure, lighted instrumentation has been used at our institution to improve visualization in the depth of the operative field. A lighted microsuction or lighted bipolar device has proven quite useful in various surgical cases involving brainstem cavernomas. These additional tools have also been useful as the senior author's surgical technique has evolved. Surgery performed without brain retraction with fixed malleable retractors is largely unnecessary for skull base approaches and exposures, including those to the brainstem (Spetzler and Sanai 2011).

These lighted instruments also provide an added ability to visualize and preserve the developmental venous anomaly often buried within the cavernous malformation. A developmental venous anomaly is often seen on MRI in conjunction with a brainstem cavernous malformation. It is critical to leave the developmental venous anomaly intact. We do not advocate its removal to prevent hemorrhages because such a risk is low compared to the complications associated with resection of a developmental venous anomaly. In fact, in our initial experience, one patient died after resection of a developmental venous anomaly (Porter et al. 1999).

The BNI Experience

Our recent experiences in both adult and pediatric patients have been reported (Abla et al. 2010a, 2011). These series included 260 adults and 40 patients younger than 18 years. The mean age was 12.3 years for the children and 41.8 years for adults. The mean size of the lesions was 2.3 cm in children and 1.8 cm in adults. The mean length of clinical follow up was 31.9 months in 36 patients with available follow up after discharge.

Six patients showed evidence of cavernoma growth in the same location at last follow up; five were suspicious for having rebled. At last follow up the mean Glasgow Outcome Scale score was 4.5 compared to 4.2 at admission and 4.05 at discharge. Preoperative symptoms and deficits improved in 16 patients. Nineteen patients had some neurologic findings after surgery, but only ten had new fixed deficits that were absent the morning before surgery. Many of these deficits were mild. The vast majority of patients did not demonstrate new hemorrhage or residual or recurrent cavernomas.

In adults, the mean length of follow up was 51 months. The mean Glasgow Outcome Scale score was 4.4 at admission, 4.2 at discharge, and 4.6 at follow up. There were 146 patients with some new or worsened neurologic findings: 21% of patients had transient findings and 35% had permanent deficits. Eighteen patients experienced 20 rehemorrhages from the surgical site. Twelve required reoperation. The postoperative risk of hemorrhage was 2.0% per year.

Overall, the pediatric patients had larger lesions and higher rates of recurrence or residual than the adults. Most patients had favourable outcomes, and treatment decreases the risk of rebleeding and the often devastating complications. In our extensive experience, temporary new cranial nerve or motor tract deficits are common but resolve. Recovery is dictated by the prevention of future hemorrhages, as in 90–95% of our patients. However, given the delicate nature of the brainstem, there is a considerable number of patients whose preoperative symptomatology may recur after surgery (which is counted as a new finding if not present the morning before surgery) and is sometimes permanent.

Summary

The approaches to brainstem cavernous malformations can be summarized as follows (Abla et al. 2010b). Midbrain lesions, including those extending into the thalamus, are approached through a supracerebellar craniotomy when located posteriorly and through the OZ approach when located in the anterior half. Ponteomesencephalic lesions are most often approached with a supracerebellar infratentorial approach, but a hybrid retrosigmoid/supracerebellar approach can be useful when the cerebellum can be retracted posteriorly as with a standard retrosigmoid approach or inferiorly as with the supracerebellar approach. Lesions of the pons are approached through a retrosigmoid approach when they are accessible from the cerebellopontine angle or middle cerebellar peduncle. Lesions in the middle cerebellar peduncle of the pons itself can be approached through a suboccipital/telovelar approach or a retrosigmoid approach, depending on how superiorly the lesion is located. Telovelar approaches are less favorable for superior and laterally situated lesions in the middle cerebellar peduncle. Although lateral approaches were once used, pontomedullary approaches are now usually accessed through a retrosigmoid craniotomy. The comfort and familiarity of the retrosigmoid approach and how well it is tolerated by patients are why we now favor it for low pontine lesions. When situated anteriorly, cervicomedullary lesions are accessed with a far-lateral approach. When the lesions are located in the posterior half of the medulla or cervicomedullary junction, a suboccipital approach is used.

References

Abla AA, Lekovic GP, Garrett M, Wilson DA, Nakaji P, Bristol R, Spetzler RF (2010a) Cavernous malformations of the brainstem presenting in childhood: surgical experience in 40 patients. Neurosurgery 67:1589–1598. doi:10.1227/NEU.0b013e3181f8d1b2; discussion 1598–1599

Abla AA, Turner JD, Mitha AP, Lekovic G, Spetzler RF (2010b) Surgical approaches to brainstem cavernous malformations. Neurosurg Focus 29:E8. doi:10.3171/2010.6.FOCUS10128

Abla AA, Lekovic GP, Turner JD, de Oliveira JG, Porter R, Spetzler RF (2011) Advances in the treatment and outcome of brainstem cavernous malformation surgery: a single-center case series of 300 surgically treated patients. Neurosurgery 68:403–414. doi:10.1227/NEU.0b013e3181ff9cde; discussion 414–415

Brown A, Thompson B, Spetzler RF (1996) The two point method: evaluating brain stem lesions. BNI Q 12:20–24

de Oliveira JG, Lekovic GP, Safavi-Abbasi S, Reis CV, Hanel RA, Porter RW, Preul MC, Spetzler RF (2010) Supracerebellar infratentorial approach to cavernous malformations of the brainstem: surgical variants and clinical experience with 45 patients. Neurosurgery 66:389–399. doi:10.1227/01.NEU.0000363702.67016.5D

Gross BA, Day AL (2010) Comments for publication for Abla AA, Lekovic GP, Garrett M, Wilson DA, Nakaji P, Bristol R and Spetzler RF (2010a) Cavernous malformations of the brainstem presenting in childhood: surgical experience in 40 patients. Neurosurgery 67:1589–1598; discussion 1598–1599. doi: 10.1227/NEU.0b013e3181f8d1b2

Kimball MM, Lewis SB, Werning JW, Mocco JD (2012) Resection of a pontine cavernous malformation via an endoscopic endonasal approach: a case report. Neurosurgery 71:186–193

Lemole GM Jr, Henn JS, Zabramski JM, Spetzler RF (2003) Modifications to the orbitozygomatic approach. Technical note. J Neurosurg 99:924–930

Porter RW, Detwiler PW, Spetzler RF, Lawton MT, Baskin JJ, Derksen PT, Zabramski JM (1999) Cavernous malformations of the brainstem: experience with 100 patients. J Neurosurg 90:50–58

Spetzler RF, Sanai N (2011) The quiet revolution: retractorless surgery for complex vascular and skull base lesions. J Neurosurg 116(2):291–300. doi:10.3171/2011.8.JNS101896

The Role of Surgical Resection for Metastatic Brain Tumors

18

Aqueel Pabaney and Steven N. Kalkanis

Contents

Abstract

Brain metastasis is the most common intracranial neoplasm in adults. The treatment of brain metastases requires a robust multidisciplinary approach. The roles of surgical resection, radiosurgery and whole brain radiation have been explored in great depths over the past several years. Although controversies exist regarding each treatment modality, it is generally agreed that the best results, in terms of both tumor response and patient survival, can be achieved by using a combination of local (surgical resection, stereotactic radiosurgery) and more generalized (whole brain radiotherapy, for example) treatments. Considering our current armamentarium of preoperative imaging, advanced intraoperative techniques followed by adjunct therapies, surgery plays an indispensible role in relieving increased intracranial pressure and significant mass effect, decreasing tumor burden as well as providing neuropathological diagnosis. Hemorrhagic and radioresistant lesions may also be treated most effectively with surgical interventions. However, important patient-specific and tumor-specific variables must be taken under consideration, and risks must be weighed against benefits to dictate patient selection and treatment. In this chapter, surgical treatment of brain metastases will be discussed with emphasis on current guidelines and controversies.

A. Pabaney • S.N. Kalkanis (✉)
Department of Neurosurgery, Henry Ford Center for Cancer Surgery, 2799 West Grand Blvd., K-11, Detroit, MI 48202, USA
e-mail: skalkan1@hfhs.org

M. Hayat (ed.), *Tumors of the Central Nervous System, Volume 13*,
DOI 10.1007/978-94-007-7602-9_18,

Introduction

Brain metastases are ten times more common than any primary brain neoplasm. Approximately 20–40% of patients suffering from systemic cancer develop symptomatic brain metastases, translating into a huge disease burden annually (Shaffrey et al. 2004). In adults, lung cancer is the most common source of brain metastases (50–60%), followed by breast cancer (15–20%) and melanoma (5–10%) (Johnson and Young 1996). The optimal management of brain metastases remains controversial. Whole brain radiotherapy (WBRT) and local treatment, including surgery and radiosurgery are considered the cornerstones of treatment. The treatment paradigm for brain metastases is rapidly evolving with emerging technologies, and multiple guidelines and controversies exist regarding each treatment modality.

Surgical treatment was first introduced as a treatment for brain metastases in 1926 (Grant 1926). Initially, the morbidity and mortality of surgical procedures was unacceptably high. Improved surgical techniques and developments in the field of neuroanesthesia, neuroimaging, neuromonitoring and postoperative care have significantly improved outcomes, and, for a select patient population with a single brain metastasis, surgical resection followed by radiotherapy is now the gold standard for the treatment of brain metastasis (Kalkanis et al. 2010). It is worth noting that most of the patients treated for brain metastases now die of extracranial disease (Khuntia et al. 2006). This is an important consideration because, although most studies have used overall survival as the main endpoint, survival is probably not the best parameter to measure the efficacy of existing brain-specific therapeutic modalities (Loeffler and Shrieve 1995).

Surgical Resection

Indications of Surgical Resection

The main aim of surgery in treating patients with brain metastases is to lengthen the survival times while improving neurological conditions and performance status (Narita and Shibui 2009). Due to recent technical advances in radiosurgery, patients treated with stereotactic radiosurgery (SRS) have outnumbered those treated with surgery. However, only surgical intervention allows rapid debulking of tumors with significant mass effect and can be considered a life saving intervention in patients with herniation syndromes and neurological deficits (Schackert et al. 2001). It also allows for restoration of CSF flow, relieving intracranial pressure and also lowers steroid dependence. Surgery can also assist in managing medically refractory seizures caused by tumors (Ranasinghe and Sheehan 2007). In most cases, the primary goal of surgery is gross-total resection of the tumor with minimal disruption of normal brain tissue. Data from a retrospective review showed that early postoperative KPS scores were improved in 59% of patients, unchanged in 32%, and worse in 9% in patients who had undergone microsurgical tumor removal. Surgery also allows for histological confirmation of the diagnosis of brain metastases (Korinth et al. 2002).

Surgical Technique

Prior to taking the patient to the operating room, outlining the preoperative surgical plan utilizing structural and functional imaging modalities is of paramount importance. Utilization of intraoperative image guidance, microsurgical techniques and perioperative neurologic monitoring reduces the possibility of surgery-related mortality and morbidity. Tumors can be resected either in a *piecemeal* fashion or *en bloc*. Data has shown that piecemeal resection of the tumor is associated with a significant increase in risk of local recurrence when compared to *en bloc* resection (Patel et al. 2010). *En bloc* resection is particularly instrumental when dealing with posterior fossa lesions; however, an eloquent location within the brain parenchyma significantly limits *en bloc* resection. In those cases, partial resection should be performed and radiosurgery can be given to the residual tumor. Despite seemingly complete resections, the reported local recurrence rate in the surgical bed is 10–34% at 1 year following surgery and radiotherapy (Sundaresan et al. 1988).

It is mostly attributable to tumor infiltration which can vary with different histological tumor types, with aggressive tumors infiltrating up to 3 mm beyond the surgical bed (Neves et al. 2001). The concept of "microscopic total resection" in which microscopic infiltrating tumor cells within a normal looking brain parenchyma are removed within a 5 mm area in the peritumoral bed region, has also been evaluated by a Korean group (Yoo et al. 2009). They were able to show that this method of resection is as effective in reducing local recurrence as gross total resection supplemented with radiotherapy.

Clinical Evidence

Evidence-Based Guidelines for the Management of Brain Metastases: AANS/CNS Initiative

There is a wealth of literature that addresses the question of optimal therapy for patients with metastatic brain tumors. Several papers attempt to make strong recommendations regarding one treatment versus the other. However, in 2010, *Journal of Neuro-oncology* published an extensive array of guidelines. This initiative was sponsored by the American Association of Neurological Surgeons (AANS), the Congress of Neurological Surgeons (CNS), and the AANS/CNS Joint Tumor Section. A multidisciplinary panel consisting of surgeons, neuro-oncologists and radiation experts was constituted that came up with eight practice guideline papers in this series after adopting a broad literature search strategy, extensive data collection, formation of writing groups and spirited discussions (Robinson et al. 2010). This set of guidelines presents the most advanced, up to date and evidence based approach to treating patients with metastatic brain tumors. Both the evidence classification and the strength of the recommendations were graded according to the criteria endorsed by the AANS/CNS (Table 18.1). Several guidelines have been mentioned in the text to follow and have been extracted from one of the papers in this series. For more information, please refer to January 2010 issue of *Journal of Neuro-oncology*.

Table 18.1 AANS/CNS evidence classes and levels of recommendation

Evidence classification
Class I
Evidence provided by one or more well-designed randomized controlled clinical trials, including overview (meta-analyses) of such trials
Class II
Evidence provided by well-designed observational studies with concurrent controls (e.g. case control and cohort studies)
Class III
Evidence provided by expert opinion, case series, case reports and studies with historical controls
Levels of recommendation
Level I
Generally accepted principles for patient management, which reflect a high degree of clinical certainty (usually this requires Class I evidence which directly addresses the clinical questions or overwhelming Class II evidence when circumstances preclude randomized clinical trials)
Level II
Recommendations for patient management which reflect clinical certainty (usually this requires Class II evidence or a strong consensus of class III evidence)
Level III
Other strategies for patient management for which the clinical utility is uncertain (inconclusive or conflicting evidence or opinion)

A concise summary of these guidelines was published by Bhangoo et al. (2011).

Surgical Resection and WBRT

There exists Level 1 evidence that surgical resection plus WBRT is a superior treatment modality than surgical resection alone in controlling local recurrence at the original site and in the brain overall (Kalkanis et al. 2010). Patchell et al. (1998) investigated the benefit of using WBRT as an adjunctive therapy following surgical tumor removal in these patients. The study randomized 95 patients to surgery alone or surgery plus WBRT. Progression of intracranial disease was fourfold greater in the surgery-alone group (70% compared with 18%, $p<0.001$), and local recurrence was also higher in this group (46% compared with 10%, $p<0.001$). Patchell et al. (1990) randomized 48 patients with single brain metastases to surgery and WBRT (25 patients) compared with WBRT alone (23 patients) and evaluated local recurrence and survival rates. In this study, the addition of surgery reduced the local recurrence in these patients from 52 to 20% ($p<0.02$) and improved median survival from 15 to 40 weeks ($p<0.001$). Patients in the surgery combined with WBRT arm also remained functionally independent for a longer period of time than those treated with WBRT alone (38 weeks compared with 8 weeks; $p<0.005$).

Surgical Resection and SRS

Stereotactic radiosurgery is increasingly being considered as the first choice of treatment in many initial and recurrent cases of metastatic brain tumors because of its non-invasiveness, high control rate and lower rates of morbidity, in addition to it being an outpatient treatment modality. However, SRS fails to relieve the signs and symptoms of increased intracranial pressure and is unable to alleviate the mass effect of the tumor in an expedited fashion. Also, SRS is not considered optimal for lesions > 3 cm in diameter. SRS can increase peritumoral edema as well. The effectiveness of SRS compared to surgical resection has not been compared within a phase III randomized trial for patients with a single brain metastasis. Approximately six studies have been published comparing SRS and surgery. There exists level 2 evidence that surgical resection plus WBRT versus SRS + WBRT both present effective treatment options, resulting in relatively equal survival rates (Gaspar et al. 2010).

Tumor Characteristics

Tumor Histopathology

Histopathology of the tumor is an important consideration when developing a treatment plan for a patient with brain metastases, because different tumors respond differently to radiation and chemotherapy options. Tumors such as sarcoma, renal cell carcinoma and melanoma are considered resistant to WBRT, but it has been shown that these tumors do respond better to radiosurgery (Brown et al. 2008). It is also deemed prudent to stage the tumor according to size, location and disease burden and define the histological grade to better estimate the prognosis. Data from several studies have shown that breast cancer metastases yield the best prognosis after surgery and postoperative radiotherapy, whereas melanoma and renal cell cancer fare the worst (Wroński et al. 1997; Sampson et al. 1998). In a recent study by Sperduto et al. (2010), a good correlation was found between diagnosis-specific graded prognosis assessment (DS-GPA) scores and outcomes for newly diagnosed brain metastases patients. Prognostic factors analyzed included Karnofsky Performance status (KPS), age, presence of extracranial metastases and number of brain lesions.

Multiple Metastases

Presence of multiple metastases has long been accepted as a partial contraindication for surgery because the patient was not expected to live long enough to realize a benefit from surgery. Patients with four or more brain tumors are usually not treated surgically, given the poor prognosis. No level I evidence defines optimal treatment of patients with more than five brain lesions. WBRT can be considered in these cases if the life expectancy is greater than 3 months based on systemic disease. There is level II evidence suggesting that SRS may be effective in up to ten brain metastases if they are smaller than 3 cm and are not associated with mass effect or significant edema (DiLuna et al. 2007). One study revealed that a highly selected subset of patients with a limited number of multiple brain metastases may benefit from resection of all lesions (Bindal et al. 1993).

Recurrent Disease

Treatment of recurrent brain metastases is a highly controversial topic. Surgery has been shown to improve survival and quality of life in patients with recurrent disease (Arbit et al. 1995). Resection of recurrent tumor also allows confirmation of histopathology and the use of local chemotherapeutic adjuncts such as BCNU wafer implants. Bindal et al. (1995) have also reported that reoperation for recurrent brain metastases after the initial resection prolonged survival and improved quality of life. Surgical resection is an option for recurrence after SRS as well. In patients with symptomatic mass effect, progressive neurological signs or symptoms, imaging evidence of tumor progression, or intractable seizures after radiosurgery, resection may become the treatment of choice.

Cerebellar Metastases

Cerebellar metastases represent a special group of brain metastases because they may cause obstructive hydrocephalus and brain stem compression, and survival of patients with cerebellar metastases has been reported as more disappointing than that reported for cerebral hemispheric metastases. However, surgical resection provides a significant benefit in cerebellar metastases. In a recent study comparing effectiveness of surgery versus radiation, 38 patients with cerebellar metastases underwent surgical resection alone; their median survival was 20.5 months. In the 27 patients who underwent surgical resection plus radiation, the median survival was 35.5 months. For 21 patients who underwent WBRT without surgical resection, the median survival was 6.5 months and for those who were treated with SRS alone, 9.1 months (Yoshida and Takahashi 2009).

Patient Selection for Surgery

As in all surgical disciplines, patient selection is of paramount importance if desirable results are to be achieved. The patient should be medically fit to undergo surgery and to withstand the recovery phase postoperatively. Several studies have evaluated variables that might make some patients a better surgical candidate than others. Factors considered favorable for surgical resection of the tumor include age less than 65 years, KPS score > 70, single tumors, tumor size < 3 cm, surgically accessible location, good control of extracranial disease and absence of leptomeningeal involvement, expected long disease – free survival and local symptomatic mass effect.

Table 18.2 Recursive Partitioning Analysis (RPA) classification for brain metastases

RPA Class I
Patients with KPS ≥ 70, <65 years of age with controlled primary and no extracranial metastases
RPA Class II
KPS < 70
RPA Class III
All others

In 1997, Gaspar et al. (1997) performed a rigorous multivariable analysis of tumor characteristics, patient profiles and treatment variables extracted from three prospective Radiation Therapy Oncology Group (RTOG) brain metastases trials. This effort was made to analyze the relative contributions of pretreatment variables to the survival of patients with brain metastases using an interactive, nonparametric statistical technique known as Recursive Partitioning Analysis (RPA), to define the influence of treatment variations on survival among patients and to identify patient subgroups or stages. Three prognostic classes were developed for patients with multiple brain metastases (Table 18.2). Recursive partitioning analysis Class I patients are considered good candidates for craniotomy and resection, whereas Class III patients are not likely to realize benefit from surgery. This method of classification was later validated in 2000 (Gaspar et al. 2000). The RPA classification was also successfully applied to surgically resected and irradiated cases of metastatic brain tumors (Agboola et al. 1998). In another large single-institution retrospective study, preoperative performance status, symptomatic response to steroid treatment, systemic tumor control, and serum lactate dehydrogenase levels were found to be independent prognostic factors in patients with brain metastases (Largerwaard et al. 1999).

Making Surgery Safe

With recent advances in image-guided surgery and with increased utilization of newer modalities like functional MRI (fMRI), intraoperative

MRI (iMRI), electrocorticography during awake craniotomy and diffusion tensor imaging (DTI), surgical resection is no longer considered "risky", but has allowed neurosurgeons to safely navigate through the parenchyma and safely resect even deeply seated lesions with an acceptable risk of neurological deficit. DTI is a form of fMRI used to delineate white matter anatomy and is based on the principle that water preferentially diffuses along the long axis of white matter tracts. Distortions of white matter architecture secondary to a tumor or the edema surrounding a tumor can be mapped in a meaningful way to provide guidance during surgical resection. fMRI is a noninvasive imaging modality that uses cortical blood flow changes as a surrogate for increased or decreased neuronal activity. Functional MR imaging maps can be matched and fused with high-resolution MR or CT images to produce neuronavigational images, a process referred to as "functional neuronavigation." The completeness of resection has also improved in recent years with increasing use of iMRI that provides instant feedback regarding residual tumor and helps overcome the problem of "brain shift", an intraoperative phenomenon in which changes in tumor volume, cerebrospinal fluid drainage, intracranial pressure, or the use of brain retractors generate intraoperative brain deformation that renders preoperative neuronavigation registration inaccurate. Stummer et al. (1998) are credited with the development of a tumor-specific fluorescent marker that allows more accurate discrimination of infiltrating tumor from normal brain parenchyma, called 5-Aminolevulinic Acid (5-ALA). Although 5-ALA was used for resection of glial neoplasms initially, groups have applied the same concept successfully to metastatic brain tumors as well (Utsuki et al. 2007). With a paradigm shift towards minimally invasive neurosurgery that has always been an attractive option, both for clinicians and patients, several treatment modalities have surfaced recently. These modalities include such techniques as Laser Interstitial Thermal Therapy (LITT), cryoablation, and radiofrequency ablation. LITT has recently seen resurgence with regard to the clinical utility of this technique secondary to advances in MRI thermometry. Carpentier et al. (2008) recently reported the successful treatment of six patients with metastatic intracranial lesions using LITT.

Conclusion

Several treatment options exist when managing a patient with metastatic brain tumor. Numerous variables are to be taken into consideration when tailoring treatment plans for individual patients. Updated guidelines regarding management of these pathologies exist and make the clinicians' job much easier in choosing the most effective treatment modality. Combining the modalities of surgery, WBRT, and radiosurgery has improved the outcome in patients with metastatic disease. With the advent of newer technologies, all the treatment modalities are not only expected to become more effective but will also develop a larger interface for overlap that will have synergistic effects towards improving patient survival. In addition, a multidisciplinary approach should be utilized for every patient to individualize care. Multispecialty interactions in the form of a "Tumor Board" allow experts to look at the pathologies from their perspective and offer evidence based opinions. Careful review of cases in tumor board conferences can potentially result in a change in the clinical management of patients presented and also serves as an excellent educational opportunity for trainees. Ultimately, it serves as a platform that brings the specialists together and yields a much enhanced treatment plan that is likely to translate into longevity and improved quality of life.

References

Agboola O, Benoit B, Cross P, Da Silva V, Esche B, Lesiuk H, Gonsalves C (1998) Prognostic factors derived from recursive partitioning analysis (RPA) of radiation therapy oncology group (RTOG) brain metastases trials applied to surgically resected and irradiated brain metastatic cases. Int J Radiat Oncol Biol Phys 42:155–159

Arbit E, Wronski M, Burt M, Galicich JH (1995) The treatment of patients with recurrent brain metastases. A retrospective analysis of 109 patients with nonsmall cell lung cancer. Cancer 76(5):765–773

Bhangoo SS, Linskey ME, Kalkanis SN, American Association of Neurologic Surgeons (AANS), Congress of Neurologic Surgeons (CNS) (2011) Evidence-based guidelines for the management of brain metastases. Neurosurg Clin N Am 22(1):97–104

Bindal RK, Sawaya R, Leavens ME, Lee JJ (1993) Surgical treatment of multiple brain metastases. J Neurosurg 79(2):210–216

Bindal RK, Sawaya R, Leavens ME, Hess KR, Taylor SH (1995) Reoperation for recurrent metastatic brain tumors. J Neurosurg 83:600–604

Brown PD, Brown CA, Pollock BE, Gorman DA, Foote RL (2008) Stereotactic radiosurgery for patients with "radioresistant" brain metastases. Neurosurgery 62(Suppl 2):790–801

Carpentier A, McNichols RJ, Stafford RJ, Itzcovitz J, Guichard JP, Reizine D, Delaloge S, Vicaut E, Payen D, Gowda A, George B (2008) Real-time magnetic resonance-guided laser thermal therapy for focal metastatic brain tumors. Neurosurgery 63:ONS21–ONS28

DiLuna ML, King JT Jr, Knisely JP, Chiang VL (2007) Prognostic factors for survival after stereotactic radiosurgery vary with the number of cerebral metastases. Cancer 109(1):135–145

Gaspar L, Scott C, Rotman M, Asbell S, Phillips T, Wasserman I, McKenna WG, Byhardt R (1997) Recursive partitioning analysis (RPA) of prognostic factors in three Radiation Therapy Oncology Groups (RTOG) brain metastases trials. Int J Radiat Oncol Biol Phys 37:745–751

Gaspar LE, Scott C, Murray K, Curran W (2000) Validation of the RTOG recursive partitioning analysis (RPA) classification for brain metastases. Int J Radiat Oncol Biol Phys 47(4):1001–1006

Gaspar LE, Mehta MP, Patchell RA, Burri SH, Robinson PD, Morris RE, Ammirati M, Andrews DW, Asher AL, Cobbs CS, Kondziolka D, Linskey ME, Loeffler JS, McDermott M, Mikkelsen T, Olson JJ, Paleologos NA, Ryken TC, Kalkanis SN (2010) The role of whole brain radiation therapy in the management of newly diagnosed brainmetastases: a systematic review and evidence-based clinical practice guideline. J Neurooncol 96(1):17–32

Grant FC (1926) Concerning intracranial malignant metastases: their frequency and the value of surgery in their treatment. Ann Surg 84(5):635–646

Johnson JD, Young B (1996) Demographics of brain metastasis. Neurosurg Clin N Am 7:337–344

Kalkanis SN, Kondziolka D, Gaspar LE, Burri SH, Asher AL, Cobbs CS, Ammirati M, Robinson PD, Andrews DW, Loeffler JS, McDermott M, Mehta MP, Mikkelsen T, Olson JJ, Paleologos NA, Patchell RA, Ryken TC, Linskey ME (2010) The role of surgical resection in the management of newly diagnosed brain metastases: a systematic review and evidence-based clinical practice guideline. J Neurooncol 96(1):33–43

Khuntia D, Brown P, Li J, Mehta M (2006) Whole brain radiotherapy in the management of brain metastases. J Clin Oncol 24:1295–1304

Korinth MC, Delonge C, Hutter BO, Gilsbach JM (2002) Prognostic factors for patients with microsurgically resected brain metastases. Onkologie 25:420–425

Largerwaard FJ, Levendag PC, Nowak PJ, Eijkenboom WM, Hanssens PE, Schmitz PI (1999) Identification of prognostic factors in patients with brain metastases: a review of 1292 patients. Int J Radiat Oncol Biol Phys 43:795–803

Loeffler JS, Shrieve D (1995) An overview of radiotherapy trials for the treatment of brain metastases. Oncology 9:1212–1216

Narita Y, Shibui S (2009) Strategy of surgery and radiation therapy for brain metastases. Int J Clin Oncol 14:275–280

Neves S, Mazal PR, Wanschitz J, Rudnay AC, Drlicek M, Czech T, Wüstinger C, Budka HI (2001) Pseudogliomatous growth pattern of anaplastic small cell carcinomas metastatic to the brain. Clin Neuropathol 20:38–42

Patchell RA, Tibbs PA, Walsh JW, Dempsey RJ, Maruyama Y, Kryscio RJ, Markesbery WR, Macdonald JS, Young B (1990) A randomized trial of surgery in the treatment of single metastases to the brain. N Engl J Med 322:494–500

Patchell RA, Tibbs PA, Regine WF, Dempsey RJ, Mohiuddin M, Kryscio RJ, Markesbery WR, Foon KA, Young B (1998) Postoperative radiotherapy on the treatment of single metastases to the brain: a randomized trial. JAMA 280:1485–1489

Patel AJ, Suki D, Hatiboglu MA, Abouassi H, Shi W, Wildrick DM, Lang FF, Sawaya R (2010) Factors influencing the risk of local recurrence after resection of a single brain metastasis. J Neurosurg 113:181–189

Ranasinghe MG, Sheehan JM (2007) Surgical management of brain metastases. Neurosurg Focus 22:E2

Robinson PD, Kalkanis SN, Linskey ME, Santaguida PL (2010) Methodology used to develop the AANS/CNS management of brain metastases evidence based clinical practice parameter guidelines. J Neurooncol 96(1):11–16

Sampson JH, Carter JH Jr, Friedman AH, Seigler HF (1998) Demographics, prognosis, and therapy in 702 patients with brain metastases from malignantmelanoma. J Neurosurg 88(1):11–20

Schackert G, Steinmetz A, Meier U, Sobottka SB (2001) Surgical management of single and multiple brain metastases: results of a retrospective study. Onkologie 24:246–255

Shaffrey ME, Mut M, Asher AL, Burri SH, Chahlavi A, Chang SM, Farace E, Fiveash JB, Lang FF, Lopes MB, Markert JM, Schiff D, Siomin V, Tatter SB, Vogelbaum MA (2004) Brain metastases. Curr Probl Surg 41(8):665–741

Sperduto PW, Chao ST, Sneed PK, Luo X, Suh J, Roberge D, Bhatt A, Jensen AW, Brown PD, Shih H, Kirkpatrick J, Schwer A, Gaspar LE, Fiveash JB, Chiang V, Knisely J, Sperduto CM, Mehta M (2010) Diagnosis-specific prognostic factors, indexes, and treatment outcomes for patients with newly diagnosed

brain metastases: a multi-institutional analysis of 4,259 patients. Int J Radiat Oncol Biol Phys 77(3):655–661
Stummer W, Stepp H, Moller G, Ehrhardt A, Leonhard M, Reulen HJ (1998) Technical principles for protoporphyrin-IX-fluorescence guided microsurgical resection of malignant glioma tissue. Acta Neurochir (Wien) 140:995–1000
Sundaresan N, Sachdev VP, DiGiacinto GV, Hughes JE (1988) Reoperation for brain metastases. J Clin Oncol 6:1625–1629
Utsuki S, Miyoshi N, Oka H, Miyajima Y, Shimizu S, Suzuki S, Fujii K (2007) Fluorescence-guided resection of metastatic brain tumors using a 5-aminolevulinic acid-induced protoporphyrin IX: pathological study. Brain Tumor Pathol 24(2):53–55
Wroński M, Arbit E, McCormick B (1997) Surgical treatment of 70 patients with brain metastases from breast carcinoma. Cancer 80(9):1746–1754
Yoo H, Kim YZ, Nam BH, Shin SH, Yang HS, Lee JS, Zo JI, Lee SH (2009) Reduced local recurrence of a single brain metastasis through microscopic total resection. J Neurosurg 110:730–736
Yoshida S, Takahashi H (2009) Cerebellar metastases in patients with cancer. Surg Neurol 71:184–187

Part V

Brain Metastasis

Factors Responsible for Local Recurrence of Brain Metastasis

19

Akash J. Patel and Raymond Sawaya

Contents

Abstract

Up to 30% of cancer patients will ultimately develop brain metastases, the treatment of which consists of surgery, radiation therapy, or a combination of the two modalities. Over the past two decades, an incredible body of work has helped us identify the factors responsible for local recurrence of brain metastasis. The factors that have been shown to affect local control can be divided into two categories: those that are intrinsic to the lesion and those that are treatment related. The histology and volume of the tumor are intrinsic factors. The primary treatment modality and adjuvant whole-brain radiotherapy have also consistently been found to affect local control. Here we review the clinical data to date.

Introduction

Brain metastases, aside from being the most common brain tumors, are one of the most feared consequences of systemic cancer because of their poor prognosis if left untreated. Although 10–30% of cancer patients ultimately develop brain metastases, the incidence is increasing owing to increased cancer survival. Treatment of brain metastasis consists of surgical resection, radiation therapy, or a combination of the two modalities. Until the advent of whole-brain radiotherapy (WBRT), the prognosis for patients who develop brain metastasis was dismal.

A.J. Patel • R. Sawaya (✉)
Department of Neurosurgery-442, The University of Texas, M. D. Anderson Cancer Center, Houston, TX 77030, USA
e-mail: rsawaya@mdanderson.org

M. Hayat (ed.), *Tumors of the Central Nervous System, Volume 13*,
DOI 10.1007/978-94-007-7602-9_19, © Springer Science+Business Media Dordrecht 2014

As imaging techniques improved, surgery began to play an increasing role in the treatment of brain metastasis; however, this remained controversial until two randomized, prospective trials confirmed the benefit of surgery (Patchell et al. 1990; Vecht et al. 1993). Surgery has since become a mainstay of therapy, but its application has become more refined with the emergence of stereotactic radiosurgery (SRS). Although controversial, data suggest that both surgery and SRS can be used to treat small brain metastases effectively, but surgery is the modality of choice for lesions ≥3 cm in maximal diameter. Additionally, SRS is useful in treating patients who are high-risk surgical candidates, i.e., those with poor clinical status and those with lesions located deep within the brain or in regions controlling eloquent brain function.

Even with treatment, the median survival time for patients ranges from 4.9 to 16.4 months depending on tumor histology (Bindal et al. 1996; Sperduto et al. 2010). Central nervous system disease progression can occur from local recurrence of a treated lesion or from distant recurrence. WBRT remains a useful adjunct to surgery or SRS to improve both local and distant tumor control. Optimizing treatment of brain metastasis by identifying factors that affect local control is crucial.

Of note, very few studies have been specifically designed to look at factors that affect local control after therapy. Thus, much of the information comes from analysis of a cumulative body of literature, large retrospective studies, or secondary endpoint analysis in prospective trials. It is also important to note that the definition of "local control" of brain metastasis differs based on primary treatment modality. For patients undergoing surgical resection, any tumor growth noted on postoperative magnetic resonance imaging, usually marked by increase in size of a contrast-enhancing lesion, is considered a recurrence. For patients undergoing SRS, recurrence has typically been defined as tumor growth ≥25% of the baseline lesion size, i.e., the lesion is permitted to increase in size up to 25% of the original pretreatment size before being considered a recurrence.

Histology

Classically, tumor histology has been used to guide treatment of brain metastasis, as some tumors respond better to specific therapies. Metastatic sarcomas are notoriously difficult to treat because of their chemo- and radioresistance as well as their proclivity for recurrence after surgery. Recurrence rates after resection of sarcoma metastatic to the brain have been reported to range from 26 to 32% (Fox et al. 2009; Patel et al. 2010). We noted that local recurrence ranged between 12 and 18% for most tumor histologies, but sarcomas were found to have a risk of recurrence that was 1.7 times greater than that for other histologies (Patel et al. 2010).

It seems intuitive that tumor histology should play a role in local control after SRS, as some tumor types are more radiosensitive than others. Flickinger et al. (1994) reported better local control rates for metastatic melanoma and renal cell carcinoma, classically thought to be radioresistant tumors when compared with "other" tumor histologies, including non-small cell lung, breast, colorectal, and other cancers. Woo et al. (2010) found contradictory results, with metastatic breast and kidney cancers showing poorer local control than non-small cell lung cancers, which showed the best response. Mehta et al. (1992) used volumetric analysis to quantify response rates to SRS and found that lymphomas, melanomas, sarcomas, and non-small cell lung cancers had a greater than 50% complete response rate. However, breast and renal cell carcinomas showed a much poorer response. These observations warrant further investigations in prospective studies with larger cohorts.

Tumor Volume

Tumor volume is another factor that has been implicated in contributing to increased risk for local recurrence, regardless of treatment modality. Mehta et al. (1992) found that the response rate decreases dramatically with increasing tumor size. A lesion with a volume of 9 cm^3 (corresponding

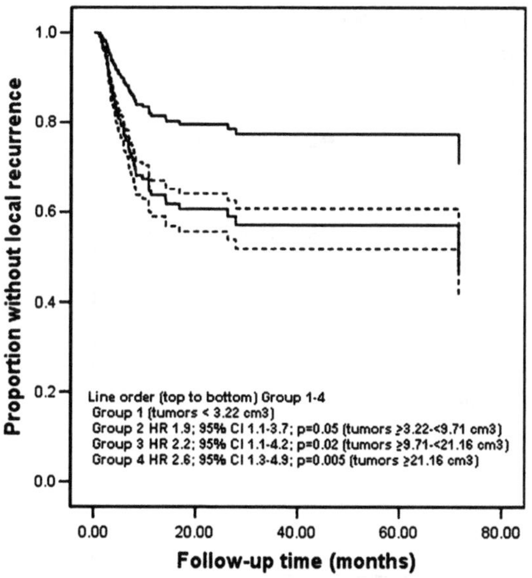

Fig. 19.1 Graph showing proportions of patients without local recurrence stratified by tumor volume. The study group was divided into four equal quartiles of volume. Plot obtained from the multivariate Cox proportional hazards model analysis (From Patel et al. (2010). Used with permission)

to a diameter of 2.58 cm) has only a 20% complete response rate. Shinoura et al. (Shinoura et al. 2002) found that lesions treated with SRS that were larger than 1 cm^3 had a higher rate of local recurrence than smaller lesions and that this difference was statistically significant in a multivariate analysis. Yang et al. (2011) sought to identify the factors that predict response of larger brain lesions to SRS. They found that patients with tumors with a volume >16 cm^3 (corresponding to a diameter of 3.1 cm in a spherical lesion) had a statistically significantly lower response rate than those with smaller lesions. This finding has been corroborated in many other studies (Kwon et al. 2009; Sheehan et al. 2002; Swinson and Friedman 2008).

In reviewing our experience with resection of single brain metastases at The University of Texas M.D. Anderson Cancer Center (M.D. Anderson), we found that tumors exceeding the median volume of 9.7 cm^3 had a significantly higher risk for local recurrence than smaller tumors (Patel et al. 2010). Figure 19.1 shows that the relationship between the increased risk of large tumors recurring locally after resection may be a result of infiltration of tumor cells beyond the border of the metastasis in more advanced stages, i.e., larger size. Neuropathological studies have shown that this "pseudogliomatous" infiltration varies with different tumor histologies, and can extend up to 3 mm from the border (Baumert et al. 2006; Neves et al. 2001).

Whole-Brain Radiotherapy

Shortly after the discovery of x-rays in 1895, radiation was used to treat cancers. Over the past 100 years, radiation therapy has evolved into a multifaceted form of treatment. For many years, WBRT was used as the primary treatment modality for brain metastasis. Patients treated with WBRT alone exhibited high rates of recurrence. In 1990, Patchell et al. (1990) demonstrated that surgical excision followed by WBRT decreased the local recurrence rate to 20% from 52% in patients treated with WBRT alone. Since then, surgery and SRS have been established as the primary treatment modalities for most brain metastases, but WBRT has played an indispensible role as an adjuvant, as it is thought to help prevent both local and distant recurrence. The landmark study by Patchell et al. (1990) prospectively demonstrated a reduction in local recurrence from 46 to 10% with postoperative WBRT (Patchell et al. 1998). Similarly, Aoyama et al. (2006) found that WBRT after SRS decreased the rate of local recurrence from 76.4 to 46.8%. The European Organization for Research and Treatment of Cancer (EORTC) conducted a randomized trial with 359 patients to evaluate whether WBRT increases intracranial tumor control after surgery or SRS in patients with one to three brain metastases (Kocher et al. 2011). After 2 years, adjuvant WBRT had reduced local recurrence from 59 to 27% postoperatively and from 42 to 23% after SRS (Kocher et al. 2011). Despite this evidence, some clinicians withhold WBRT after resection of single brain metastases. However, McPherson et al. (2010) conducted a retrospective review of 358 patients with newly diagnosed single brain metastases who were treated by resection with or without adjuvant WBRT. They found that adjuvant

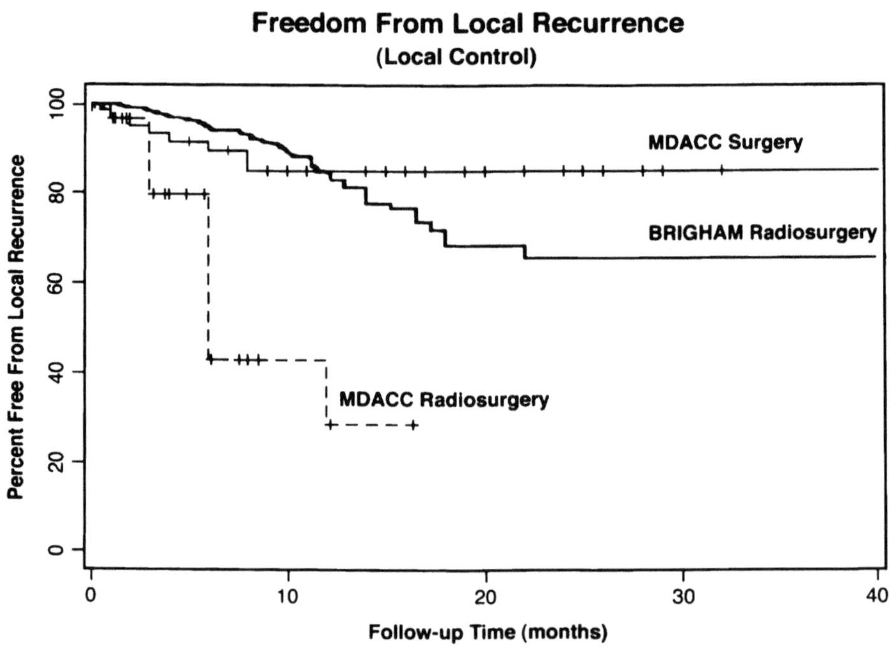

Fig. 19.2 Freedom from local recurrence (local control) of brain metastases with radiosurgery and with surgery. Curve showing time from radiosurgical treatment to local failure (BRIGHAM Radiosurgery) in 42 patients treated at Brigham and Women's hospital (Mehta et al. 1992) superimposed on curves showing time from surgical resection (MDACC Surgery) to local recurrence in 62 patients and time to local failure in 31 patients radiosurgically treated (MDACC Radiosurgery) at The University of Texas M.D. Anderson Cancer Center (Bindal et al. 1996) (From Sawaya 1999. Used with permission)

WBRT significantly reduced both local and distant recurrence, with this difference being particularly significant in lesions ≥3 cm in maximal diameter (Fig. 19.2).

Primary Treatment Modality

Both surgery and SRS are used to treat brain metastases, usually with adjuvant WBRT. The ideal treatment regimen varies based on the tumor histology, size of the lesion, patient's clinical status and presentation. For example, small cell lung cancer is so radiosensitive that it is commonly treated with WBRT. Germ cell tumors are treated with chemotherapy because of their exquisite chemosensitivity. Surgery is favored when treating radioresistant tumors, such as melanoma and renal cell carcinoma.

In addition to providing histologic diagnosis and immediate relief of symptoms from mass effect, surgical excision has been shown to provide excellent local control, especially when compared with WBRT alone, as in the study by Patchell et al. (1990). Since then, surgery has played a central role in the treatment of brain metastasis. The rate of local recurrence after surgery alone has been reported to range from 15 to 46% without adjuvant therapy (Nieder et al. 2007; Patchell et al. 1998; Patel et al. 2010; Vecht et al. 1993). When reviewing our experience in patients who underwent surgery alone for a single, previously untreated brain metastasis, we identified two factors that independently affected local recurrence (Patel et al. 2010). In addition to tumor volume, the method of resection significantly influenced local recurrence. Patients undergoing *piecemeal* resection had a risk of developing local recurrence that was 1.7 times greater than those with tumors excised *en bloc*. Based on the multivariate analysis, this risk was negated in tumors larger than 9.71 cm^3; however, for patients with smaller tumors, those with tumors resected by *piecemeal* methodology had a

2.7 times greater risk for recurrence than those with tumors resected *en bloc*.

SRS was first developed by Lars Leksell at the Karolisnka Institute (Leksell 1951). Subsequently, it has emerged as an alternative to surgery in the treatment of brain metastasis. A multi-institutional, randomized trial compared patients with one to three brain metastases smaller than 4 cm in maximal diameter that were treated either by WBRT alone or by WBRT followed by an SRS boost and demonstrated a reduction in local recurrence from 29 to 18% in patients receiving the SRS boost (Andrews et al. 2004).

Although there is no question that surgery or SRS in addition to WBRT reduces local recurrence compared with WBRT alone, whether surgery or SRS is the optimal treatment modality remains controversial. It is accepted that SRS is the treatment of choice for patients whose medical condition precludes them from being surgical candidates or who decline surgery. Lesions that are less than 1 cm in maximal diameter and surgically inaccessible (e.g., in deep cortex that controls eloquent functions) are also amenable to SRS. Similarly, surgery is usually the clear choice for patients diagnosed with symptoms of mass effect. Historically, the remainder of the lesions have been stratified as those either larger or smaller than 3 cm in maximal diameter. As outlined above, for large lesions, SRS is associated with a higher rate of recurrence than surgery. Chang et al. (2003) retrospectively reviewed the records of 135 patients who underwent SRS and found that lesions that were greater than 1 cm in maximal diameter had only a 56% local control rate. Thus, the 3 cm cutoff that is commonly used requires further study in a prospective manner.

Shinoura et al. (2002) compared patients who underwent surgery plus postoperative WBRT with those who underwent SRS alone and found that the mean times to recurrence were 25 and 7.2 months, respectively. Muacevic et al. (2008) published the results of a randomized trial (prematurely terminated owing to poor patient accrual) comparing surgery plus WBRT with SRS alone. They reported no difference in local control between these two groups. But both of these studies had approximately 30 patients in each arm and did not directly compare surgery and SRS because the SRS group did not receive WBRT. A prospective study at M.D. Anderson with both randomized and nonrandomized arms sought to compare patients receiving surgery or SRS alone for single brain metatastasis (Lang et al. 2008). The randomized arm had 30 patients in the surgical group and 29 patients in the SRS group; the nonrandomized arm had 89 and 66 patients, respectively. A multivariate analysis to eliminate confounding covariables demonstrated that patients in the SRS-alone group had significantly more local recurrences than those undergoing surgical excision. As expected, distant recurrence rates were the same in both groups.

It is difficult to compare the two treatments because of differences in the way recurrence is defined. Published values of local tumor control after SRS permit an increase in tumor size of up to 25%. It is important to keep this in mind when reviewing the available data and comparing these two treatment modalities, as this can overstate the level of local tumor control in patients receiving SRS.

Conclusion

There have been very few studies specifically addressing what factors affect the risk of local recurrence of brain metastasis; thus, most of the aforementioned information has been assembled from the secondary endpoints of various prospective and retrospective studies. Nevertheless, these studies represent an incredible body of work compiled over the past few decades. Table 19.1 summarizes the published class I studies to date. The factors that have been shown to affect local control can be divided into two categories: those that are intrinsic to the lesion and those that are treatment related.

Tumor volume and histology are intrinsic factors. For lesions treated by SRS, the radiosensitivity of the histologic tumor type directly affects the level of local control. Thus, metastastic small cell lung cancer is very responsive to SRS, whereas metastatic breast and renal cell carcinomas are not. Metastatic sarcoma, which is notable

Table 19.1 Summary of all class I studies evaluating treatment of brain metastasis

Study	Groups	Local control	Distant control	Median survival (months)
Patchell et al. (1990)	WBRT (n=23)	48%[a]	87%[a]	3.5[a]
	Surgery+WBRT (n=25)	80%[a]	80%[a]	9.2[a]
Vecht et al. (1993)	WBRT (n=31)	N/A	N/A	6[a]
	Surgery + WBRT (n=32)	N/A	N/A	10[a]
Mintz et al. (1996)	WBRT (n=43)	N/A	N/A	6.3
	Surgery+WBRT (n=41)	N/A	N/A	5.6
Patchell et al. (1998)	Surgery (n=46)	54%[a]	30%[a]	9.9
	Surgery + WBRT (n=49)	90%[a]	82%[a]	11.1
Kondziolka et al. (1999)	WBRT (n=14)	0%[a]	N/A	7.5
	SRS+WBRT (n=13)	92%[a]	N/A	11
Andrews et al. (2004)	WBRT (n=167)	71%[a]	N/A	6.5
	SRS+WBRT (n=164)	82%[a]	N/A	5.7
Aoyama et al. (2006)	SRS (n=67)	73%	36%	8
	SRS+WBRT (n=65)	89%	58%	7.5
Muacevic et al. (2008)	SRS (n=31)	97%	74%	10.3
	Surgery+WBRT (n=33)	82%	97%	9.5

N/A data not available
[a]Difference is statistically significant

for its relentless course, seems to recur locally after surgery more often than other histologic types. Increased tumor volume, regardless of treatment modality, has been shown to independently increase the risk of local recurrence. In the case of larger lesions treated by SRS, the radiation dose cannot be concomitantly increased for fear of neurotoxicity to the surrounding tissue.

We can gather information about how treatment modality affects local recurrence from a number of class I studies. Regardless of the primary treatment modality, adjuvant WBRT has been shown to decrease both local and distant recurrence. Surgery and SRS are both more effective at achieving local control than WBRT alone. In our experience, the *en bloc* method of resection results in lower rates of local recurrence than *piecemeal* resection. However, for larger lesions, the level of local control is identical regardless of method of resection. Additionally, as tumor size increases, surgical excision confers a higher rate of local control than SRS. This may also be true for smaller tumors but awaits confirmation via randomized trial.

Acknowledgment We thank David M. Wildrick, Ph.D., for editorial assistance with the manuscript.

References

Andrews DW, Scott CB, Sperduto PW, Flanders AE, Gaspar LE, Schell MC, Werner-Wasik M, Demas W, Ryu J, Bahary JP, Souhami L, Rotman M, Mehta MP, Curran WJ Jr (2004) Whole brain radiation therapy with or without stereotactic radiosurgery boost for patients with one to three brain metastases: phase III results of the RTOG 9508 randomised trial. Lancet 363:1665–1672

Aoyama H, Shirato H, Tago M, Nakagawa K, Toyoda T, Hatano K, Kenjyo M, Oya N, Hirota S, Shioura H, Kunieda E, Inomata T, Hayakawa K, Katoh N, Kobashi G (2006) Stereotactic radiosurgery plus whole-brain radiation therapy vs stereotactic radiosurgery alone for treatment of brain metastases: a randomized controlled trial. JAMA 295:2483–2491

Baumert BG, Rutten I, Dehing-Oberije C, Twijnstra A, Dirx MJ, Debougnoux-Huppertz RM, Lambin P, Kubat B (2006) A pathology-based substrate for target definition in radiosurgery of brain metastases. Int J Radiat Oncol Biol Phys 66:187–194

Bindal AK, Bindal RK, Hess KR, Shiu A, Hassenbusch SJ, Shi WM, Sawaya R (1996) Surgery versus radiosurgery in the treatment of brain metastasis. J Neurosurg 84:748–754

Chang EL, Hassenbusch SJ 3rd, Shiu AS, Lang FF, Allen PK, Sawaya R, Maor MH (2003) The role of tumor size in the radiosurgical management of patients with ambiguous brain metastases. Neurosurgery 53:272–280; discussion 280–281

Flickinger JC, Kondziolka D, Lunsford LD, Coffey RJ, Goodman ML, Shaw EG, Hudgins WR, Weiner R,

Harsh GR, Sneed PK, Larson DA (1994) A multi-institutional experience with stereotactic radiosurgery for solitary brain metastasis. Int J Radiat Oncol Biol Phys 28:797–802

Fox BD, Patel A, Suki D, Rao G (2009) Surgical management of metastatic sarcoma to the brain. J Neurosurg 110:181–186

Kocher M, Soffietti R, Abacioglu U, Villa S, Fauchon F, Baumert BG, Fariselli L, Tzuk-Shina T, Kortmann RD, Carrie C, Ben Hassel M, Kouri M, Valeinis E, van den Berge D, Collette S, Collette L, Mueller RP (2011) Adjuvant whole-brain radiotherapy versus observation after radiosurgery or surgical resection of one to three cerebral metastases: results of the EORTC 22952-26001 study. J Clin Oncol 29:134–141

Kondziolka D, Patel A, Lunsford LD, Kassam A, Flickinger JC (1999) Stereotactic radiosurgery plus whole brain radiotherapy versus radiotherapy alone for patients with multiple brain metastases. Int J Radiat Oncol Biol Phys 45:427–434

Kwon AK, Dibiase SJ, Wang B, Hughes SL, Milcarek B, Zhu Y (2009) Hypofractionated stereotactic radiotherapy for the treatment of brain metastases. Cancer 115:890–898

Lang FF, Suki D, Maor M, Chang E, Hess KR, Graham S, Hentschel S, Sawaya R (2008) Conventional surgery versus stereotactic radiosurgery in the treatment of single brain metastasis: a prospective study with both randomized and nonrandomized arms. Am Assoc Neurol Surg Meet [abstract]. Article ID: 48938

Leksell L (1951) The stereotaxic method and radiosurgery of the brain. Acta Chir Scand 102:316–319

McPherson CM, Suki D, Feiz-Erfan I, Mahajan A, Chang E, Sawaya R, Lang FF (2010) Adjuvant whole-brain radiation therapy after surgical resection of single brain metastases. Neuro Oncol 12:711–719

Mehta MP, Rozental JM, Levin AB, Mackie TR, Kubsad SS, Gehring MA, Kinsella TJ (1992) Defining the role of radiosurgery in the management of brain metastases. Int J Radiat Oncol Biol Phys 24:619–625

Mintz AH, Kestle J, Rathbone MP, Gaspar L, Hugenholtz H, Fisher B, Duncan G, Skingley P, Foster G, Levine M (1996) A randomized trial to assess the efficacy of surgery in addition to radiotherapy in patients with a single cerebral metastasis. Cancer 78:1470–1476

Muacevic A, Wowra B, Siefert A, Tonn JC, Steiger HJ, Kreth FW (2008) Microsurgery plus whole brain irradiation versus Gamma Knife surgery alone for treatment of single metastases to the brain: a randomized controlled multicentre phase III trial. J Neurooncol 87:299–307

Neves S, Mazal PR, Wanschitz J, Rudnay AC, Drlicek M, Czech T, Wustinger C, Budka H (2001) Pseudogliomatous growth pattern of anaplastic small cell carcinomas metastatic to the brain. Clin Neuropathol 20:38–42

Nieder C, Astner ST, Grosu AL, Andratschke NH, Molls M (2007) The role of postoperative radiotherapy after resection of a single brain metastasis. Combined analysis of 643 patients. Strahlenther Onkol 183:576–580

Patchell RA, Tibbs PA, Walsh JW, Dempsey RJ, Maruyama Y, Kryscio RJ, Markesbery WR, Macdonald JS, Young B (1990) A randomized trial of surgery in the treatment of single metastases to the brain. N Engl J Med 322:494–500

Patchell RA, Tibbs PA, Regine WF, Dempsey RJ, Mohiuddin M, Kryscio RJ, Markesbery WR, Foon KA, Young B (1998) Postoperative radiotherapy in the treatment of single metastases to the brain: a randomized trial. JAMA 280:1485–1489

Patel AJ, Suki D, Hatiboglu MA, Abouassi H, Shi W, Wildrick DM, Lang FF, Sawaya R (2010) Factors influencing the risk of local recurrence after resection of a single brain metastasis. J Neurosurg 113:181–189

Sheehan JP, Sun MH, Kondziolka D, Flickinger J, Lunsford LD (2002) Radiosurgery for non-small cell lung carcinoma metastatic to the brain: long-term outcomes and prognostic factors influencing patient survival time and local tumor control. J Neurosurg 97:1276–1281

Shinoura N, Yamada R, Okamoto K, Nakamura O, Shitara N (2002) Local recurrence of metastatic brain tumor after stereotactic radiosurgery or surgery plus radiation. J Neurooncol 60:71–77

Sperduto PW, Chao ST, Sneed PK, Luo X, Suh J, Roberge D, Bhatt A, Jensen AW, Brown PD, Shih H, Kirkpatrick J, Schwer A, Gaspar LE, Fiveash JB, Chiang V, Knisely J, Sperduto CM, Mehta M (2010) Diagnosis-specific prognostic factors, indexes, and treatment outcomes for patients with newly diagnosed brain metastases: a multi-institutional analysis of 4,259 patients. Int J Radiat Oncol Biol Phys 77:655–661

Sawaya R (1999) Surgical treatment of brain metastases. Clin Neurosurg 45:41–47

Swinson BM, Friedman WA (2008) Linear accelerator stereotactic radiosurgery for metastatic brain tumors: 17 years of experience at the University of Florida. Neurosurgery 62:1018–1031; discussion 1031–1032

Vecht CJ, Haaxma-Reiche H, Noordijk EM, Padberg GW, Voormolen JH, Hoekstra FH, Tans JT, Lambooij N, Metsaars JA, Wattendorff AR et al (1993) Treatment of single brain metastasis: radiotherapy alone or combined with neurosurgery? Ann Neurol 33:583–590

Woo HJ, Hwang SK, Park SH, Hwang JH, Hamm IS (2010) Factors related to the local treatment failure of gamma knife surgery for metastatic brain tumors. Acta Neurochir (Wien) 152:1909–1914

Yang HC, Kano H, Lunsford LD, Niranjan A, Flickinger JC, Kondziolka D (2011) What factors predict the response of larger brain metastases to radiosurgery? Neurosurgery 68:682–690; discussion 690

Role of MMP2 in Brain Metastasis

20

George Stoica and Gina Lungu

Contents

G. Stoica (✉) • G. Lungu
Veterinary Pathobiology, Texas A&M University,
College Station, TX 77843-4467, USA
e-mail: GSTOICA@cvm.tamu.edu

Abstract

Matrix metalloproteinase 2 (MMP2) is important in breast cancer (BC) invasion and metastasis.

In order to study the expression of MMP2, in breast cancer brain metastasis, we used a syngeneic rat model of distant metastasis of ENU1564, a carcinogen-induced mammary adenocarcinoma cell line. At 6 weeks post left-ventricle inoculation we observed development of micro-metastasis in the brain. Immunohistochemistry (IHC) and Western blotting (WB) analyses showed that MMP2 protein expressions were significantly higher in neoplastic brain tissue compared to normal brain tissue. These results were confirmed by RT-PCR and gelatin zymography increased in MMP2 and activity in brain metastasis. The MMP2 mechanism of action in the brain is still under intense scrutiny. To study the role of MMP2 in the development of BC brain metastasis we transfected ENU1564 rat mammary adenocarcinoma cells with tissue inhibitor of MMP2 (TIMP2). Animals inoculated with ENU1564-TIMP2 cells had decreased orthotopic tumor growth, decreased orthotopic metastastic behavior and did not develop brain metastases. Mitogen activated protein kinase (MAPK) pathway components, such as ERK1/2, have been correlated to MMP expression and/or astrocyte activity. We found that BC brain metastases have peripheral astrocyte reactivity and higher expression of glial fibrillary acidic protein (GFAP) and phosphorylated-ERK1/2 (p-ERK1/2). Blockage of ERK1/2

M. Hayat (ed.), *Tumors of the Central Nervous System, Volume 13*,
DOI 10.1007/978-94-007-7602-9_20, © Springer Science+Business Media Dordrecht 2014

phosphorylation by treatment with MEK inhibitor (PD98059) decreased the expression of MMP2 in cancer cells grown in rat astrocyte-conditioned media. Our results are highly suggestive that MMP2 plays a role in the development of BC metastases, in particular to the brain. Furthermore, our results suggest that astrocyte factors and the ERK1/2 signaling pathway may be associated with BC brain metastasis development; and that ERK1/2 may regulate MMP2 in a way that is modifiable by astrocyte factors.

Introduction

The metastatic process of breast cancer (BC) has been the subject of intense scrutiny. The brain is one of the most common organs affected in the spread of BC that ultimately results in fatal overcome of the disease. Brain metastasis is an increasingly common complication in breast cancer patients. Approximately 15–30% of breast cancer patients develop brain metastasis (Somerville et al. 2003). A suitable specific environment is important to the development of tumor cells (Hall and Stoica 1994). The exact role of the brain environment to the development of the metastatic process has yet to be clarified. Many theories have been developed to study and understand metastatic behavior. Factors such as neoplastic cell molecular and genetic characteristics and biological environment are thought to be determinants in the metastatic process.

Matrix Metalloproteinases (MMPs) are a broad family of zinc-dependent proteinases that play a key role in extracellular matrix degradation, implicated in numerous pathogenic processes (Leppa et al. 2004). Tumor cells are thought to secrete these matrix-degrading enzymes and/or induce host cells to elaborate them (Stetler-Stevenson et al. 1993). MMPs have been associated with pathology within the central nervous system in neoplastic disease, such as glioma and melanoma brain metastasis (Fujimake et al. 1996) and non-neoplastic disease, such as trauma, ischemia and immune-mediated disease (Nie and Pei 2003). Physiologically these enzymes play a role in normal tissue remodeling as well as in angiogenesis and mammary gland involution. They belong to a family of 23 gene products, which encode for zinc-dependent and calcium dependent proteases that are endopeptidases (Somerville et al. 2003). There are also two other large families of proteases that have major roles in extracellular proteolysis, the ADAM family (a disintegrin and metalloproteinase domain, with about 33 members in humans) and the ADAMTS family (a disintegrin-like and metalloproteinase domain (reprolysin type) with thrombospondin type I repeats, with about 19 members in humans). Numerous classifications of MMPs can be made. Based on their solubility they can be divided in two major groups: (1) Soluble type MMPs; include collagenases, stromelysins, gelatinases and matrilysins. (2) Membrane-anchored metalloproteinases; include Type II and type II types. They can also be classified according to their substrates. They are known to degrade a large array of substrates such as Collagens (C), Fibronectin (FN), Cartilage oligomeric protein (COMP), Laminin (LN) and Proteoglycan (PG).

MMP regulation occurs at multiple levels that include transcription, activation of zymogen forms, and activity of extracellular inhibitors (Somerville et al. 2003). There is moderate variation in the protein structure of MMPs. Metalloproteinases are composed by a pre-catalytic, a procatalytic domain, a fibronectin-like domain, a domain for binding to zinc and a homeopexin domain. These enzymes depend on zinc for catalytic activity. The presence of the prodomain keeps the enzyme inactive. In order to be activated a cystein residue that inactivates ligand binding to the zinc catalytic site must be removed. This can be done by conformational change or proteolysis accomplished by plasmin or other MMPs. MMP zymogens can be activated by themselves. MT1-MMP activates MMP2 and requires TIMP2 binding to its active site in order to do so (Lafleur et al. 2003). Inactivation of MMPs can occur by direct interaction with tissue inhibitors of MMPs (TIMPs), alpha 2 macroglobolin and other molecules such as pro-collagen C-proteinase enhancer (Lebeau et al. 1999).

Different animal models have been used for in vivo study of the role of MMPs in the development of cancer. Most of these studies describe lung, bone and/or node metastasis and are usually concurrent with the studies conducted on human patients that correlated MMPs with increased tumor invasion and metastatic behavior (Tester et al. 2004; Yoneda 2000). Most utilize immunocompromised animals, such as nude mice, that usually develop metastasis in the bone, lung and liver (Ohshiba et al. 2003). These models do not consider the importance of the immune system in cancer development and its relevance to the development of metastatic disease. We use a syngeneic rat model to study distant metastasis of breast cancer. The ENU1564 cell line used in our study is a highly metastatic breast cancer cell line originated from a N-ethyl-N-nitrosourea (ENU) induced mammary adenocarcinoma in a female Berlin-Druckrey IV (BD-IV) rat.

MMP2 is thought to play an important role in breast cancer invasion, metastasis and tumor angiogenesis (Hynes 2003). MMP2 over-expression and activation have been associated with the invasive potential of human tumors. Active MMP2 is detected more frequently in malignant than benign carcinomas. Some reports, however, do not correlate MMP2 immunohistochemical staining with the presence of metastases at the time of diagnosis or with disease outcome (La Rocca et al. 2004). Absence of distinct positive immunoreactivity for MMP2 has been observed in both invasive and non-invasive tumor cells without apparent differences in the staining intensity (Talvensaari-Mattila et al. 2003). In this regard, additional in vivo studies that characterize MMP 2 expression in brain metastasis are needed.

It is important to determine if MMP2 have different effects/roles in the development of metastasis to the brain because this may help to understand why BC cells metastasize to preferential organs. We focused on the metastatic process of BC to the brain in a rodent rat model. Understanding the mechanisms of BC brain metastasis could be utilized in the development of the current therapeutic approach to metastatic cancer.

Matrix Metalloproteinase 2 Overview

Matrix Metalloproteinase 2 Localization in Tumors

The morphological intratumoral localization of MMP2 in breast cancer has been the subject of numerous studies. Different studies have sometime-contradictory data on the location of MMP2. Some co-localize MMP2 with neoplastic epithelial cells whereas others associated them with different components of the neoplastic stroma. Therefore, according to some reports, MMP2 can be observed in stromal tumor fibroblasts and well differentiated invasive cancer cells (Heppner et al. 1996) in the neoplastic cell plasma membrane in peritumor stromal cells (Caudroy et al. 1999) and/or angiogenic blood vessels (Jones et al. 1999). In our studies we observed staining for MMP2 in the cytoplasm of neoplastic epithelial cells in all neoplastic sites evaluated (brain, mammary gland, lung, pancreas, heart and kidney). MMP2 also stained stromal fibroblasts. Staining was also in normal epithelium and macrophages within metastatic foci.

Matrix Metalloproteinase 2 and Tumor-Stroma Interaction

Tumor environment is very important for expression and activities of MMPs. For instance IL12, a cytokine observed in the extracellular matrix, can enhance the activity of MMPs (Scott et al. 2000). In the tumor cell-cell interactions, pericellular environment and products of degradation of the extracellular matrix are important for MMP production and activation.

Tumor cells also interact with stromal cells or cell-bound factors that stimulate the production of MMPs. MMP2 is expressed predominately in peritumoral fibroblasts (Heppner et al. 1996). MT-MMP1 is produced in fibroblasts and is a major activator of MMP2 and this suggests that the stroma component is fundamental for MMP2 production. MT1-MMP is anchored to the cell surface and acts as a receptor for TIMP2 that binds to MT1-MMP through his N-terminal domain. This binary complex acts then as a receptor for pro-MMP2. TIMP2 C-terminal binds to pro-MMP2 and MT1-MMP cleaves then pro-MMP2 causing the formation of an intermediate species. Stromal fibroblasts at the tumor invasion front are thought to produce the bulk of MMP2. Tumor cells usually express low constitutive levels of MMP2. Stromal cells have strong but short induction of MMP2 (Stetler-Stevenson et al. 1993). This very high and complex regulation of the expression of MMPs represents a host response to the tumor and neoplastic cell interaction with the tumor stromal component is fundamental for cancer invasion and metastasis. Although the exact mechanism of activation and function of MMPs in brain disease is still under intense scrutiny, it is reported that astrocyte co-culture with glioma cells increases activation of MMP2 (Ohshiba et al. 2003). In addition, astrocytes can produce and/or regulate MMP2 production (Dzwonek et al. 2004). Astrocytes also have been implicated in the mechanism of activation of other MMPs in the CNS (Lee et al. 2003). Astrocyte factors such as interleukin 6 (IL6), fibroblast growth factor-b (FGFb), and insulin-like growth factor (IGF) receptor are up regulated in breast cancer brain metastases.

Matrix Metalloproteinase 2 and Tissue Inhibitors

The activities of MMPs are in part influenced by the presence of tissue inhibitors of metalloproteinases (TIMPs). An increase in the amount of TIMPs relative to MMPs could function to block tumor cell invasion and metastasis. In fact, tumor cell invasion and metastasis can be inhibited by up-regulation of TIMP expression or by an exogenous supply of TIMPs. Tissue inhibitors of metalloproteinases 2 (TIMP2) is reported to be a physiologic inhibitor of MMP2 (Danilewicz et al. 2003). An increase in the amount of TIMP2 relative to MMP2 may decrease MMP2 activity and block tumor cell invasion and metastases (Li et al. 2001). However, the role of TIMP2 in cancer development is still under investigation. Although TIMP2 was at first considered a suppressor of invasion and metastases, the complexity of TIMP2/MMP2 interactions led to reconsideration of the role of TIMP2 in BC development (Gakiopoulou et al. 2003). TIMP2 expression in BC patients has been correlated with advanced disease, decrease in survival time, increased tumor size, node positive status, and tumor recurrence (Gakiopoulou et al. 2003; Remacle et al. 2000). Paradoxically, genetic manipulation of cancer cells has correlated experimental TIMP2 over expression with decreased metastatic behavior (Li et al. 2001).

Matrix Metalloproteinase 2 and MAPK

The mitogen activated protein kinase (MAPK) pathway is one of most important transduction signaling pathways that is related to numerous pathogenic processes, including neoplasia. Genes that codify for molecules in this family, such as MKK4 and MAPKK4 are considered to be important in the metastatic cascade (Alessandrini 2002). Moreover components of these pathways have been correlated with cancer invasion, and development. They have also been linked with MMP2 expression. Activation state of the ERK pathway in tumor cells correlates with the invasive phenotype, which was determined by the ability of cells to invade through reconstituted extracellular matrix. Our data showed that MMP2 mRNA expression, protein expression and gelatinolytic activity are correlated with ERK phosphorylative activity. MEK inhibitor PD98059 inhibits MMP2 promoter activity and Sp1 phosphorylation. Overexpression of constitutively active MEK1 stimulates Sp1 phosphorylation and MMP2 promoter activity (Pan and Hung 2002).

Matrix Metalloproteinase 2 in Brain Metastasis

MMPs are expressed physiologically in central nervous system cells. Matrix metalloproteinases have been extensively correlated with neoplastic pathological processes within the central nervous system (CNS) (Liuzzi et al. 2004; Massengale et al. 2002). They have been involved in the degradation of extracellular periphery of brain tumors such as glioma as well as in glioma invasion and vascularization (Nagashima et al. 2002). Although the exact mechanism of activation and function of MMP2 in brain disease is still under intense scrutiny, it is reported that astrocyte co-culture with glioma cells increases activation of MMP2 (Le et al. 2003). Astrocytes can produce and/or regulate MMP2 production (Leveque et al. 2004; Muir et al. 2002). Additionally, astrocytes may be involved in the development of BC brain metastases (Dzwonek et al. 2004). Among several signaling transduction pathways, mitogen activated protein kinase (MAPK) pathway components, such as ERK1/2, have been correlated to MMP2 activation and expression; associated with astrocyte activity; and are considered genes related with tumor metastases.

Matrix Metalloproteinase 2 Expression in Breast Cancer Brain Metastasis

In order to to evaluate the expression and activity of MMP2 in metastatic foci of BC in the brain. We used an in vivo model that consistently produces brain metastasis (Hall and Stoica 1994). We found that levels of MMP2 mRNA and protein, in BC brain metastasis are higher than those of normal brain tissue (Fig. 20.1a, b). To determine if the higher expression of MMP2 was correlated with intra-tumoral increased enzyme activity, we performed gelatin zymography studies. Gel zymography showed that there was a significant increase of MMP2 activity in metastatic foci than in non-affected brains (Fig. 20.2a, b). Immunohistochemistry was performed in order to characterize MMP2-protein expression within the metastatic brain foci. Immunohistochemical staining for MMP2 showed immunolabeling with moderate intensity in the cytoplasm of neoplastic cells within the brain metastatic foci (Fig. 20.1c). To confirm IHC and WB results on MMP2 protein expression, we extracted total RNA from frozen specimens. Semi-quantitative RT-PCR analysis of MMP2 mRNA of brain metastatic foci of breast cancer was compared with mRNA obtained from age-matched non-inoculated rats. The comparison revealed that the amounts of MMP2 mRNAs of brain metastasis foci of breast cancer were higher than those of control tissues (Fig. 20.1d).

Matrix Metalloproteinase 2 Role in the Development of Breast Cancer Brain Metastasis

We further investigated whether MMP2 has an effect in tumor growth and metastasis development (specifically brain metastasis) in a syngeneic animal model by over-expression of TIMP2 in rat mammary adenocarcinoma -ENU1564-cells. In addition, we investigated if astrocytes are associated with BC brain metastasis development and MMP2 expression, and if this association has a relationship with MAPK pathway components.

Because we did not observe development of brain metastases (Fig. 20.3a) in any of the animals inoculated with ENU1564-TIMP2 cells, we used material collected from orthotopic mammary tumors developed from inoculation of both ENU1564 and ENU1564-TIMP2 cells to evaluate in vivo change in expression and activity of MMP2.

WB evaluation revealed non-significant difference in levels of MMP2 protein ($p>0.1$). Gel zymography evaluation revealed significantly higher MMP2 activity in samples obtained from animals inoculated with ENU1564 cells when compared with animals in the control group (Fig. 20.3b). Gross evaluation did not reveal any significant changes in the central nervous system. Upon histological evaluation 44.4% of the animals in the control group had brain metastases. Brain metastases were not detected in any of the animals in the group inoculated with ENU1564-TIMP2 cells ($p<0.001$).

To evaluate if TIMP2 over expression has an effect on in vivo tumor growth and metastasis development, we examined tumor progression by orthotopic inoculation of ENU1564-TIMP2 cells. We observed, at two different time points, that tumors derived from ENU1564-TIMP2 cells

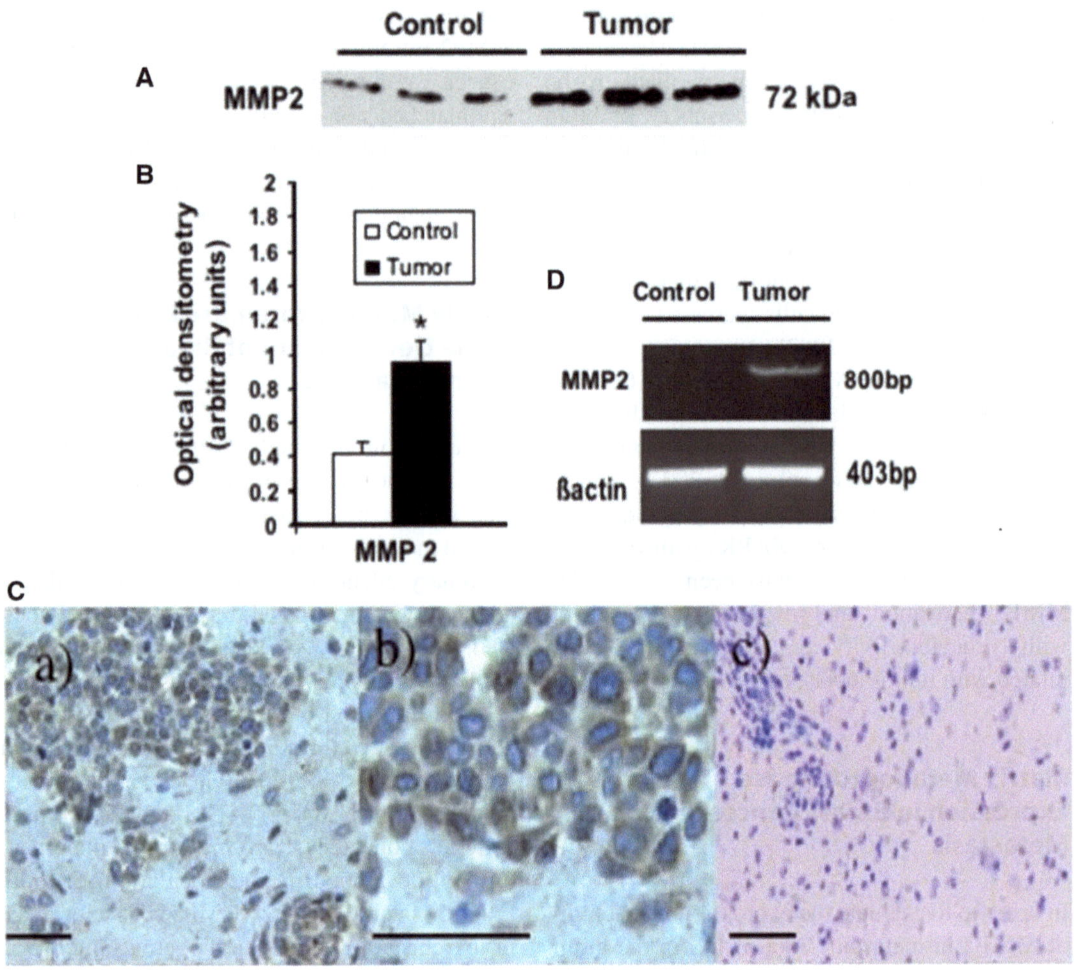

Fig. 20.1 Increased expression of MMP2 protein in the metastatic brain foci. (**a**) Evaluation of protein expression by Western blotting. The membranes were stripped and re-probed with β-actin antibody to confirm equal loading. (**b**) Quantitative analysis of MMP2 expression was determined by densitometry. The results shown in the histogram are the mean ± standard deviation from three control and three tumor samples. (*) for statistically significant when P was ≤0.05). (**c**) Localization of MMP2 in the brain metastatic foci. Immunohistochemical staining (*brown*) of MMP2 (**A, B**) protein in the brain metastatic foci revealed positivity within neoplastic cell cytoplasm. Negative controls for MMP2 (**C**). Note that glial cells are also positive. Bars indicate 100 μm. (**d**) Increased expression of MMP2 mRNA in the metastatic brain foci. Semi-quantitative RT-PCR analysis was used to detect MMP2 and β-actin in total RNAs from normal brain and metastatic brain foci. β-actin was used as an internal control

were significantly smaller than tumors derived from ENU1564 cells ($P<0.001$ and <0.05) in the two different time points, days 32 and 42 post-inoculation. Tumor weights were obtained at the time of sacrifice (day 42 post-inoculation). Additionally, post-mortem evaluation revealed that only animals from the control group developed metastases.

Breast cancer brain metastases induce a marked astrocyte response (Nishizuka and Ishikawa 2002). We reported previously that prominent astrocyte reaction is associated with BC brain metastatic foci (Mendes et al. 2005). Therefore, we investigated if astrocytes play a role in the development of BC brain metastases in our model. To examine the astrocyte response, we

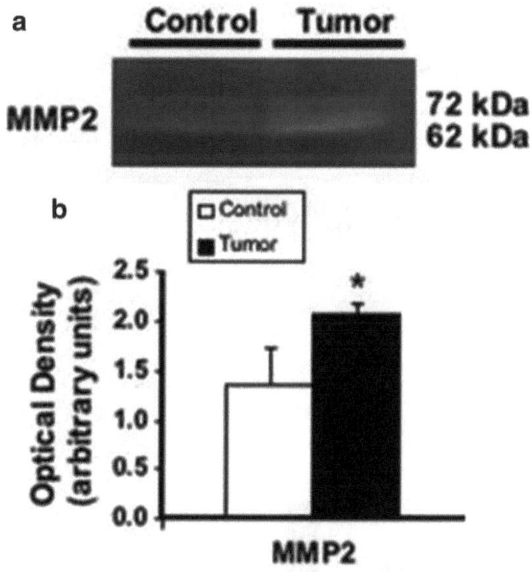

Fig. 20.2 Increased MMP2 enzymatic activities in the metastatic brain foci. (**a**) Evaluation of MMP2 activity by gel zymography. (**b**) Quantitative analysis of MMP2 activity was determined by densitometry of respective active bands (62 kDa). The results shown in the histogram are the mean ± standard deviation from three control and three tumor samples. (*) for statistically significant when $p \leq 0.05$

evaluated GFAP (glial fibrillary protein-astrocyte marker) reactivity by IHC. We observed a marked increase of immunohistochemical staining of GFAP in reactive astrocytes around brain metastatic foci when compared with normal brain. To confirm the IHC results, we performed WB analysis for GFAP protein in samples obtained from dissected frozen brain specimens. We observed that tumor tissues express higher levels of GFAP protein when compared with control brains of animals free of tumor.

To further test the hypothesis that astrocytes have a role in MMP2 expression in ENU1564 cells, we used 48-h-rat astrocyte-conditioned media. ENU1564 cells revealed an increase in vitro invasive behavior of ENU1564 cells when compared with ENU1564 cells without astrocyte-conditioned media treatment. This invasive behavior was also reduced in ENU1564-TIMP2 transfected cells ($P<0.001$) (Fig. 20.3c). Conversely, incubation with astrocyte-conditioned media increased expression of MMP2 protein when compared with MMP2 expression in ENU1564 without astrocyte conditioned media.

Our studies showed that astrocyte factors increase in vitro invasiveness of ENU1564 cells and MMP2 expression and activity are increased in BC brain metastases in our model. Since the MAPK pathway has been linked with the metastatic cascade, MMP2 activity and/or astrocyte factors we hypothesized that BC brain metastasis development and astrocyte factor-related MMP2 expression could be associated with MAPK pathway components. To test this hypothesis in our model, we performed IHC analysis for p-ERK1/2, p38, and JNK. Neoplastic epithelial cells were positive for p-ERK1/2 and staining was more intense at the periphery of the neoplastic lesions. Neoplastic cells were not stained with phosphorylated-p38 and JNK antibodies. To confirm the IHC results, we performed WB analysis for p-ERK1/2 protein in samples obtained from dissected frozen brain specimens. We observed that tumor tissue expresses higher levels of p-ERK1/2 protein when compared with control brains of tumor-free suggesting that ERK1/2 may be associated with BC brain metastasis development. Further, to evaluate the relationship between ERK1/2, MMP2 protein expression and astrocyte factors, we examined whether PD98059, a potent ERK1/2-MAPK inhibitor, can affect the expression of MMP2 in ENU1564 cells treated with 48 h astrocyte-conditioned media (Fig. 20.3d–f). We observed that p-ERK1/2 and MMP2 protein expression were increased in ENU1564 cells treated with astrocyte-conditioned media ($P<0.05$) when compared with ENU1564 cells without astrocyte-conditioned media. There was no significant difference in the levels of non-phosphorilated ERK in ENU1564 cells when compared to ENU1564 cells treated with astrocyte-conditioned media. Treatment with PD98059 induced a significant decrease of MMP2 protein expression ($P<0.05$) as well as the expected decrease in p-ERK1/2 expression in ENU1564 cells and ENU1564 cells treated with rat astrocyte-conditioned media ($P<0.001$).

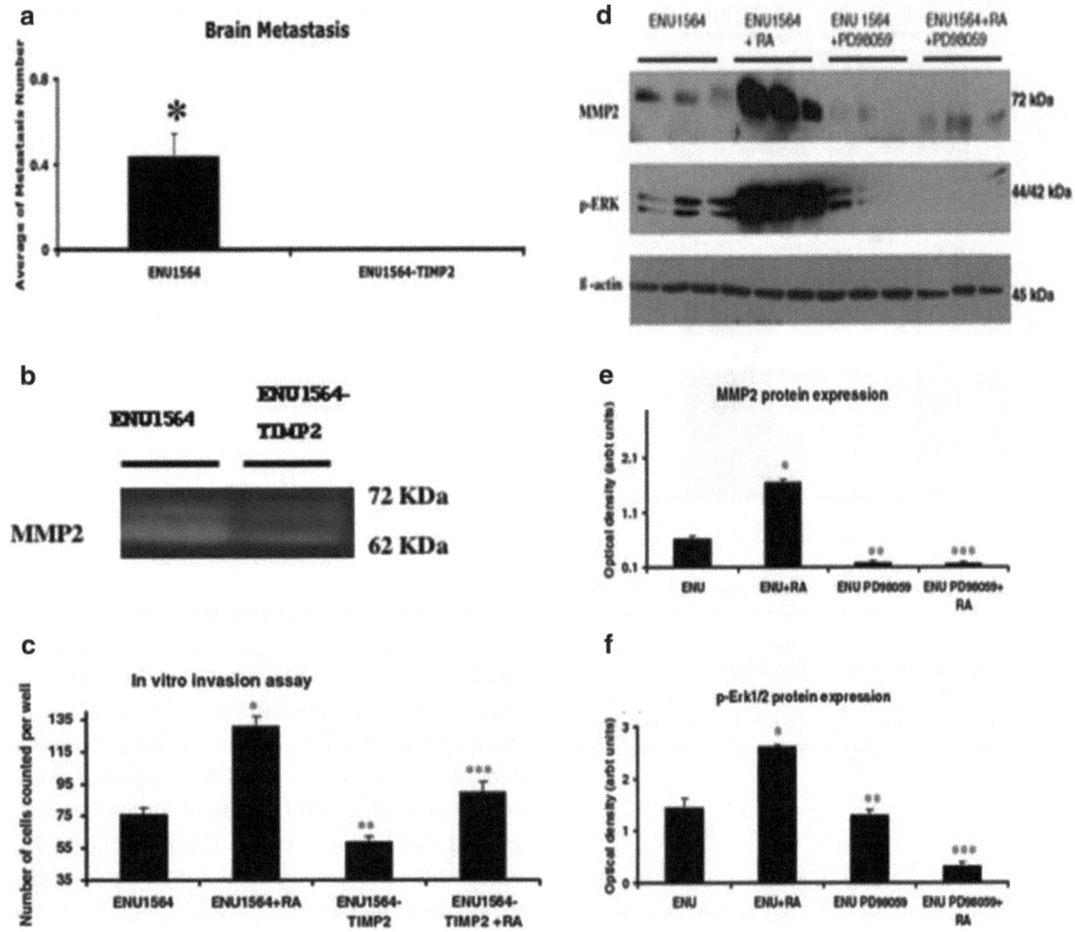

Fig. 20.3 Metastatic potential of ENU1564 vs ENU1564-TIMP2 and MMP2 enzymatic activity in orthotopic tumors originated from inoculation with ENU1564-TIMP2 vs ENU1564 control cells and ERK1/2 pathway. (**a**) Effects of TIMP2 overexpression in brain metastatic tumor development. (*) for statistically significant when $p<0.05$. (**b**) Evaluation of MMP2, activity by gel zymography. (**c**) In vitro invasion chamber assay for ENU1564 cells was performed and the results shown in the histogram are the mean ± SD of two individual experiments run in triplicate. *RA* rat astrocyte-conditioned media. *Asterisk*, statistically significantly different from ENU1564 ($P<0.001$). (**d**) Evaluation of phosphorylated-ERK1/2 and MMP2 protein expression by WB. ENU15164 cells were treated with PD98059 (a potent ERK1/2-MAPK inhibitor) with or without the presence of rat astrocyte-conditioned media (*RA*). The membranes were striped and re-probed with β-actin antibody to confirm equal protein loading and transfer. (**e**) Quantitative analysis of phosphorylated-ERK1/2 expression was determined by densitometry. (**f**) Quantitative analysis MMP2 expression was determined by densitometry. The results shown in the histograms are the mean ± SD from three individual experiments run in triplicate. *Single asterisk*, statistically significantly different from ENU1564 ($P<0.05$). *Double asterisk*, statistically significantly different from ENU1564 ($P<0.05$). *Triple asterisk*, statistically significantly different from ENU1564-RA (rat astrocyte-conditioned media treated-ENU1564) ($P<0.001$)

Discussion

We have determined, using a rat syngeneic model for BC brain metastases that BC brain metastases significantly express higher levels of MMP2. We also determined that there is a correlation between MMP expression and enzymatic activity within the neoplastic foci. Our results suggest that this proteinase may play a role in the development of BC brain metastases. Matrix metalloproteinase 2 is believed to play an important role in BC invasion and metastases. MMP2 over expression and activity have been associated with the invasive potential of human tumors. Active MMP2 was detected more frequently in malignant than benign breast carcinomas (Hanemaaijer et al. 2000). The tissue physiological inhibitors of MMP2 are thought to influence MMP2 activity. In human patients affected with brain cancer metastasis over-expression of MMP2 tissue inhibitor TIMP2 has been associated with poor prognosis. We observed that transfection of ENU1564 cells with TIMP2 causes in vitro TIMP2 over-expression associated with decreased in vitro invasive behavior of TIMP2-ENU1564 cancer cells. Concurrently, when these transfected cells were inoculated in vivo, there was a marked increase in TIMP2 expression. In order to characterize the effect of TIMP2 over expression in in vivo expression and activity of MMP2, we analyzed material collected from orthotopic mammary tumors developed from animals inoculated with ENU1564 and ENU1564-TIMP2 cells. As expected, no significant variation was observed in the levels of MMP2 protein expression because TIMP2 has no reported influence on MMP2 protein expression. However, tumors derived from animals inoculated with ENU1564 cells had higher MMP2 activity when compared with tumors originated from animals inoculated with ENU1564-TIMP2 cells. Additionally, TIMP2 over expression decreased in vivo orthotopic tumor growth, size, and weight; and also influenced the metastatic behavior of orthotopic tumors. None of the animals inoculated with ENU1564-TIMP2 cells developed metastases, compared with development of lung metastases in all animals inoculated with ENU1564 cells. Moreover, and in concurrence with these results, none of the animals inoculated with ENU1564-TIMP2 cells developed brain metastases. Additionally, in contrast with animals inoculated with ENU1564-TIMP1 and ENU1564-TIMP4, only animals inoculated with ENU1564-TIMP2 had statistically significant less brain metastasis than controls. Taken together, these data suggest that TIMP2 over-expression decreases the metastatic brain behavior of BC cancer cells in this model. Additionally, TIMP2 over-expression effectively decreases MMP2 activity. This suggests that MMP2 is, not only important in the development of orthotopic BC but also, in the biological brain metastatic behavior of these cancer cells. Cell transfection with TIMP2 in in vivo models decreases not only tumor growth, but also metastatic potential. Experimental TIMP over-expression relates to decreased node and pulmonary metastases in bladder cancer.

Astrocytes may play an important role in the development of brain metastases, as they have been shown to respond to extracellular stimuli by producing many cytokines and growth factors that can modulate tumor cell proliferation, growth and/or metastases. Cytokines produced by glial cells in vivo (such as IL6, tumor necrosis factor alpha and IGF 1) may contribute, in a paracrine manner, to the development of brain metastases by BC cells. Astrocytes have also been reported to produce MMP2 and/or regulate its production. We reported previously that there is prominent astrocyte reaction associated with BC brain metastatic foci. Therefore, we investigated whether astrocytes play a role in the development of BC brain metastases in our model and if this role is related to MMP2. Increased IHC stain of GFAP around brain metastatic foci when compared to brain tissue non-infiltrated by neoplasia suggests that astrocytes are associated with the development of BC brain metastases in our model. Additionally, tumor samples (that include brain tissue around tumor foci) express higher levels of GFAP protein when compared to controls. ENU1564 cells express very low levels of MMP2 in vitro, making the study of MMP2/TIMP2 in vitro difficult and of limited value. Rat astrocyte-conditioned media causes increased in vitro expression of MMP2 protein in ENU1564 cells but has no effect in in vitro MMP3 and/or MMP9 protein expression. Additionally, astrocyte-conditioned media also increases the invasive behavior of ENU1564. These results are concurrent with current literature (Leveque et al. 2004; Sierra et al. 1997) and suggest that astrocyte factors are associated with MMP2 protein expression and that this association could influence cancer cell invasive phenotype in our model. We demonstrated that expression and activity of MMP2 are increased in BC brain metastases and showed that factors derived from astrocytes could be involved in increased MMP2 protein expression in ENU1564 cells and increased ENU1564 invasive behavior. These results prompted us to investigate what signaling pathway is involved in the regulation of MMP2 expression in ENU1564 cells. MAPK pathway has been related to MMP activation and expression. MMP2 expression and activity is correlated with ERK phosphorylation. Blockade of the ERK pathway by treatment with PD184352, a specific powerful inhibitor of MAPK/ERK kinase (MEK) suppressed expression of MMPs. In addition, activation of the ERK pathway in tumor cells is well correlated with cancer cell invasive and metastatic phenotype. To determine if there is an in vivo correlation between brain metastases and the main components of the MAPK pathway, we performed IHC analysis for several components of the MAPK pathway p-ERK1/2, -p38 and-JNK. Neoplastic epithelial cells were positive for p-ERK1/2 and staining was more intense at the periphery of the brain metastatic foci. Since these peripheral tumor cells correspond to the invasive front of the tumor foci, these results suggest that p-ERK1/2 may be involved in brain metastases development. No staining was observed with phosphorylated-p38 and -JNK (data not shown). Additionally, we observed that tumor tissue expresses higher levels of p-ERK1/2 protein when compared with control brains of tumor-free animals. In addition, we conducted in vitro experiments to evaluate if astrocyte factor-associated MMP2 expression could be related to the ERK pathway. We observed that, although there was no significant difference in the levels of non-phosphorylated ERK in ENU1564 cells when compared to ENU1564 cells treated with astrocyte-conditioned media, an increase of p-ERK1/2 protein in ENU1564 cells after treatment with astrocyte-conditioned media was present and parallel with increased expression of MMP2 in these cells. When we treated ENU1564 cells (with or without astrocyte-conditioned media) in the presence of an ERK1/2 inhibitor (PD98059), we observed a significant decrease in MMP2 protein expression, as well as the expected decrease in p-ERK1/2 expression. These results strongly suggest that MMP2 expression is regulated by ERK1/2 phosphorylation and activation and, at least in part, by factors produced by astrocytes. Future in vitro and in vivo studies will be needed to determine the specific astrocyte factors that are associated with the exact mechanism of MMP2 BC cell expression within the CNS.

In conclusion, the results of our study indicate that MMP2 plays a role in the process of establishment of BC brain metastases. Additionally, we suggest that astrocytes and the ERK1/2 pathway can be important factors within the BC brain metastatic cascade and may regulate MMP2 expression in that process.

References

Alessandrini A (2002) The roles of map kinases in controlling cancer metastasis. In: Welch DR (ed) Cancer metastasis-related genes, V. 3, vol 20. Kluwer Academic, Boston, pp 35–51

Caudroy S, Polette M, Tournier JM, Burlet H, Toole B, Zucker S, Birembaut P (1999) Expression of the extracellular matrix metalloproteinase inducer (EMMPRIN) and the matrix metalloproteinase-2 in bronchopulmonary and breast lesions. J Histochem Cytochem 47: 1575–1580

Danilewicz M, Sikorska B, Wagrowska-Danilewicz M (2003) Prognostic significance of the immunoexpression of matrix metalloproteinase MMP2 and its inhibitor TIMP2 in laryngeal cancer. Med Sci Monit 9:MT42–MT47

Dzwonek J, Rylski M, Kaczmarek L (2004) Matrix metalloproteinases and their endogenous inhibitors in neuronal physiology of the adult brain. FEBS Lett 567:129–135

Fujimake T, Price JE, Fan D, Bucana CD, Itoh K, Kirino T, Fidler IJ (1996) Selective growth of human melanoma cells in the brain parenchyma of nude mice. Melanoma Res 6:363–371

Gakiopoulou H, Nakopoulou L, Siatelis A (2003) Tissue inhibitor of metalloproteinase-2 as a multifunctional molecule of which the expression is associated with adverse prognosis of patients with urothelial bladder carcinomas. Clin Cancer Res 9:5573–5581

Hall DG, Stoica G (1994) Characterization of brain and bone-metastasizing clones selected from an ethylnitrosurea-induced rat mammary carcinoma. Clin Exp Metastasis 12:283–295

Hanemaaijer R, Verheijen JH, Maguire TM (2000) Increased gelatinase-A and gelatinase-B activities in malignant vs. benign breast tumors. Int J Cancer 86:204–207

Heppner KJ, Matrisian LM, Jensen RA, Rodgers WH (1996) Expression of most matrix metalloproteinase

family members in breast cancer represents a tumor-induced host response. Am J Pathol 149:273–282
Hynes RO (2003) Metastatic potential: generic predisposition of the primary tumor or rare, metastatic variants-or both? Cell 113:821–823
Jones JL, Glynn P, Walker RA (1999) Expression of MMP-2 and MMP-9, their inhibitors, and the activator MT1-MMP in primary breast carcinomas. J Pathol 189:161–168
La Rocca G, Pucci-Minafra I, Marrazzo A, Taormina P, Minafra S (2004) Zymographic detection and clinical correlations of MMP-2 and MMP-9 in breast cancer sera. Br J Cancer 88:1318–1326
Lafleur MA, Tester AM, Thompson EW (2003) Selective involvement of TIMP-2 in the second activational cleavage of pro-MMP-2: refinement of the pro-MMP-2 activation mechanism. FEBS Lett 553:457–463
Le DM, Besson A, Fogg DK (2003) Exploitation of astrocytes by glioma cells to facilitate invasiveness: a mechanism involving matrix metalloproteinase-2 and the urokinase-type plasminogen activator-plasmin cascade. J Neurosci 23:4034–4043
Lebeau A, Nerlich AG, Sauer U, Lichtinghagen R, Lohrs U (1999) Tissue distribution of major matrix metalloproteinases and their transcripts in human breast carcinomas. Anticancer Res 19:4257–4264
Lee WJ, Shin CY, Yoo BK, Ryu JR, Choi EY, Cheong JH, Ryu JH, Ko KH (2003) Induction of matrix metalloproteinase-9 (MMP-9) in lipopolysaccharide- stimulated primary astrocytes is mediated by extracellular signal-regulated protein kinase 1/2 (Erk1/2). Glia 41: 15–24
Leppa S, Saarto T, Vehmanen L, Blomqvist C, Elomaa I (2004) A high serum matrix metalloproteinase-2 level is associated with an adverse prognosis in node-positive breast carcinoma. Clin Cancer Res 10:1057–1063
Leveque T, Le Pavec G, Boutet A (2004) Differential regulation of gelatinase A and B and TIMP-1 and -2 by TNFalpha and HIV virions in astrocytes. Microbes Infect 6:157–163
Li H, Lindenmeyer F, Grenet C (2001) AdTIMP-2 inhibits tumor growth, angiogenesis, and metastases, and prolongs survival in mice. Hum Gene Ther 5:515–526
Liuzzi GM, Mastroianni CM, Latronico T (2004) Anti-HIV drugs decrease the expression of matrix metalloproteinases in astrocytes and microglia. Brain 127:398–407
Massengale JL, Gasche Y, Chan PH (2002) Carbohydrate source influences gelatinase production by mouse astrocytes in vitro. Glia 38:240–245
Mendes O, Kim HT, Stoica G (2005) Expression of MMP2, MMP9 and MMP3 in breast cancer brain metastasis in a rat model. Clin Exp Metastasis 22: 237–246
Muir EM, Adcock KH, Morgenstern DA (2002) Matrix metalloproteases and their inhibitors are produced by overlapping populations of activated astrocytes. Brain Res Mol Brain Res 100:103–117
Nagashima G, Suzuki R, Asai J (2002) Immunohistochemical analysis of reactive astrocytes around glioblastoma: an immunohistochemical study of postmortem glioblastoma cases. Clin Neurol Neurosurg 104:125–131
Nie J, Pei D (2003) Direct activation of pro-matrix metalloproteinase-2 by leukolysin/membrane-type 6 matrix metalloproteinase/matrix metalloproteinase 25 at the asn(109)-Tyr bond. Cancer Res 63:6758–6762
Nishizuka I, Ishikawa T (2002) Analysis of gene expression involved in brain metastases from breast cancer using cDNA microarray. Breast Cancer 9:26–32
Ohshiba T, Miyaura C, Inada M, Ito A (2003) Role of RANKL-induced osteoclast formation and MMP-dependent matrix degradation in bone destruction by breast cancer metastasis. Cancer 90:1414–1421
Pan MR, Hung WC (2002) Nonsteroidal anti-inflammatory drugs inhibit matrix metalloproteinase-2 via suppression of the ERK/Sp1-mediated transcription. J Biol Chem 277:32775–32780
Remacle A, McCarthy K, Noel A (2000) High levels of TIMP-2 correlate with adverse prognosis in breast cancer. Int J Cancer 89:118–121
Scott KA, Holdsworth H, Balkwill FR, Dias S (2000) Exploiting changes in the tumour microenvironment with sequential cytokine and matrix metalloprotease inhibitor treatment 133 in a murine breast cancer model. Br J Cancer 83:1538–1543
Sierra A, Price JE, Garcia-Ramirez M (1997) Astrocyte-derived cytokines contribute to the metastatic brain specificity of breast cancer cells. Lab Invest 77: 357–368
Somerville RP, Oblander SA, Apte SS (2003) Matrix metalloproteinases: old dogs with new tricks. Genome Biol 4:205–216
Stetler-Stevenson WG, Aznavoorian S, Liotta LA (1993) Tumor cell interactions with the extracellular matrix during invasion and metastasis. Annu Rev Cell Biol 9:541–573
Talvensaari-Mattila A, Paakko P, Turpeenniemi-Hujanen T (2003) Matrix metalloproteinase-2 (MMP-2) is associated with survival in breast carcinoma. Br J Cancer 89:1270–1275
Tester AM, Waltham M, Oh SJ, Bae SN, Bills MM, Walker EC, Kern FG, Stetler- Stevenson WG, Lippman ME, Thompson EW (2004) Pro-matrix metalloproteinase-2 transfection increases orthotopic primary growth and experimental metastasis of MDA-MB-231 human breast cancer cells in nude mice. Cancer Res 64:652–658
Yoneda T (2000) Cellular and molecular basis of preferential metastasis of breast cancer to bone. J Orthop Sci 5:75–81

21 Differentiating Choroid Plexus Tumors from Metastatic Carcinomas: Use of Inwardly Rectifying K+ Channel KIR7.1 and Excitatory Amino Acid Transporter-1

Rudi Beschorner

Contents

R. Beschorner (✉)
Institute of Pathology and Neuropathology, Department of Neuropathology, University Hospital Tübingen, Tübingen, Germany
e-mail: Rudi.Beschorner@med.uni-tuebingen.de

Abstract

The choroid plexus is a villous structure within the ventricular system of the brain and consists of a fibrovascular stroma with fenestrated blood vessels, which is covered by the choroid plexus epithelium. The choroid plexus epithelium is of neuroectodermal origin and bears both features of glial and epithelial cells. Tumors of the choroid plexus (CPT) include choroid plexus papilloma (CPP), atypical CPP and choroid plexus carcinoma (CPC). CPP and atypical CPP are typically papillary tumors that may contain areas of solid growth. CPC are typically pleomorphic tumors with predominantly solid growth. Thus, histological distinction of CPT may be difficult from both, metastasis from well differentiated papillary adenocarcinomas and from less differentiated carcinomas. CPTs are rare neoplasm's most frequently affecting children. In contrast, brain metastases from adenocarcinomas are rather frequent but usually arise in adult and elderly patients. The overall frequency of solitary metastases to the choroid plexus or its direct vicinity outnumbers primary CPT. Thus, the differential diagnosis of CPT and brain metastasis should always be considered and additional immunostainings are required, especially in adult patients. Under this point of view, the value and the limits of several antibodies are summarized and the most significant role of Kir7.1 and EAAT1 is sustained.

M. Hayat (ed.), *Tumors of the Central Nervous System, Volume 13*,
DOI 10.1007/978-94-007-7602-9_21, © Springer Science+Business Media Dordrecht 2014

Introduction

Choroid plexus tumors (CPT) are rare neoplasm's accounting for 0.3–0.6% of all brain tumors and arise most frequently in the first decade of life but may occur at any age (Paulus and Brandner 2007). CPT may occur at any site where the choroid plexus is physiologically present, i.e., within any of the four cerebral ventricles or at the cerebellopontine angle. CPT may spread through the cerebrospinal fluid. According to the World Health Organization (WHO) classification of tumors of the CNS, primary tumors of the choroid plexus (CPT) are classified as choroid plexus papilloma (CPP, grade I WHO), atypical CPP (grade II WHO) and choroid plexus carcinoma (CPC, grade III WHO) (Paulus and Brandner 2007). Besides a typical papillary growth pattern closely resembling normal choroid plexus, CPTs may rarely show unusual histological features, such as tubular glandular architecture or mucinous degeneration or a solid growth pattern. Thus, the histological distinction of CPTs from metastatic (adeno-) carcinomas may be difficult. CPP can usually be cured by surgery and even for CPC a 10-year survival rate of ~35% can be expected. In contrast, overall survival of patients with brain metastases is usually <1 year. Thus, prognosis and therapeutic consequences differ between primary CPT and metastasis to the choroid plexus.

In contrast to CPT, metastatic tumors are much more common CNS neoplasms, and carcinomas are the most frequent source of brain metastases. Brain metastases usually occur in adult and elderly patients, an age group in which CPT are less common. Metastatic carcinomas can also be located within the choroid plexus or at sites where choroid plexus tissue might physiologically be present (intraventricular, cerebellopontine angle) (Kohno et al. 1996; Cha et al. 2000; Gopal et al. 2008). The incidence of solitary metastasis to the choroid plexus accounts for approximately 0.14% of all brain metastasis (Gopal et al. 2008). Taking together these data, the overall incidence of primary CPT is approximately only two to fourfold higher than the incidence of solitary metastasis to the choroid plexus. Consequently, in adult and elderly patients, the incidence of metastatic adenocarcinomas at these locations is probably equally or even higher than primary CPT. Thus, especially in adult and elderly cases, additional immunohistochemical examinations are needed to make an appropriate diagnosis. A suggestion of useful immunostainings to prove choroid plexus origin is given in this chapter. If initially a brain metastasis is expected, the panel of helpful antibodies depends on histology and clinical data, especially if a primary carcinoma is not reported in the files.

Immunoreactivity of Normal and Neoplastic Choroid Plexus Epithelium

Both, non-neoplastic choroid plexus epithelium and CPT show immunohistochemically a variable expression pattern, and there is no single immunohistochemical marker proved to be specific for choroid plexus. Normal choroid plexus epithelium and CPTs are consistently immunoreactive to antibodies against cytokeratin (CK) in the majority of cells. Among CPT subtypes that express CK in normal and neoplastic choroid plexus, CK7 is more frequently found than CK20. The most frequent combination of CK7 and CK20 observed in CPT is CK7+/CK20-, but any other combination may also be present in CPT (Gyure and Morrison 2000). CK8 and CK18 have also been noted in normal and neoplastic choroid plexus (Gianella-Borradori et al. 1992; Gertz et al. 1990).

The carcinoembryonic antigen (CEA) is usually negative in CPP but may be expressed in a subset of CPC (Ang et al. 1990; Coffin et al. 1986). However, expression of CEA may point to metastatic adenocarcinomas than CPT (Paulus and Janisch 1990).

The inwardly rectifying potassium channel Kir7.1 (Nakamura et al. 1999) has been reported to be expressed in 34/35 (97%) samples from normal choroid plexus and in 17/23 (74%) of CPT, including 5/5 CPC (Hasselblatt et al. 2006b). In our experience, all samples from normal (n=5) and neoplastic (n = 75) choroid plexus are strongly Kir7.1 positive, including five CPCs.

The excitatory amino acid transporter-1 (EAAT1, homologous to rodent glutamate/aspartate transporter/GLAST) is a glial glutamate transporter that is predominantly expressed in astrocytes (Williams et al. 2005). In normal choroid plexus, membranous expression of EAAT1 is exceptional and significantly decreases with age. Expression levels of EAAT1 in the choroid plexus decrease throughout fetal life, and is usually absent in choroid plexus after birth. In normal choroid plexus, only a minority of pediatric samples may show few EAAT1 positive cells, whereas in adult cases EAAT1 immunostaining is constantly negative (Beschorner et al. 2009). In contrast, the majority of CPT (81%) is immunoreactive for EAAT1, and there is a significant increase in EAAT1 expression with age. CPTs from adult and elderly patients express EAAT1 in 90% of cases. This makes EAAT1 the sole marker to distinguish between normal and neoplastic choroid plexus, especially in adult cases (Beschorner et al. 2009).

S100 and transthyretin (prealbumin/TTR) are also frequently expressed in normal and neoplastic choroid plexus (Gaudio et al. 1998; Jacobsen et al. 1982; Redzic et al. 2005). Similarly, E-cadherin and aquaporin-1 are frequently expressed in both, normal and neoplastic choroid plexus (Figarella-Branger et al. 1995; Rickert and Paulus 2001). Immunoreactivity for vimentin and podoplanin (D2-40) may be found in normal and neoplastic choroid plexus (Nakamura et al. 2006).

Expression of Antibodies Applied in Diagnosis of Choroid Plexus Tumors, in Adenocarcinomas and Other Brain Tumors

Antibodies that are frequently applied in the diagnosis of CPT are also expressed in a variable frequency in adenocarcinomas of different primary sites. Depending on the origin of a metastatic carcinoma, besides pan-cytokeratin, cytokeratin subtypes such as CK7 and/or CK20 may be expressed, and the expected combination of CK7 and CK20 expression depends mainly on the anatomic origin of the tumor and may vary among histological subtypes. Thus, as in CPT, any combination of CK7/CK20 expression may be present in a brain metastasis. Expression of CK8 and CK18, which has also been noted in normal and neoplastic choroid plexus (Gianella-Borradori et al. 1992; Gertz et al. 1990), is not uncommon in metastatic adenocarcinomas of various origins. Furthermore, other primary brain tumors may be cytokeratin-positive (e. g., papillary tumor of the pineal region, atypical teratoid-rhabdoid tumor). The CEA is also not specific for CPT and is frequently expressed in gastrointestinal carcinomas. However, one study pointed out that expression of CEA may indicate metastatic adenocarcinoma instead of CPT (Paulus and Janisch 1990).

The inwardly rectifying K+ channel Kir7.1 is highly expressed in thyroid follicular cells and intestinal epithelial cells from rats (Nakamura et al. 1999), and has been reported at the DNA level in human small intestine, stomach, and kidney (Partiseti et al. 1998). Presently, there are no reports on immunohistochemical expression of Kir7.1 in humans outside the CNS. In a series of 45 cerebral metastases of carcinomas, all cases were found to be negative for Kir7.1. Thereby, metastases from different organs were investigated, including kidney, lung, breast, thyroid, colon, ovary, stomach, and testis (Hasselblatt et al. 2006b). Thus, if Kir7.1 is expressed in human epithelial organs, its expression level might be to low for immunohistochemical detection. Membranous immunoreactivity for Kir7.1 has also been reported in a minority of papillary tumors of the pineal region (Hasselblatt et al. 2006a), and may be found in a subset (~15–20%) of atypical teratoid-rhabdoid tumors.

In cell cultures, prostate cancer cell lines PC-3 and LNCaP both expressed EAAT1 at the mRNA level (Pissimissis et al. 2009). But, immunohistochemically an expression of the membranous glutamate transporter EAAT1 has not been observed in metastatic adenocarcinomas. In a series of 64 metastases from adenocarcinomas derived from different organs, including prostate, breast, lung, colon, kidney, and thyroid, all tumors were EAAT1-negative (Beschorner et al. 2006). Metastases from urothelial carcinomas were also EAAT1-negative. However, other primary brain tumors such as astrocytomas and ependymomas as well as atypical teratoid-rhabdoid tumors and papillary tumors of the pineal region are positive for EAAT1 in a variable frequency. Several adenocarcinomas of different organs may be vimentin-positive, including carcinomas from kidney (including clear cell carcinoma and papillary carcinoma), thyroid, ovary, and uterus.

S100 and transthyretin (prealbumin/TTR) are also expressed in variable numbers of adenocarcinomas from different sites (up to 75%), including thyroid, kidney and breast (for review see Beschorner et al. 2006). E-cadherin may also be expressed in adenocarcinomas of different sites, including lung and kidney. Aquaporin-1 and podoplanin (D2-40) can be found in metastatic carcinomas especially with primary sites in ovary, and breast, and in the kidney, respectively (Longatti et al. 2006b; Hoque et al. 2006; Rivera et al. 2010). Aquaporin-1 and podoplanin immunoreactivity also occurs in hemangioblastomas and glioblastomas (Longatti et al. 2006a; Boon et al. 2004) and in ependymomas, glioblastomas and meningiomas, respectively (Ishizawa et al. 2009; Nakamura et al. 2006).

In conclusion, the differential diagnosis of primary CPT and a metastasis of adenocarcinomas is challenging and appropriate tools for routine diagnosis, especially in adult cases in which brain metastases outnumber CPT. However, there is no marker known to be specific for choroid plexus epithelium. Kir7.1 is consistently expressed in normal and neoplastic choroid plexus epithelium, and is, therefore, the most suitable antibody to prove choroid plexus origin. Kir7.1 is very useful in distinguishing CPT from brain metastases, as adenocarcinomas are Kir7.1 negative. However, Kir7.1 does not differentiate between normal and neoplastic choroid plexus epithelium, and is not commercially available. EAAT1 is commercially available and also significantly distinguishes CPT from brain metastases. Furthermore, EAAT1 expression significantly distinguishes CPT (EAAT1-positive) from nonneoplastic (EAAT1-negative) choroid plexus. Thus, if available, a combination of Kir7.1 and EAAT1 immunostaining is recommended in achieving the diagnosis of a CPT. If the applied immunostainings point against a CPT, a panel of additional antibodies should be selected on the basis of histology and clinical data to obtain brain metastasis diagnosis.

References

Ang LC, Taylor AR, Bergin D, Kaufmann JC (1990) An immunohistochemical study of papillary tumors in the central nervous system. Cancer 65:2712–2719

Beschorner R, Schittenhelm J, Schimmel H, Iglesias-Rozas JR, Herberts T, Schlaszus H, Meyermann R, Wehrmann M (2006) Choroid plexus tumors differ from metastatic carcinomas by expression of the excitatory amino acid transporter-1. Hum Pathol 37:854–860

Beschorner R, Pantazis G, Jeibmann A, Boy J, Meyermann R, Mittelbronn M, Schittenhelm J (2009) Expression of EAAT-1 distinguishes choroid plexus tumors from normal and reactive choroid plexus epithelium. Acta Neuropathol 117:667–675

Boon K, Edwards JB, Eberhart CG, Riggins GJ (2004) Identification of astrocytoma associated genes including cell surface markers. BMC Cancer 4:39

Cha ST, Jarrahy R, Mathiesen RA, Suh R, Shahinian HK (2000) Cerebellopontine angle metastasis from papillary carcinoma of the thyroid: case report and literature review. Surg Neurol 54:320–326

Coffin CM, Wick MR, Braun JT, Dehner LP (1986) Choroid plexus neoplasms. Clinicopathologic and immunohistochemical studies. Am J Surg Pathol 10:394–404

Figarella-Branger D, Lepidi H, Poncet C, Gambarelli D, Bianco N, Rougon G, Pellissier JF (1995) Differential expression of cell adhesion molecules (CAM), neural CAM and epithelial cadherin in ependymomas and choroid plexus tumors. Acta Neuropathol 89:248–257

Gaudio RM, Tacconi L, Rossi ML (1998) Pathology of choroid plexus papillomas: a review. Clin Neurol Neurosurg 100:165–186

Gertz HJ, Unger M, Lobeck H, Stoltenburg-Didinger G (1990) Histological and immunocytochemical findings in a case of fetal choroid plexus papilloma. Zentralbl Allg Pathol 136:719–722

Gianella-Borradori A, Zeltzer PM, Bodey B, Nelson M, Britton H, Marlin A (1992) Choroid plexus tumors in childhood. Response to chemotherapy and immunophenotypic profile using a panel of monoclonal antibodies. Cancer 69:809–816

Gopal P, Parker JR, Debski R, Parker JC Jr (2008) Choroid plexus carcinoma. Arch Pathol Lab Med 132:1350–1354

Gyure KA, Morrison AL (2000) Cytokeratin 7 and 20 expression in choroid plexus tumors: utility in differentiating these neoplasms from metastatic carcinomas. Mod Pathol 13:638–643

Hasselblatt M, Blumcke I, Jeibmann A, Rickert CH, Jouvet A, van de Nes JA, Kuchelmeister K, Brunn A, Fevre-Montange M, Paulus W (2006a) Immunohistochemical profile and chromosomal imbalances in papillary tumours of the pineal region. Neuropathol Appl Neurobiol 32:278–283

Hasselblatt M, Bohm C, Tatenhorst L, Dinh V, Newrzella D, Keyvani K, Jeibmann A, Buerger H, Rickert CH, Paulus W (2006b) Identification of novel diagnostic markers for choroid plexus tumors: a microarray-based approach. Am J Surg Pathol 30:66–74

Hoque MO, Soria JC, Woo J, Lee T, Lee J, Jang SJ, Upadhyay S, Trink B, Monitto C, Desmaze C, Mao L, Sidransky D, Moon C (2006) Aquaporin 1 is overexpressed in lung cancer and stimulates NIH-3T3 cell proliferation and anchorage-independent growth. Am J Pathol 168:1345–1353

Ishizawa K, Komori T, Shimada S, Hirose T (2009) Podoplanin is a potential marker for the diagnosis of ependymoma: a comparative study with epithelial membrane antigen (EMA). Clin Neuropathol 28:373–378

Jacobsen M, Jacobsen GK, Clausen PP, Saunders NR, Mollgard K (1982) Intracellular plasma proteins in human fetal choroid plexus during development. II. The distribution of prealbumin, albumin, alpha-fetoprotein, transferrin, IgG, IgA, IgM, and alpha 1-antitrypsin. Brain Res 255:251–262

Kohno M, Matsutani M, Sasaki T, Takakura K (1996) Solitary metastasis to the choroid plexus of the lateral ventricle. Report of three cases and a review of the literature. J Neuro-Oncol 27:47–52

Longatti P, Basaldella L, Orvieto E, Dei Tos AP, Martinuzzi A (2006a) Aquaporin 1 expression in cystic hemangioblastomas. Neurosci Lett 392:178–180

Longatti P, Basaldella L, Orvieto E, Dei TA, Martinuzzi A (2006b) Aquaporin(s) expression in choroid plexus tumours. Pediatr Neurosurg 42:228–233

Nakamura N, Suzuki Y, Sakuta H, Ookata K, Kawahara K, Hirose S (1999) Inwardly rectifying K+ channel Kir7.1 is highly expressed in thyroid follicular cells, intestinal epithelial cells and choroid plexus epithelial cells: implication for a functional coupling with Na+, K+-ATPase. Biochem J 342(Pt 2):329–336

Nakamura Y, Kanemura Y, Yamada T, Sugita Y, Higaki K, Yamamoto M, Takahashi M, Yamasaki M (2006) D2-40 antibody immunoreactivity in developing human brain, brain tumors and cultured neural cells. Mod Pathol 19:974–985

Partiseti M, Collura V, Agnel M, Culouscou JM, Graham D (1998) Cloning and characterization of a novel human inwardly rectifying potassium channel predominantly expressed in small intestine. FEBS Lett 434:171–176

Paulus W, Brandner S (2007) Choroid plexus tumours. In: Louis DN, Ohgaki H, Wiestler OD, Webster KC (eds) WHO classification of tumours of the central nervous system. WHO Press, Geneva, pp 81–86

Paulus W, Janisch W (1990) Clinicopathologic correlations in epithelial choroid plexus neoplasms: a study of 52 cases. Acta Neuropathol 80:635–641

Pissimissis N, Papageorgiou E, Lembessis P, Armakolas A, Koutsilieris M (2009) The glutamatergic system expression in human PC-3 and LNCaP prostate cancer cells. Anticancer Res 29:371–377

Redzic ZB, Preston JE, Duncan JA, Chodobski A, Szmydynger-Chodobska J (2005) The choroid plexus-cerebrospinal fluid system: from development to aging. Curr Top Dev Biol 71:1–52

Rickert CH, Paulus W (2001) Tumors of the choroid plexus. Microsc Res Tech 52:104–111

Rivera AL, Takei H, Zhai J, Shen SS, Ro JY, Powell SZ (2010) Useful immunohistochemical markers in differentiating hemangioblastoma versus metastatic renal cell carcinoma. Neuropathology 30:580–585

Williams SM, Sullivan RK, Scott HL, Finkelstein DI, Colditz PB, Lingwood BE, Dodd PR, Pow DV (2005) Glial glutamate transporter expression patterns in brains from multiple mammalian species. Glia 49:520–541

Part VI

General CNS Diseases

Alexander Disease: Role of Glial Fibrillary Acidic Protein

22

Tomokatsu Yoshida and Masanori Nakagawa

Contents

Abstract

Alexander disease is a rare neurodegenerative disease that develops as a result of mutations in the gene encoding glial fibrillary acidic protein (GFAP). Alexander disease pathology is characterized by the existence of aggregates consisting mainly of GFAP in the cytoplasm of astrocytes. We found that the *GFAP* mutations, V87G, R88C, and R416W, induced structural alterations in GFAP and were associated with dysfunctional cell proliferation, as observed in a morphological study. We performed a migration assay to examine the dynamic functional effects of these mutant GFAPs in astrocytes. Finally, we carried out time-lapse recording of two *GFAP* mutant cells, R239C and R416W, located in the rod and tail domains, respectively, to examine the dynamic process of aggregation between the different domains of GFAP. One-third of R239C cells first appeared as aggregates, and clusters of these aggregates in the cytoplasm tended to move inward and form amorphous aggregates. Eighty percent of R416W cells showed disrupted GFAP, with a bubble- or ring-like structure. However, most cells maintained their structure and were capable of cell division. Our study indicated that R239C and R416W differed in their processes of inducing aggregate formation. Recent data from other studies suggest a complex mechanism of mutant GFAP accumulation; oligomeric forms of mutant GFAP located in the helical rod domain inhibit proteasome activity.

T. Yoshida (✉) • M. Nakagawa
Department of Neurology, Graduate School of Medical Science, Kyoto Prefectural University of Medicine, Kyoto, Japan
e-mail: toyoshid@koto.kpu-m.ac.jp

M. Hayat (ed.), *Tumors of the Central Nervous System, Volume 13*,
DOI 10.1007/978-94-007-7602-9_22, © Springer Science+Business Media Dordrecht 2014

Proteasome inhibition induces the activation of several stress-response pathways that cause toxicity or apoptosis within astrocytes, and also activates αB-crystallin, which can reinstate proteasomal degradation. GFAP aggregation caused by mutant GFAP in the tail domain induces the association of small heat shock proteins and ubiquitin, and is an initiating event in Alexander disease and contributes to aggregate maturation and astrocyte malfunction. GFAP-δ/ε, an alternatively spliced GFAP isoform, may play a modulating role in aggregate formation.

Introduction

Alexander disease is a rare neurodegenerative disorder characterized by white matter degeneration and the formation of cytoplasmic inclusions called Rosenthal fibers, which are demonstrated in astrocytes by pathological studies (Alexander 1949). Rosenthal fibers, which particularly accumulate in astrocyte end-feet in the subpial and perivascular zones, consist of glial fibrillary acidic protein (GFAP), heat shock protein 27, and αB-crystallin (Iwaki et al. 1989; Johnson and Bettica 1989; Tomokane et al. 1991). Clinically, Alexander disease is classified into the following three subtypes: the infantile, juvenile, and adult forms, according to the age of disease onset (Li et al. 2005). The infantile form is the most common and severe, it usually presents when the patient is under 2 years of age and causes developmental delay, megalocephaly, spasticity, and seizures. The adult form is milder and presents with spastic paresis and ataxia, with or without palatal myoclonus.

Genetically, heterozygous *GFAP* mutations have been identified in patients with Alexander disease (Brenner et al. 2001). The corresponding domain is composed of the head, alpha-helical rod, and tail domains. Most *GFAP* mutations are located in the rod domain and fewer reside in the tail and head domains (Li et al. 2005), although the genotype-phenotype correlation is currently unknown.

Alexander disease is considered a disorder of astrocytes associated with protein misfolding and aggregation, because *GFAP* transgenic mice over expressing human wild-type *GFAP* showed characteristic pathological changes and clinical features (Messing et al. 1998). The mechanisms leading to aggregate formation, which are considered models of Rosenthal fibers, were investigated in recent studies.

We prepared vectors expressing three different *GFAP* mutations and investigated the expression pattern of mutant GFAPs in astrocytoma-derived cells, and performed a migration assay to examine the dynamic functional effects of these mutant GFAPs in astrocytes (Yoshida et al. 2007). Furthermore, to examine the dynamic process of aggregation between different domains of GFAP, we performed time-lapse recording of R239C and R416W GFAP mutants, which are localized in the rod and tail domains, respectively (Yoshida et al. 2009).

Morphological and Functional Studies Using Cell Models

Morphological Analysis

Human astrocytoma-derived cells (mtU251) were used in the morphological study and the migration assay, wherein GFAP was no longer spontaneously expressed during cell culture. The cells were grown in RPMI 1640 medium (Nikken Biochemical Laboratory, Kyoto, Japan) supplemented with 10% fetal bovine serum (FBS) and amphotericin B (0.125 μg/ml).

The transiently transfected cells; wild-type, V87G, R88C, and R416W cells, were blocked for 20 min with 1% normal goat serum in Tris-buffered saline (TBS) at 25 °C. We used anti-cow GFAP as the primary antibody and Alexa Fluor® 488-conjugated goat anti-rabbit IgG (Molecular Probes, OR, USA) as the secondary antibody.

Following GFAP immunostaining, we examined whether the transfected cells had impaired proliferation by performing immunostaining of proliferating cell nuclear antigen (PCNA). The cells were rehydrated with 1× phosphate buffered saline (PBS) at room temperature for 5 min. Then, the cells were blocked with 3% hydrogen peroxide with ethanol for 15 min. We used PC10 antibody (mouse monoclonal anti-PCNA) (Nichirei, Tokyo, Japan) as the primary antibody, and Alexa Fluor 555®-conjugated goat anti-mouse IgG antibody (Molecular Probes, OR, USA) as the secondary antibody at a dilution of 1/2000. All immunocytostaining procedures were performed at 25 °C. Cells were incubated for 60 min with the primary antibody. After washing with PBS, cells were incubated for 30 min with the secondary antibody. Stained cells were viewed under a fluorescence microscope at a magnification of 20×.

The transfected cells were classified into three patterns according to GFAP expression: filamentous pattern, polarized distribution pattern and aggregated or amorphous pattern. The filamentous pattern appeared to be a fundamentally normal GFAP structure in both the wild-type and mutant GFAPs. The polarized distribution pattern was also observed in both the wild-type and mutant GFAPs. This pattern showed a bipolar GFAP distribution in mitotic process with filaments between poles when viewed under a confocal microscope, and a local distribution after mitosis. In this pattern, the bipolar distribution was stained with PCNA, which plays a fundamental role in DNA replication and repair. The aggregated or amorphous patterns showed irregular GFAP aggregation with few filamentous structures, or appeared to have destroyed filament networks. The cells with an aggregated or amorphous pattern were not stained by PCNA. The most critical findings were that the aggregated or amorphous pattern was observed significantly more frequently in cells transfected with mutant GFAPs than in those transfected with wild-type GFAP ($p<0.01$). However, there was no great difference in the proportion of aggregated and amorphous patterns among mutant GFAPs (2–4%). The patterns observed 24–48 h after transfection were similar to those observed 72 h after transfection.

Migration Assay

The mtU251 cells were also used in migration assay. The GFP-tagged GFAP cells, including wild-type, V87G, R88C, and R416W, were placed in serum-free RPMI 1640 medium inside the culture inserts, which had an 8 μm-pore fluorescence-blocking microporous membrane (HTS FluoroBlok™ Inserts, Becton Dickinson, Franklin Lakes, NJ, USA). The culture insert was set in a 24-well plate containing 10% FBS, and was incubated for 8 h in a CO_2 incubator at 37 °C. The invading cells were visualized with an Olympus IX70 microscope at a magnification of 100×. Up to four different fields for each condition were counted after recording every well with a resolution of 1,392×1,024 pixels per channel. Individual migrating cells per field recognized over time were compared using a regression analysis carried out using the JMP® software. The Tukey-Kramer HSD test was used to compare the velocity of migration between individual cells expressing different plasmids.

All transfected cells migrated in a time-dependent manner, and the relationship between the velocity of migration and time was linear. The same experiment was repeated four times and the velocity of the migration among these cells was compared. The velocities of migration of the V87G and R88C cells were significantly higher than that of the wild-type and R416W cells ($p<0.01$). The migration velocity differences between the V87G and R88C cells or between the wild-type and R416W cells were not found to be statistically significant.

Time-Lapse Recording

The mtU251 cells were assayed for the expression of the transfected gene after 48 h. Just before the time-lapse recording, cell nuclei were stained with Hoechst 33342 dye at 37 °C for 30 min. A plate for real-time acquisition was placed in the incubator at 37 °C with 5% CO_2/95% air on the stage of a fluorescence microscope (Biorevo BZ-9000, Keyence). Time-lapse images were acquired every 20 min for 24 h at 10× or 40× magnification.

Real-time images of cells expressing either GFP-wild-type GFAP (GFP-Wt) or GFP-mutant GFAP showed two initial phenotype patterns: an apparently normal filamentous network or an aggregated phenotype. Overall, 86.1% of GFP-Wt cells (n=79) first appeared to have a normal filamentous network and 95.6% (n=65) maintained the filamentous network and were capable of cell division. The remaining 13.9% of GFP-Wt cells (n=11) first appeared to have aggregates. Almost all of these cells (90.9%, n=10) were unchanged and were incapable of cell division. In GFP-R239C cells (n=293), 32.4% (n=95) first appeared as aggregates. In these cells, a cluster of aggregates in the cytoplasm tended to move inward and form amorphous aggregates. Cells with aggregates appeared to be incapable of cell division. Overall, 82.0% of GFP-R416W cells (n=73) first appeared as a filament network when viewed at 10× magnification; however, when examined at 40× magnification these cells showed that the filaments were constructed of a bubble – or ring-like structure; 79.5% of these cells (n=58) maintained their structure and were capable of cell division, whereas 20.5% of cells that appeared with an apparently filamentous network (n=15) were induced to aggregates.

Pathomechanism of Alexander Disease

Morphologically, the transfected cells could be classified into three patterns: the filamentous pattern, polarized distribution pattern and aggregated or amorphous pattern. The filamentous pattern could not be distinguished from normal GFAP filamentous architecture when viewed at a low magnification. However, R416W cells showed an apparently filamentous pattern, and the cell filaments appeared to be constructed as a bubble-or ring-like structure when viewed at a high magnification on the time-lapse recording, and approximately 20% of the cells exhibited aggregated filament formation. The polarized distribution pattern was considered to perform a function during cell division. On time-lapse recording, this pattern was present in a small population of cells with aggregates which recovered the filamentous network morphology during the time-lapse recording. The aggregated or amorphous pattern in mutant GFAPs were considered to be "abnormal" structures and appeared to be incapable of cell division, because the cells with these patterns were not stained with PCNA, which shows immunoreactivity in the proliferative compartment of cells (Hall et al. 1990). On time-lapse recording, more than 90% of the cells that first appeared as aggregates were unchanged or moved inward and formed larger amorphous aggregates. These cells did not undergo cell division. The proportions of "abnormal" structure in the three kinds of mutant GFAPs were significantly larger than that in the wild-type GFAP cells. These findings strongly suggest that *GFAP* mutations lead to an altered GFAP structure and disrupt cell proliferation.

Findings from the present functional study, migration assay, and time-lapse recording study also indicated that the mutant GFAP residing in the helical rod domain in *GFAP* can alter the function of astrocytes, and that the process of inducing these aggregates is different from the R416W mutant, which is located in the tail domain in *GFAP*.

V87, R88, and R239 are located in the helical rod domain or central helical domain in *GFAP* (Fig. 22.1), which is considered to be important for interfilament network formation, filament assembly, and the stabilization of subunits (Fuchs 1996).The elevated proportions of an "abnormal" GFAP structure in V87G and R88C mutants of GFAP suggest that *GFAP* mutations in the helical domains contribute to alterations in the filamentous architecture of GFAP to varying degrees. Furthermore, functional alteration, which was confirmed by the migration assay, may influence the interaction with other extracellular components and might therefore affect the formation of the blood-brain barrier or the development and differentiation of GFAP, or may induce astrocytosis, leading to pathological changes that determine the phenotypic severity. A previous time-lapse recording study showed that aggregates of R236H, which were also located in the

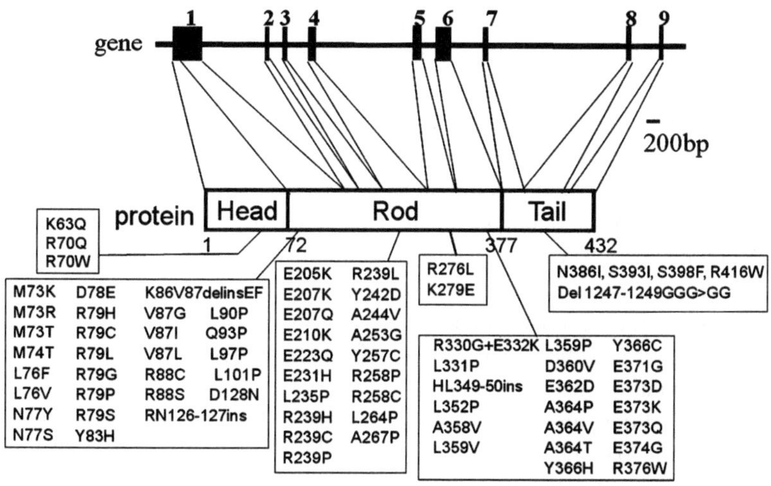

Fig. 22.1 Schematic depicting the *GFAP* gene, the corresponding protein and localization of *GFAP* mutations identified in patients with Alexander disease

rod domain of *GFAP*, either disappeared or were associated with cell survival, or coalesced in a huge juxtanuclear structure associated with cell death (Mignot et al. 2007). Mutant GFAP in the rod domain might therefore be unable to maintain the fundamental filamentous architecture of GFAP and be induced to aggregate, suggesting that the degree of severity of GFAP mutations in the rod domain depend on the degree of disruption of the fundamental structure, which may have a dominant effect over wild-type GFAP.

The R416W *GFAP* mutation is located in the tail domain (Fig. 22.1), which is conserved between all type III intermediate filament proteins and is thought to play a role in stabilizing protofibrillar interactions and filament diameter (Fuchs 1996). Furthermore, R416W GFAP would be expected to affect interactions with other cytoskeletal elements (Quinlan 2001) and be a cause of Alexander disease, as well as a residue of the point mutation in the tail domain of other intermediate filament, desmin, which causes idiopathic dilated cardiomyopathy (Cary and Klymkowsky 1995; Li et al. 1999). Recently, Perng et al. demonstrated that the R416W GFAP mutant disrupted normal filament assembly in vitro, and that protein chaperones such as αB-crystallin and HSP27 were specifically associated with GFAP aggregates in R416W GFAP cells (Perng et al. 2006). In our study, the proportion of abnormal GFAP structures in R416W cells was also significantly higher than that in the wild-type cells, while the velocity of R416W cell migration was not significantly different from the wild-type cells. The time-lapse recording findings revealed that many cells expressing R416W GFAP were able to undergo cell division, although the filamentous structure of these cells was apparently disrupted. These results suggest that intrafilament disruption could cause morphological alteration, and that the residue of point mutation in the tail domain per se may not be a direct determinant of the interaction with extracellular components. Our results suggested that these cells could maintain the fundamental structure of GFAP, and that the alteration of R416W GFAP function in astrocytes depended on other proteins that interact with GFAP, suggesting that the phenotype of mutant *GFAP* in the tail domain has a variety of clinical features of Alexander disease with varying severity (Brenner et al. 2001; Kinoshita et al. 2003; Li et al. 2005).

The fact that the same mutation can cause different phenotypes suggests that the phenotype is influenced not only by structural and functional changes induced by *GFAP* mutations, but is also affected by other components, including genetic, environmental, embryological, and GFAP-regulating factors. The current data suggest that a combination of events contribute to Alexander disease. Mitochondrial abnormalities are one consideration, because few reports have described a relationship between Alexander disease and mitochondrial abnormalities. Gingold et al. (1999) reported a female patient who was clinically diagnosed initially with Leigh disease, then later was pathologically confirmed as having Alexander disease due to the finding of a massive deposition of Rosenthal fibers. Schuelke et al. (1999) reported a female patient who showed the phenotype of Alexander disease and an *NDUFV1* homozygous mutation, which encodes a component of the mitochondrial complex I. A patient with an R88C *GFAP* mutation also showed mitochondrial DNA abnormalities, an A-to-G homoplasmic transition at np8291 and a 9-base pair deletion between np8272-8280 (Nobuhara et al. 2004). Plectin, which is a member of the plakin family of cytolinker proteins and is localized at the cytoskeleton-plasma membrane interface, is another potential candidate contributing to Alexander disease development. Tian et al. (2006) reported that the reduced levels of total plectin in response to mutant GFAP expression in astrocytes can promote the abnormal organization and aggregation of GFAP into Rosenthal fibers. Among the minor isoforms of GFAP, GFAP-δ/ε alters the binding of αB-crystallin to GFAP filaments (Perng et al. 2008). GFAP-δ/ε mRNA expression is the result of alternative splicing of intron 7, with variable use of exon 7a, and skipping exons 8 and 9 (Nielsen et al. 2002). GFAP aggregates induced by R239C mutation aggravate the effects of GFAP accumulation in inducing the intracellular stress response (Tang et al. 2008). Most recently, Tang et al. (2010) have reported that GFAP aggregates induced by R239C mutation inhibited the proteasomal system, and that αB-crystallin reinstated proteasomal degradation and inhibited the accumulation of GFAP.

In summary, *GFAP* point mutations lead to structural changes, and the mechanism of GFAP aggregation depends on the domain in which the point mutation is located and may explain the differences in clinical features observed in these patients. Because many clinical features of Alexander disease of varying severities have been attributed to each mutation, other factors that modify the clinical features of Alexander disease must be present. The identification of the factors will suggest new considerations for reducing the severity of Alexander disease, and may eventually lead to a therapy for the treatment of patients with Alexander disease. A different approach to Alexander disease therapy should be considered according to changes in the GFAP domains.

References

Alexander WS (1949) Progressive fibrinoid degeneration of fibrillary astrocytes associated with mental retardation in a hydrocephalic infant. Brain 72:373–381

Brenner M, Johnson AB, Boespflug-Tanguy O, Rodriguez D, Goldman JE, Messing A (2001) Mutations in GFAP, encoding glial fibrillary acidic protein, are associated with Alexander disease. Nat Genet 27:277–286

Cary RB, Klymkowsky MW (1995) Disruption of intermediate filament organization leads to structural defects at the intersomite junction in xenopus myotomal muscle. Development 121:1041–1052

Fuchs E (1996) The cytoskeleton and disease: genetic disorders of intermediate filaments. Ann Rev Genet 30:197–231

Gingold MK, Bodensteiner JB, Schochet SS, Jaynes M (1999) Alexander's disease: unique presentation. J Child Neurol 14:325–329

Hall PA, Levison DA, Woods AL, Yu CW, Kellock DB, Watkins JA, Barnes DM, Gillett CE, Camplejohn R, Dover R, Waseem NH, Lane DP (1990) Proliferating cell nuclear antigen (PCNA) immunolocalization in paraffin sections: an index of cell proliferation with evidence of deregulated expression in some neoplasms. J Pathol 162:285–294

Iwaki T, Kume-Iwaki A, Liem RK, Goldman JE (1989) Alpha-B-crystallin is expressed in non-lenticular tissues and accumulates in Alexander's disease brain. Cell 57:71–78

Johnson AB, Bettica A (1989) On-grid immunogold labeling of glial intermediate filaments in epoxy-embedded tissue. Am J Anat 185:335–341

Kinoshita T, Imaizumi T, Miura Y, Fujimoto H, Ayabe M, Shoji H, Okamoto Y, Takashima H, Osame M, Nakagawa M (2003) A case of adult-onset alexander disease with Arg416Trp human glial fibrillary acidic protein gene mutation. Neurosci Lett 350:169–172

Li D, Tapscoft T, Gonzalez O, Burch PE, Quiñones MA, Zoghbi WA, Hill R, Bachinski LL, Mann DL, Roberts R (1999) Desmin mutation responsible for idiopathic dilated cardiomyopathy. Circulation 100:461–464

Li R, Johnson AB, Salomons G, Goldman JE, Naidu S, Quinlan R, Cree B, Ruyle SZ, Banwell B, D'Hooghe M, Siebert JR, Rolf CM, Cox H, Reddy A, Gutiérrez-Solana LG, Collins A, Weller RO, Messing A, van der

Knaap MS, Brenner M (2005) Glial fibrillary acidic protein mutations in infantile, juvenile, and adult forms of Alexander disease. Ann Neurol 57:310–326

Messing A, Head MW, Galles K, Galbreath EJ, Goldman JE, Brenner M (1998) Fatal encephalopathy with astrocyte inclusions in GFAP transgenic mice. Am J Pathol 152:391–398

Mignot C, Delarasse C, Escaich S, Gaspera BD, Noé E, Colucci-GuyoOn E, Babinet C, Pekny M, Vicart P, Boespflug-Tanguy O, Dautigny A, Rodriguez D, Pham-Dinh D (2007) Dynamics of mutated GFAP aggregates revealed by real-time imaging of an astrocyte model of Alexander disease. Exp Cell Res 313:2766–2779

Nielsen AL, Holm IE, Johansen M, Bonven B, Jørgensen AL (2002) A new splice variant of glial fibrillary acidic protein, GFAPε, interacts with the presenilin proteins. J Biol Chem 277:29983–29991

Nobuhara Y, Nakahara K, Higuchi I, Yoshida T, Fushiki S, Osame M, Arimura K, Nakagawa M (2004) Juvenile form of Alexander disease with GFAP mutation and mitochondrial abnormality. Neurology 63:1302–1304

Perng MD, Su M, Wen SF, Li R, Gibbon T, Prescott AR, Brenner M, Quinlan RA (2006) The Alexander disease-causing glial fibrillary acidic protein mutant, R416W, accumulates into Rosenthal fibers by a pathway that involves filament aggregation and the association of αB-crystallin and HSP27. Am J Hum Genet 79:197–213

Perng MD, Wen SF, Gibbon T, Middeldorp J, Sluijs J, Hol EM, Quinlan RA (2008) Glial fibrillary acidic protein filaments can tolerate the incorporation of assembly-compromised GFAP-δ, but with consequences for filament organization and αB-crystallin association. Mol Biol Cell 19:4521–4533

Quinlan R (2001) Cytoskeletal catastrophe causes brain degeneration. Nat Genet 27:10–11

Schuelke M, Smeitink J, Mariman E, Loeffen J, Plecko B, Trijbels F, Stöckler-Ipsiroglu S, van den Heuvel L (1999) Mutant NDUFV1 subunit of mitochondrial complex I causes leukodystrophy and myoclonic epilepsy. Nat Genet 21:260–261

Tang G, Yue Z, Talloczy Z, Hagemann T, Cho W, Messing A, Sulzer DL, Goldman JE (2008) Autophagy induced by Alexander disease-mutant GFAP accumulation is regulated by p38/MAPK and mTOR signaling pathways. Hum Mol Genet 17:1540–1555

Tang G, Perng MD, Wilk S, Quinlan R, Goldman JE (2010) Oligomers of mutant glial fibrillary acidic protein (GFAP) inhibit the proteasome system in Alexander disease astrocytes, and the small heat shock protein αB-crystallin reverses the inhibition. J Biol Chem 285:10527–10537

Tian R, Gregor M, Wiche G, Goldman JE (2006) Plectin regulates the organization of glial fibrillary acidic protein in Alexander disease. Am J Pathol 168:888–897

Tomokane N, Iwaki T, Tateishi J, Goldman JE (1991) Rosenthal fibers share epitopes with alpha-B-crystallin, glial fibrillary acidic protein, and ubiquitin, but not with vimentin: immunoelectron microscopy with colloidal gold. Am J Pathol 138:875–885

Yoshida T, Tomozawa Y, Arisato T, Okamoto Y, Hirano H, Nakagawa M (2007) The functional alteration of mutant GFAP depends on the location of the domain: morphological and functional studies using astrocytoma-derived cells. J Hum Genet 52:362–369

Yoshida T, Sasayama H, Nakagawa M (2009) The process of inducing GFAP aggregates in astrocytoma-derived cells is different between R239C and R416W mutant GFAP. A time-lapse recording study. Neurosci Lett 458:11–14

Lipoma: An Overview

23

María Elena Erro Aguirre and
Elena Hernández M. de Lapiscina

Contents

Abstract

Intracranial lipomas are uncommon benign mesenchymal tumors. They are usually found near the midline and the interhemispheric fissure is the most common location. Different malformations of the central nervous system are associated with lipomas specially the agenesis of the corpus callosum. Although asymptomatic, they can sometimes trigger neurological symptoms, specifically epileptic seizures. The radiological diagnostic clue of lipomas is a well delineated lobulated extra axial fatty mass. On computed tomography lipomas demarcate areas of marked hypodensity and on magnetic resonance they appear as hyperintense lesions on T1-weighted sequences and iso to hypointense on T2-weighted images. The majority of lipomas are incidental findings and do not cause life threatening symptoms. Surgical management proves to be challenging due to the high vascularity and adherence to the lesion to the surrounding parenchyma and should be only pursued in cases of hydrocephalus or in sylvian fissure lipomas when epilepsy can not be controlled with medication.

Introduction

Lipomas of the central nervous system (CNS) are very rare lesions. They were originally considered to be neoplasms of mesodermal origin and today lipomas are known to be a type of benign, slow-growing, congenital hamartomatous

M.E. Erro Aguirre (✉) • E. Hernández M. de Lapiscina
Department of Neurology, Hospital Complex of Navarra, Pamplona, Spain
e-mail: elena.erro.aguirre@cfnavarra.es

M. Hayat (ed.), *Tumors of the Central Nervous System, Volume 13*,
DOI 10.1007/978-94-007-7602-9_23, © Springer Science+Business Media Dordrecht 2014

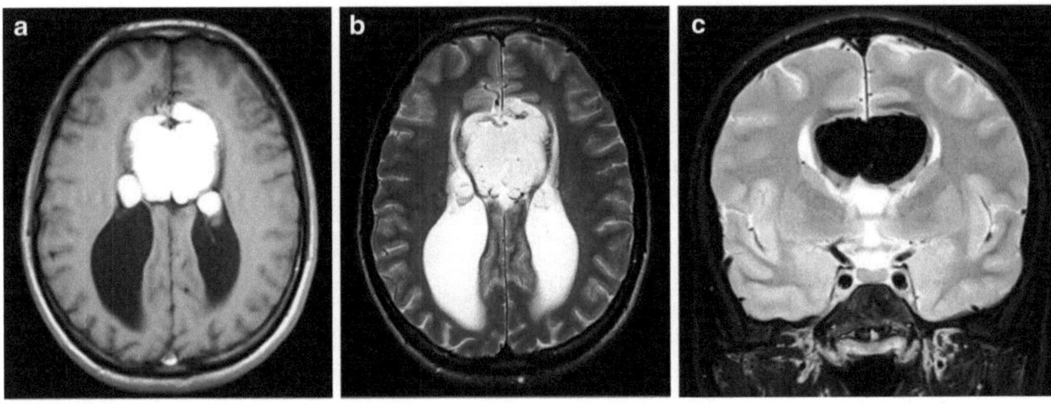

Fig. 23.1 (**a**) Axial T1-weighted cranial magnetic resonance (MR) imaging showing the hyperintense appearance of a giant interhemispheric lipoma; (**b**) Axial MR T2-weighted images showing the same lesion with reduced density; (**c**) Coronal MR fat saturation pulse sequence of the same lipoma, agenesis of the corpus callosum can be appreciated

condition rather than a true neoplasm. Since the original description of these lesions in 1856 by (Rokitansky 1856), ~200 cases have been reported (Loddenkemper et al. 2006). These masses originate from a developmental disorder in mesodermal germ plaque beneath leptomeninxs and neural tissue during the early phase of pregnancy and they are frequently associated with maldevelopment of various nervous system structures.

Lipomas of the CNS can be located inside the cranium (intracranial lipomas) or inside the spinal canal. According to the largest dataset of intracranial lipomas performed by Truwit and Barkowich (1990) they are usually found in the medial line in the interhemispheric fissure (40–50%), the suprasellar region (15–20%), the pineal region (25%) and in other uncommon regions like the lateral cerebral fissure (5%). Interhemispheric fissure lipomas are often placed over the corpus callosum (Figs. 23.1 and 23.2) and may extend into the lateral ventricles or choroid plexus. Intracranial lipomas in the suprasellar region usually attach to the infundibulum or hypothalamus (Fig. 23.3) and those in the pineal region usually attach to the quadrigeminal lamina. Lipomas can also be detected within the subarachnoid cisterns including the ambiens and interpeduncular cisterns, cerebellopontine angle and occasionally in the yugular foramen or foramen magnum. Spinal intradural lipomas are frequently found in the lumbosacral area as components of spinal dysraphism, whereas lipomas of the cervical and thoracic cord are quite rare (Vila Mengual et al. 2009). In particular, cervical location with intracranial extension is extremely rare (Şanh et al. 2010).

Epidemiology

Intracranial lipomas are usually silent and this fact explains the difficulty of a proper statistical assessment, therefore no robust data concerning the prevalence of intracranial lipomas are available. The first descriptions of intracranial lipomas were achieved mainly through incidental findings at autopsy (Jeffers et al. 2009). The expanded use of neuroimaging techniques has allowed increasing the diagnosis of lipomas. Intracranial lipoma accounts for 0.46–1% of all intracranial tumors (Donati et al. 1992). Prevalence of intracranial lipomas detected on cranial magnetic resonance (MR) in the patient population of a hospital was 0.045% (17/38,000) (Kemmling et al. 2008). Spinal intradural lipoma is a rare condition and occurs in approximately 1% of all primary spinal cord tumors (Vila Mengual et al. 2009). Lipomas of the spinal cord are frequently associated with spina bifida and are more commonly located in lumbosacral region. The thoracic spinal cord is the second most common region of involvement followed by the cervicothoracic and cervical regions. Intradural lipomas of the spinal cord with intracranial extension are very rare (Şanh et al. 2010).

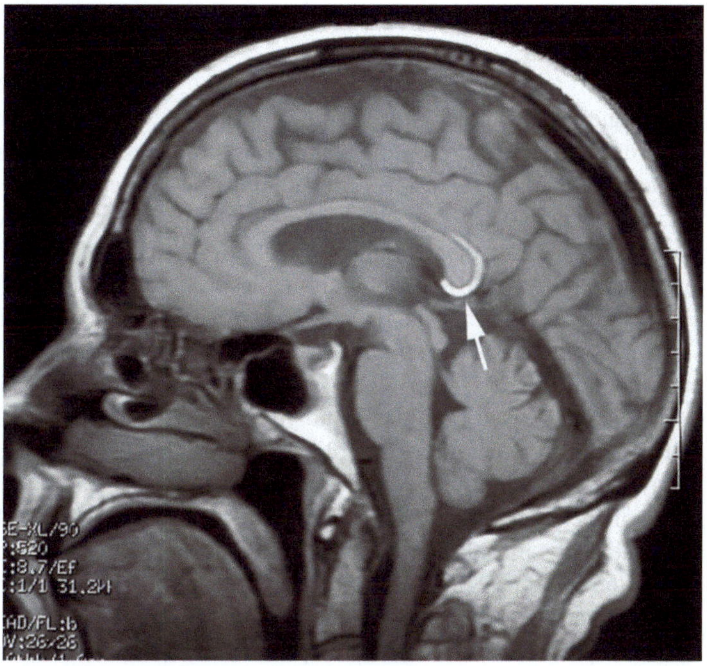

Fig. 23.2 Sagital T1-weighted cranial magnetic resonance (MR) imaging showing the hyperintense appearance of a small curvilinear lipoma of the interhemispheric fissure

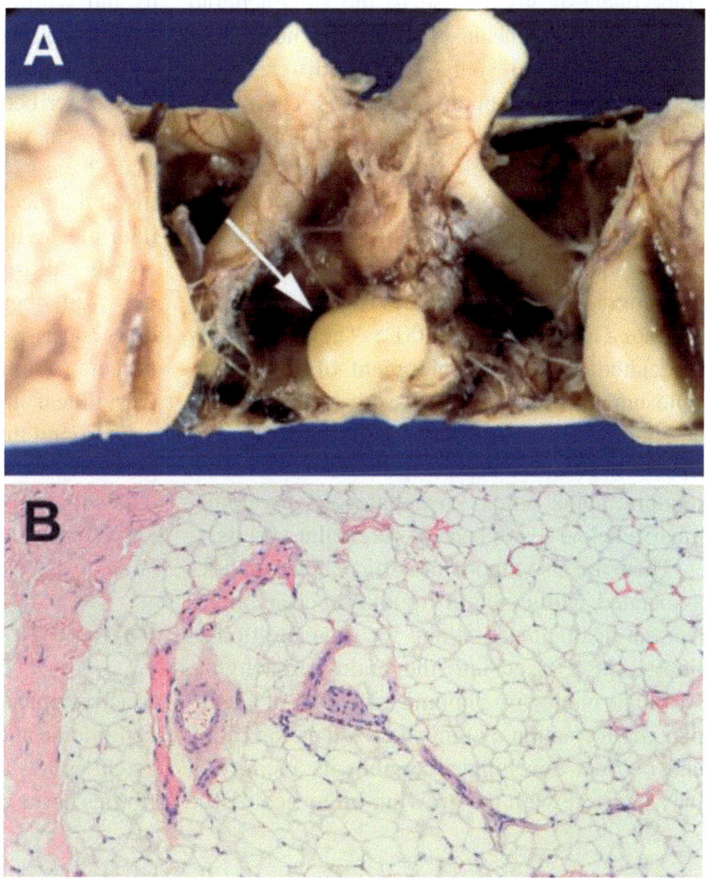

Fig. 23.3 (**a**) Macroscopic appearance of a small lipoma of the suprasellar region attached to the hypothalamus. (**b**) Photomicrograph of a lipoma composed of mature adipose cells

Pathogenesis

Truwit and Barkowich (1990) have reviewed the numerous theories regarding the pathogenesis of intracranial lipomas. The precise etiopathology of intracranial lipomas has been a topic of discussion. Theories regarding the hystogenesis of these lesions include: (1) hypertrophy from the preexisting fatty tissue of the meninges, (2) metaplasia of meningeal connective tissue, (3) heterotopic malformation of dermal origin, (4) tumor like malformation derived from the primitive meninx and (5) fatty degeneration of proliferated glia. Today intracranial lipomas are accepted to be the result of meningeal maldifferentiation and they are genetically linked to other midline defects due to the improper closure of the neural tube. The presence of lipomas is explained by the abnormal persistence of the meninx primitive, a mesenchymal derivative of the neural crest that is usually reabsorbed in an orderly fashion during embryogenesis giving rise to the subarachnoid cisterns. When the meninx primitive does not regress into subarachnoid space and maldifferentiates into adipose tissue a lipoma originates. This theory explains the cisternal locations of lipomas and the intralesional locations of blood vessels and cranial nerves.

The malformative mechanism of lipomas is supported by its association with midline malformations of the cns in up to 40% of cases (Gómez-Gonsálvez et al. 2003). They are mainly associated with maldevelopment of the corpus callosum in the form of agenesis/dysgenesis. Other anomalies of the CNS associated with lipomas are the absence of septum pellucidum, cranium bifidum, spina bifida, encephalocele, myelomeningocele and hypoplasia of cerebellar vermis.

There are also reports of cortical abnormalities associated with cerebral lipomas such as architectural disorganization and focal penetration of fibroadipose tissue into the brain parenchyma. As the formation of a lipoma takes part of a complex malformation that involves sulcus formation and cortical development within its vicinity, it may interfere with the growing of cortical tissue during the ongoing formation of the sylvian fissure, resulting in cortical dysplasia of the vicinity (Kakita et al. 2005).

In rare occasions intracranial lipoma is associated with subcutaneous lipoma. In this case different communication patterns can be observed between intracranial and extracranial component of the lipoma. They may have no connection (Tubbs et al. 2007), may connect to each other by a fibrous lipomatous stalk (Yamashita et al. 2005) or may have direct continuity with each other through cranium bifidum (Sethi et al. 2008).

The development of a lipoma may also involve vascular abnormalities and a variety of vascular abnormalities have been described in association with intracranial lipoma including dilatation, tortuosity or narrowing of feedings arteries and veins, engulfment of the cerebral arteries, arteriovenous malformations and aneurysms and malformations of venous sinus (Saatci et al. 2000). Hypervascularization has been observed adjacent to and within a lipoma located in the Sylvian fissure (Kakita et al. 2005).

Two groups of interhemispheric lipomas have been described (Yildiz et al. 2006). Tubulonodular type is characterized by nodular lesions smaller than 2 cm. They are a result of a more severe insult that occurs at an early embryonic stage and interferes with the normal development of the corpus callosum (Fig. 23.1). It is located anteriorly with the epicentre in the genu in 83% of cases and associated with a high incidence of cranial defects, frontal masses and encephaloceles. Curvilinear lipomas are thin and located posteriorly around the splenium (Fig. 23.2). This type is generally associated with a normal corpus callosum and has a low incidence of associated anomalies.

Pathology

Macroscopically lipomas vary in size from subcentimeter nodules to large masses. They have a bright yellow appearance and a soft, lobulated and fibrous tissue consistency not easily fragmentably.

Microscopic examination after hematoxylin and eosin staining reveals mature adipose tissue surrounded by a fibrous capsule with varied amounts of collagen and blood vessels (Fig. 23.3). The capsule and surrounding parenchyma frequently contain calcifications (Feldman et al. 2001). An exceptional myelomatous change in a lipoma has been described by Suri et al. (2008). These authors consider that metaplastic differentiation in the lipoma gave origin to myelolipoma which contain hematopoietic elements including erytrocytes, myeloid cells, megakaryocytes and focal lymphoid aggregate formation.

Clinical Manifestations

Although the prevalence of symptomatic lipomas remains controversial, epilepsy, headache, psychomotor retardation and cranial nerve paralysis may occur (Venkatesh et al. 2003; Yilmaz et al. 2006). Most intracranial lipomas are considered to be asymptomatic and frequently they are an incidental finding in patients undergoing a cranial computerized tomography (CT) or MR after a cranial trauma (Lin et al. 2009).

Epilepsy is the most common symptom associated with lipomas. A 5% of cranial lipomas present with epilepsy (Gómez-Gonsálvez et al. 2003). Epileptic seizures have been noted in a large proportion of patients with sylvian lipoma presumably due to irritation of the mesiotemporal cortex or to the variety of associated neocortical abnormalities (Saatci et al. 2000; Feldman et al. 2001; Vela-Yebra et al. 2002; Yildiz et al. 2006). However the association of interhemispheric lipomas and epilepsy remains controversial (Martínez-Lapiscina et al. 2010). Some authors have suggested that the symptomatic nature of corpus callosum lipomas is dependent on the interruption of the callosal fibers, which are replaced by the neoplasm. Such disconnection is responsible for each hemisphere to develop epileptic discharges. However, few case reports reported clinical and electroencephalographic characteristics congruent with cranial MR or CT which could allow attributing the etiology of epilepsy to the lipoma. Loddenkemper et al. (2006) reviewed 3,500 epilepsy patients for the presence of intracranial lipomas; only five cases were found and epilepsy could be linked to the lipoma in only a single patient. Therefore, these authors suggested that intracranial lipomas were incidental findings in this population. Yilmaz et al. (2006) reported a prevalence of epilepsy of 20% in an adult case series and Gómez-Gonsálvez et al. (2003) reported a prevalence of 5% in a similar pediatric case series.

Patients with lipomas may present with symptoms that depend on its location. Lipomas located in the quadrigeminal cistern often present with diplopia secondary to brainstem compression and with signs of intracranial pressure due to hydrocephalus secondary to obstruction of the cerebrospinal fluid pathway through the sylvian aqueduct (Truwit and Barkovich 1990). Tubbs et al. (2007) reported an unusual case of a giant lipoma that due to its size compressed the foramina of Monro resulting in hydrocephalus. Cerebellopontine angle lipomas may present with vestibulocochlear dysfunction such as hearing loss, tinnitus, dizziness and vertigo or with hemifacial spasm by compressing the seventh cranial nerve (Romero-Blanco and Monteiro-Santos 2004). Fandiño et al. (2005) reported a case of calcarin fissure lipoma presenting with cuadrantanopsia.

Patients with intradural spinal lipomas present with severe neurological dysfunction as the result of spinal cord compression and produce a progressive paraparesis o tetraparesis depending on the spinal level involved by the lipoma (Şanh et al. 2010).

Radiologic Characteristics

Jabot et al. (2009) have reviewed the neuroimaging appearance of intracranial lipomas. The best diagnostic clue of intracranial lipomas is a well delineated lobulated extra axial fatty mass. On cranial CT lipomas demarcate areas of marked hypodensity, which do not show enhancement after administration of intravenous contrast. They usually have a Houmsfield unit range between −50 UH and −100 UH. Calcification is often present in interhemispheric lipomas most commonly within the fibrous capsule surrounding the lesion. The calcification may be curvilinear, extending around the periphery of the lipoma or it may be nodular within the center of the lesion. Nearly one-half of the suprasellar and interpeduncular lipomas ossify. The magnetic resonance (MR) appearance of the lipoma is that of a homogeneous hyperintense lesion on T1-weighted sequences and iso to hypointense on T2-weighted images. On fat saturation pulse sequences lipomas are isointense to gray matter. Vascular imaging may show arterial abnormalities.

Radiologic differential diagnosis should be established with dermoid tumor and teratomas. Dermoids tumors are radiologically similar as they tend to occur adjacent to midline, but they appear round or lobulated on CT and usually have a slight mass effect and calcification foci with no contrast enhancement or surrounding edema. Dermoids and lipomas display measurably different densities on CT imaging. A dermoid will demonstrate a Houmsfield unit range from −20 to −40 HU. In addition dermoids are more heterogeneous. They have high signal intensity on T1-weighted images due to their lipid content and a heterogeneous signal on T2-weighted images due to the mixed composition of the tumor. Teratomas develop in the same location as lipomas and may have a more heterogeneous appearance with more foci of contrast enhancement. Small intracranial lipomas close to a cerebral artery are hyperintense on time-to-flight MR angiography and could be mistaken with a partially thrombosed aneurysm. A defining characteristic of lipomas on time-to-flight MR angiography results from the out-of-phase India ink artifact. This dark fringe in the periphery of the lesion is characteristic and helps to avoid potential diagnostic pitfalls (Kemmling et al. 2008). In some cases of head trauma injuries, low-density attenuation image of lipoma on brain CT has been misdiagnosed as pneumocranium (Lin et al. 2009).

Treatment and Prognosis

Surgical intervention is generally unnecessary for stable or asymptomatic intracranial lipomas because they grow very slowly, do not involve mass effect on brain tissue and malignant differentiation has never been reported. Surgery should be considered if the lipoma causes compressive effect and in cases of hydrocephalus (Spallone et al. 2004). In cases in which the lipoma causes hydrocephalus, placement of a ventricular shunt will provide adequate treatment. The majority of lipomas do not cause life threatening symptoms and epilepsy can be controlled with medication. In most occasions the risk of surgical intervention outweighs the potential benefit. Radical removal of sylvian fissure lipomas is very difficult if not impossible as well as very dangerous (Feldman et al. 2001). If surgery is necessary, partial resection was recommended by most authors because of the deep location of the lesion, its strong adherence to the sylvian cortex as well as the intricate involvement of the median cerebral artery or its branches. Attempts at radical excision increase the risk of brain injury but some authors reported that some sylvian fissure lipomas could be removed totally without complication and that symptomatic improvement may result (Chao et al. 2008).

Arresting the progression of symptoms is the principal surgical goal for intradural spinal lipomas of any location. The tendency of most neurosurgeons is to be conservative as complete surgical removal is difficult without the risk of great morbidity. Some authors perform a subtotal removal of the tumor with a decompressive laminectomy (Vila Mengual et al. 2009; Şanh et al. 2010).

References

Chao S-C, Shen C-C, Cheng W-Y (2008) Microsurgical removal of sylvian fissure lipoma with pterion keyhole approach-case report and review of the literature. Surg Neurol 70:85–90

Donati F, Vassella F, Kaiser G, Blumberg A (1992) Intracranial lipomas. Neuropediatrics 23:32–38

Fandiño J, Bermúdez J, Arán E (2005) Quadrigeminal cistern and calcarine fissure lipoma: case report and review of the literature. Neurocirugía 16:173–176

Feldman RP, Marcovici A, LaSala PA (2001) Intracranial lipoma of the sylvian fissure. Case report and review of the literature. J Neurosurg 94:515–519

Gómez-Gonsálvez FA, Menor-Serrano F, Téllez de Meneses-Lorenzo M, Aleu Pérez-Gramunt M, Sala-Sánchez AG, Rubio-Soriano A, Carbonell-Nadal J, Mulas F (2003) Intracranial lipomas in pediatrics: a

retrospective study of 20 patients. Rev Neurol 37:515–521
Jabot G, Stoquart-Elsankari S, Saliou G, Toussaint P, Deramond H, Lehmann P (2009) Intracranial lipomas: clinical appearance on neuroimaging and clinical significance. J Neurol 256:851–855
Jeffers SK, Bourne TD, Lopes MBS (2009) A 58 year old woman with a corpus callosum nodule at autopsy. Brain Pathol 19:743–744
Kakita A, Inenaga C, Kameyama S, Masuda H, Ueno T, Honma J, Takahashi H, Shimohata M (2005) Cerebral lipoma and the underlying cortex of the temporal lobe: pathological features associated with the malformation. Acta Neuropathol 109:339–345
Kemmling A, Noelte I, Gerigk L, Singer S, Groden C, Scharf J (2008) A diagnostic pitfall for intracranial aneurysms in time-of-flight MR angiography: small intracranial lipomas. Am J Roentgenol 190:W62–W67
Lin Y-F, Hsi S-C, Chen Y-Q, Long W-Z (2009) Interhemispheric lipoma masquerading as pneumocranium in a patient with head injury. Am J Emerg Med 27:516.e1–516.e3
Loddenkemper T, Morris HH III, Diehl B, Lachhwani DK (2006) Intracranial lipomas and epilepsy. J Neurol 253:590–593
Martínez-Lapiscina EH, Moreno García MP, Bujanda Alegría M (2010) Epileptic seizure of lipoma of corpus callosum: cause or incidental finding. Neurologia 25:331–332
Rokitansky C (1856) Lehrbuch der Pathologischen Anatomie. Braumuller, Vienna, pp 468–478
Romero-Blanco M, Monteiro-Santos E (2004) Cerebellopontine angle lipoma: a case report. Rev Neurol 39:238–240
Saatci I, Aslan C, Renda Y, Besim A (2000) Parietal lipoma associated with cortical dysplasia and abnormal vasculature: case report and review of the literature. Am J Neuroradiol 21:1718–1721
Şanh AM, Türkoğlu E, Kahveci R, Şekerci Z (2010) Intradural lipoma of the cervicothoracic spinal cord with intracranial extension. Child Nerv Syst 26:847–852
Sethi PK, Sethi NK, Torgovnick J, Arsura E (2008) Neuroimage: giant intracranial lipoma with extracranial extension. Eur Neurol 60:49–50
Spallone A, Pitskhelauri DI (2004) Lipomas of the pineal region. Surg Neurol 62:52–59
Suri V, Sharma MC, Suri A, Karak AK, Garg A, Sarkar C, Jain D (2008) Myelolipomatous change in an interhemispheric lipoma associated with corpus callosum agenesis: case report. Neurosurgery 62:E745
Truwit CL, Barkovich AJ (1990) Pathogenesis of intracranial lipomas: an MR study in 42 patients. Am J Roentgenol 155:855–864
Tubbs RS, Louis RG Jr, Loukas M, Shoja MM, Blount JP (2007) Giant intracranial lipoma. Folia Neuropathol 45:247–249
Vela-Yebra R, Pastor-Pons E, Altuzarra-Corral A, García Del Moral Garrido R, Hervás-Natividad R, Sánchez-Álvarez JC (2002) Lipoma of the cerebral convexity and refractory focal epilepsy. Rev Neurol 34:742–745
Venkatesh SK, Phadke RV, Kumar S, Mishra UK (2003) MR appearance of interpeduncular lipoma. Singapore Med J 44:39–41
Vila Mengual M, Miranda Lloret P, López González A, Simal JA, Alvarez Garijo JA (2009) Spinal cord lipoma without dysraphism in the infancy that extends intracranially. Case report and review of the literature. Surg Neurol 71:613–615
Yamashita S, Kunishio K, Tamiya T, Nakamura T, Ogawa D, Igawa HH, Kuroda Y, Nagao S (2005) Parietal lipomeningocele. Case report. Neurol Med Chi (Tokyo) 45:112–115
Yildiz H, Hakyemez B, Koroglu M, Yesildag A, Baykal B (2006) Intracranial lipomas: importance of localization. J Neurol 48:1–7
Yilmaz N, Unal O, Kiymaz N, Yilmaz C, Etlik O (2006) Intracranial lipomas – a clinical study. Clin Neurol Neurosurg 108:363–368

Tumefactive Demyelination

24

Adrian Häne and Ulrich Roelcke

Contents

Abstract

The term tumefactive demyelination (TD) describes intra-axial space occupying lesions of the brain which histopathologically are characterized by perivascular lymphocytic infiltrates, macrophages, reactive astrocytes and myelin loss. Most lesions are located within the brain. As there is no reliable clinical, imaging or laboratory hint to diagnose TD biopsy is mandatory. Whereas several patients initially presenting with TD develop multiple sclerosis or central nervous system lymphoma a subset of TD patients remains stable without further disease activity. TD in these patients may be coined 'idiopathic'. Mechanisms which induce TD are not known. Assuming an underlying auto-immune process treatment includes steroids, plasma exchange and immunosuppressive agents.

Introduction

A precise definition of tumefactive demyelination (TD) does not exist. The term TD describes a brain lesion which is characterized by a mass effect on brain imaging and demyelination on histopathological examination. At the time of first presentation the differential diagnosis includes onset of multiple sclerosis (MS) particularly in younger patients, and primary or secondary brain tumor at any age. Some patients who initially present with a biopsy proven space occupying demyelination develop MS or central

A. Häne (✉) • U. Roelcke
Department of Neurology and Brain Tumor Center, Cantonal Hospital Aarau, Aarau, CH-5001, Switzerland
e-mail: adi.haene@bluewin.ch

M. Hayat (ed.), *Tumors of the Central Nervous System, Volume 13*,
DOI 10.1007/978-94-007-7602-9_24,

nervous system (CNS) lymphoma during the further course. It is still debated whether TD represents an own entity. Assuming an own entity the etiology is still unknown and may be considered 'idiopathic'. Kepes (1993) reported on a patient who received a flu vaccine 10 days before the onset of symptoms which may indicate that the stimulation of the immune system is required to induce TD. Others describe TD as a distinct variant of multiple sclerosis (Poser et al. 1992; Lucchinetti et al. 2008). As TD may show relapses it was also considered to represent recurrent disseminated encephalomyelitis (Brinar 2004). In the following we illustrate the clinical course of two TD patients treated at our institution. In addition the differential diagnosis, treatment options and the clinical course will be discussed.

Clinical Presentation

Patient 1. A 70 year old man presented with subacute confusion, headache and neglect. Magnetic resonance imaging (MRI) of the brain revealed a single right parieto-occipital lesion (Fig. 24.1a, d; Häne et al. 2011). Search for a primary neoplasm including computed tomography (CT) of the chest and abdomen as well as whole body F-18 fluoro-deoxyglucose positron emission tomography (FDG PET) was negative. The cerebrospinal fluid (CSF) showed mild lymphocytic pleocytosis (27 cells/μl) but no tumor cells. Biopsy of the cerebral lesion showed foamy macrophages, perivascular lymphocytic cuffs and loss of myelin with sparing of axons which is indicative of demyelination (Fig. 24.1g–j). There was no evidence of a neoplasm. The patient rapidly improved on dexamethasone starting with 12 mg/day. Eight, 23 and 29 months later the patient developed again single symptomatic lesions, which were located in the left pontine, left frontal and right parietal region (Fig. 24.1b, e, c, f). Again dexamethasone led to rapid clinic improvement which was paralleled by a decrease of lesion volume and contrast enhancement on MRI. Repeated search for a primary tumor was negative. The CSF showed mildly elevated protein (up to 0.71 g/l) and cell count (up to 27 cells/μl), but no oligoclonal bands. Within 2 years from onset of TD the patient developed myelodyplastic syndrome (MDS) and metastatic renal cancer which was the cause of death 34 months after the first manifestation of TD. Brain autopsy yielded demyelination but no tumor.

Patient 2. A 56 year old man presented with headaches, speech difficulties and seizures. MRI of the brain showed a non-enhancing single left parietal lesion (Fig. 24.2a). The lesion was moderately positive on amino acid PET (F-18 ethyl-fluoro-tyrosine) which is compatible with low-grade glioma. Blood cell and CSF analysis were normal. No biopsy was made at this stage and anticonvulsive treatment with phenytoin was started. The follow up MRI 6 weeks later showed spontaneous regression of the lesion. Two months later the patient was again hospitalized with confusion, right-sided hemihypesthesia and seizures. The MRI showed progression of the left parietal lesion as well as several new lesions in the left frontal and right fronto-temporal white matter (Fig. 24.2b, c). Biopsy of the frontal lesion revealed inflammation with perivascular CD20 positive cells which was compatible with demyelination. No tumor cells were present. Treatment with dexamethasone was started (16 mg/day). The patient clinically improved and in a follow-up MRI 3 months later there was substantial regression of all lesions. Twenty-six months after the first presentation the patient suffered from speech difficulties. MRI revealed a new lesion in the right insular region. Dexamethasone therapy

Fig. 24.1 MRI of patient #1 (*upper row* T2-weighted, *lower row* gadolinium-enhanced T1 image): onset with a single right parieto-occipital lesion (**a**, **d**). Subsequent lesions at 23 months (**b**, **e**) and at 29 months (**c**, **f**). (**g–j**) Histology (biopsy): the white matter shows focal hypercellularity and perivascular lymphocytic cuffs (*arrows* in **g**). Abundant macrophages are visible in the hypercellular areas (**h**). There is loss of myelin (**i**) with relative sparing of axons (*arrowheads* in **j**). Stainings: **g**, **h** hematoxylin and eosin, (**i**) luxol fast blue/PAS, (**j**) immunohistochemistry for neurofilament. Magnifications: **g** ×100, **h–j** ×400 (Published in *J Neurol* (Häne et al. 2011), with kind permission)

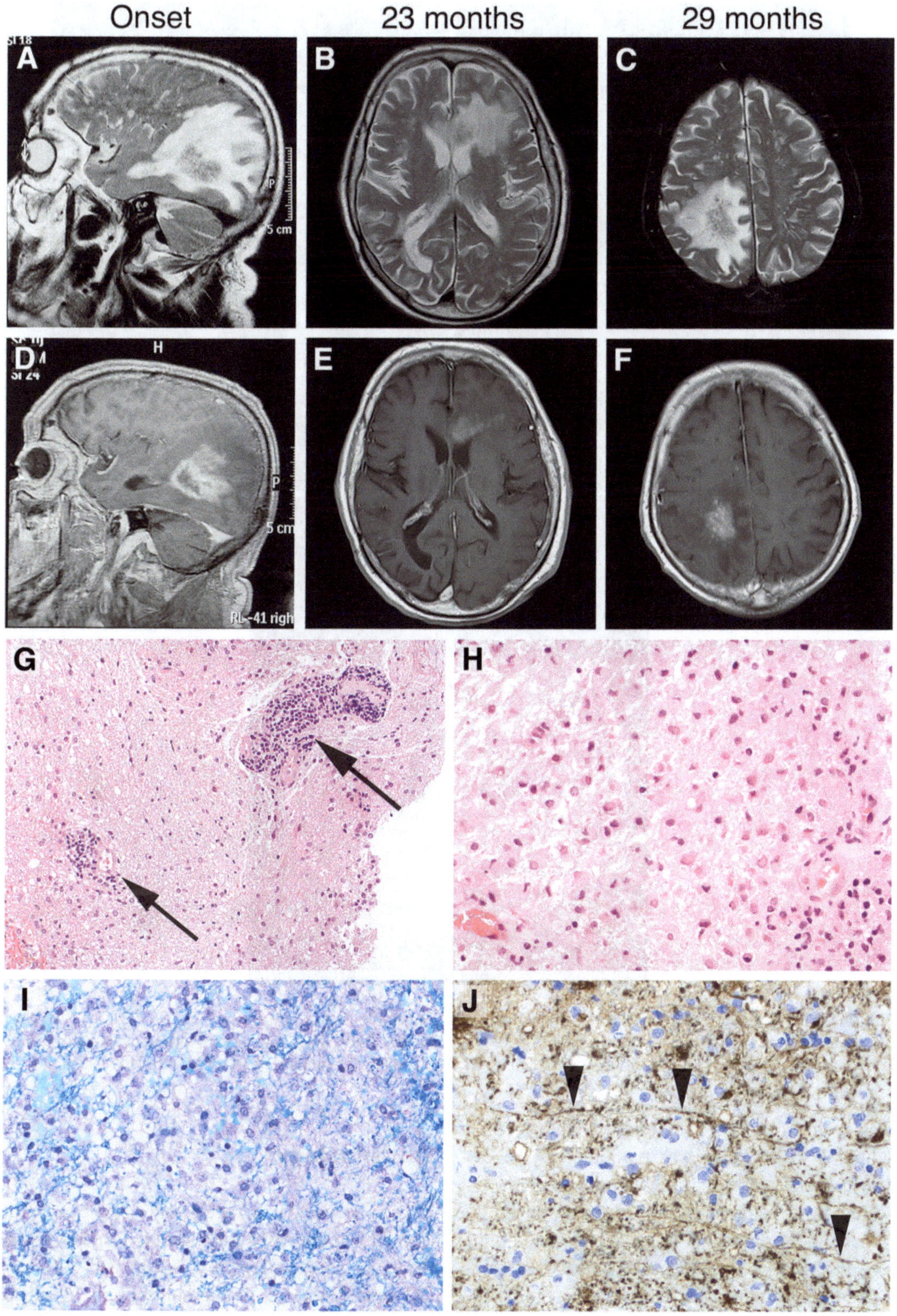
Onset
23 months
29 months
A
B
C
D
E
F
G
H
I
J
5 cm
5 cm
RL -41 righ
H
P
P

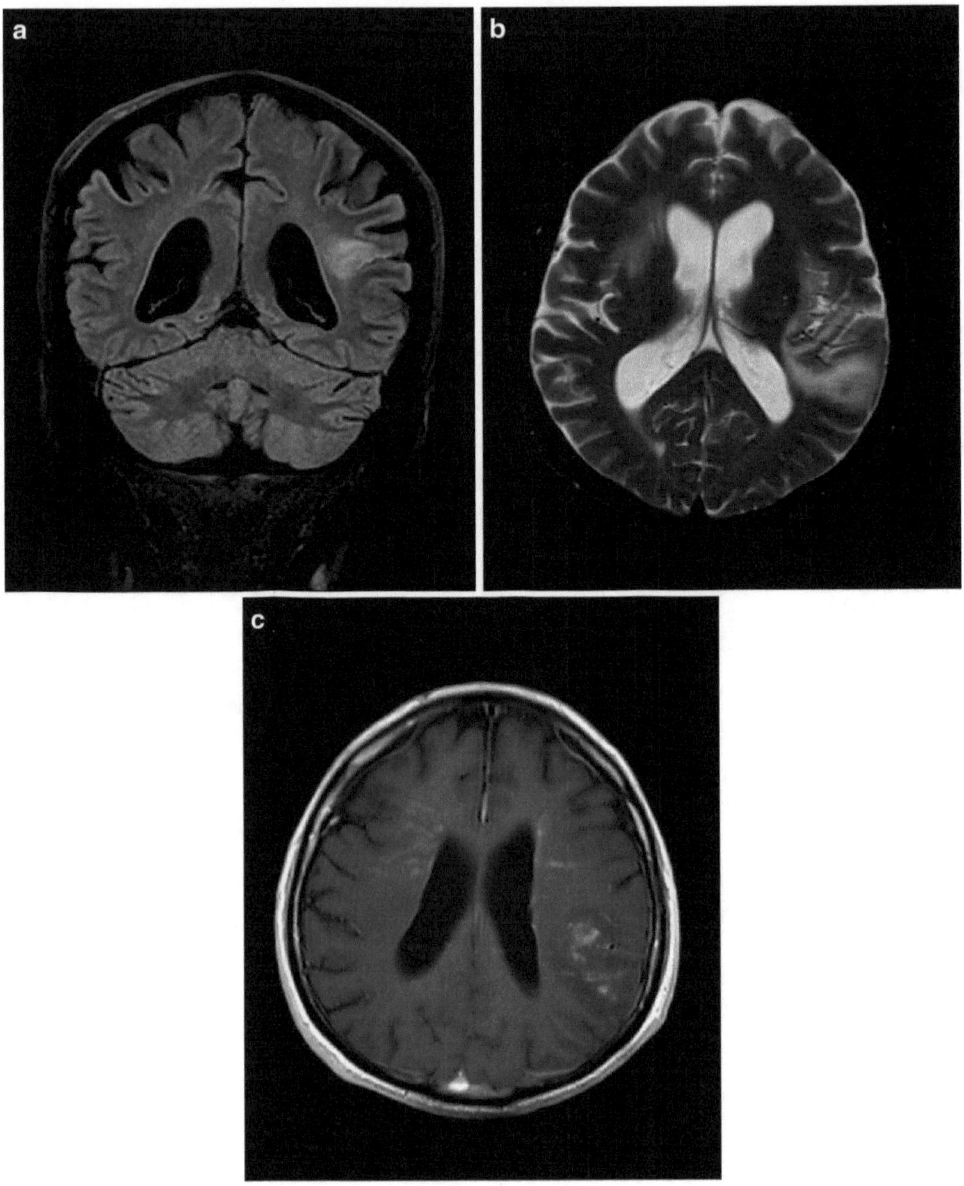

Fig. 24.2 MRI of patient #2: A single left parietal lesion was present at the time of first manifestation of TD (**a**); coronal FLAIR image). At the time of first recurrence bilateral space occupying lesions were observed (**b**), axial T2-weighted image). The lesions now showed mild contrast-enhancement (**c**), axial gadolinium-enhanced T1-weighted image)

was resumed and led to clinical improvement within 3 weeks. This was paralleled by regression of the lesion as shown on MRI. At last follow-up 44 months after disease onset the patient showed residual mild aphasia and cognitive dysfunction. There was no clinical or imaging evidence of tumor in and outside the brain. The clinical course, imaging and biopsy findings in this patient are compatible with TD.

Epidemiology

TD can be found in all age groups. However, more than 50% occur in the third to fifth decade. Overall the incidence of 'idiopathic' TD appears to be low. TD represents the first manifestation of newly diagnosed multiple sclerosis in 1–2 per 1,000 cases (Poser et al. 1992). An accurate number on the incidence of 'idiopathic' TD can not be given also since the term TD is heterogeneously used.

Clinical Course

Most TD patients present with an acute to subacute onset. As illustrated in our cases single or multiple lesions can develop at various sites of the brain over time. The majority of lesions are located within the brain. Rarely do they grow in the spinal cord (Xia et al. 2009). Recurrences are frequently observed. In most patients there is no preceding viral infection or vaccination. The clinical course is variable. In the series presented by Kepes 31 patients were observed up to 12 years (Kepes 1993). The diagnosis of TD was established in these patients by biopsy. Twenty-three of them developed no additional lesions, but showed remissions over many years, and may be considered as 'idiopathic' TD. Lucchinetti et al. (2008) reported on a group of 168 patients with biopsy confirmed lesions classified as 'CNS inflammatory demyelinating disease'. After a median follow-up of 3.9 years 14% remained with an 'isolated syndrome' which however was not further specified, whereas 70% developed definite MS (Lucchinetti et al. 2008). In other cases CNS lymphoma occurred within months (Alderson et al. 1996) to several years (Ng et al. 2007) from the onset of TD.

Diagnosis/Differential Diagnosis

The differentiation of TD from MS, glioma, CNS lymphoma and infection can be challenging and is essential in terms of treatment strategies. In the following we will briefly summarize diagnostic tools applied to evaluate the above mentioned differential diagnoses.

Imaging

Anatomical MRI using T1 and T2 weighted, contrast enhanced and FLAIR sequences is not capable to derive parameters specific for TD. Central necrosis, cystic degeneration and contrast enhancement may occur in TD but represent also typical features of high-grade glioma and lymphoma. Diffusion weighted MRI measures the restriction of water diffusion and allows to differentiate brain abscess, which shows high diffusion restriction in the lesion center, from gliomas. In accordance it may be suitable to differentiate abscess from TD (Abou Zeid et al. 2012). Perfusion CT can estimate the intra-lesional blood volume and may assist to differentiate TD from high grade glioma where an increased tumor blood volume is present due to neo-angiogenesis in most cases. Although mean blood volume values are significantly higher in high-grade gliomas compared with TD (Jain et al. 2010) there is considerable overlap between these groups (TD: 0.71–1.23 ml/100 g, high-grade glioma: 0.71–4.49 ml/100 g). No criteria exist for other imaging modalities. Magnetic resonance spectroscopy can detect patterns of abnormal choline, lactate, and lipid peaks in MS, however, substantial overlap with TD can be expected (Cianfoni et al. 2007; Saini et al. 2011). Diffusion tensor imaging (DTI) which provides information on the integrity of fiber tracts could aid in the differential diagnosis of TD, however, no validated data have been reported yet. Positron emission tomography (PET) and single photon emission computed tomography (SPECT) may reveal alterations of glucose metabolism and amino acid uptake in TD. However, these findings are non-specific and are usually also present in gliomas and lymphomas.

Laboratory Parameters Including CSF

Blood cell analysis does not reveal specific results. Mild pleocytosis and elevated protein can be found in the CSF. The absence of intrathecal oligoclonal bands in TD does not exclude the diagnosis of MS.

Histopathology

Biopsy remains the procedure to establish the diagnosis of TD. Histopathological findings include loss of myelin as detected by luxol fast blue staining and preservation of axons. Perivascular chronic inflammatory infiltrates can be found. They are often surrounded by foamy macrophages and reactive astrocytes (Fig. 24.1g–j, Häne et al. 2011). Circumscribed mitotic figures can lead to an incorrect diagnosis of a glial tumor (Tan et al. 2004).

Treatment and Clinical Follow Up

The use of steroids is the treatment of choice (Xia et al. 2009; Nadkar et al. 2008). Concerning the steroid schedule there are no guidelines. In our experience initial methylprednisolone doses of up to 1 g/day, or dexamethasone up to 16 mg/day, appear appropriate. After 7–10 days dose tapering over weeks has to be adapted to the clinical response. In unresponsive cases plasma exchange can be effective (Mao-Draayer et al. 2002). However, there exist no recommendations on how long to treat with plasma exchange. They used up to seven courses given in 2–3 day intervals. In patients with imminent herniation decompressive surgery is advisable (Nilsson et al. 2009). In order to prevent relapses immunosuppressive therapy may be applied (e.g., using azathioprine, cyclophosphamide, cyclosporine). As there are no data from large TD patients series treatment regimen may be adapted to recommendations used in patients with autoimmune disorders of the nervous system.

The clinical follow up has to be scheduled in order to monitor the residual neurological syndrome and to plan further MRI scans according to the individual clinical course. As stated above the cause of TD is not known. The assumption of an underlying auto-immune dysregulation which leads to demyelination remains speculative. It also remains open whether TD confers a risk for associated disorders. Kepes (1993) reported on TD patients also suffering from chronic myelomonocytic leukemia, seminoma, and immunoblastic sarcoma. One of our patients developed myelodysplastic syndrome and renal cell carcinoma (Häne et al. 2011). Of note, these disorders are possibly related to a disturbance of the immune system (Sloand and Rezvani 2008; Derweesh et al. 2003). Although these observations cannot serve as a proof of a causal relationship between TD and malignancies we monitor TD patients closely with regard to concomitant disorders.

References

Abou Zeid N, Pirko I, Erickson B, Weigand SD, Thomsen KM, Scheithauer B, Parisi JE, Giannini C, Linbo L, Lucchinetti CF (2012) Diffusion-weighted imaging characteristics of biopsy-proven demyelinating brain lesions. Neurology 78:1655–1662

Alderson L, Fetell MR, Sisti M, Hochberg F, Cohen M, Louis DN (1996) Sentinel lesions of primary CNS lymphoma. J Neurol Neurosurg Psychiatry 60:102–105

Brinar VV (2004) Non-MS recurrent demyelinating diseases. Clin Neurol Neurosurg 106:197–210

Cianfoni A, Niku S, Imbesi SG (2007) Metabolite findings in tumefactive demyelinating lesions utilizing short echo time proton magnetic resonance spectroscopy. AJNR Am J Neuroradiol 28:272–277

Derweesh IH, Tannenbaum CS, Rayman PA, Finke JH (2003) Mechanisms of immune dysfunction in renal cell carcinoma. Cancer Treat Res 116:29–51

Häne A, Bargetzi M, Hewer E, Bruehlmeier M, Khamis A, Roelcke U (2011) Recurrent tumefactive demyelination without evidence of multiple sclerosis or brain tumour. J Neurol 258:318–320

Jain R, Ellika S, Lehman NL, Scarpace L, Schultz LR, Rock JP, Rosenblum M, Mikkelsen T (2010) Can permeability measurements add to blood volume measurements in differentiating tumefactive demyelinating lesions from high grade gliomas using perfusion CT? J Neurooncol 97:383–388

Kepes JJ (1993) Large focal tumor-like demyelinating lesions of the brain: intermediate entity between multiple sclerosis and acute disseminated encephalomyelitis? A study of 31 patients. Ann Neurol 33:18–27

Lucchinetti CF, Gavrilova RH, Metz I, Parisi JE, Scheithauer BW, Weigand S, Thomsen K, Mandrekar J, Altintas A, Erickson BJ, König F, Giannini C, Lassmann H, Linbo L, Pittock SJ, Brück W (2008) Clinical and radiographic spectrum of pathologically confirmed tumefactive multiple sclerosis. Brain 131:1759–1775

Mao-Draayer Y, Braff S, Pendlebury W, Panitch H (2002) Treatment of steroid-unresponsive tumefactive demyelinating disease with plasma exchange. Neurology 59:1074–1077

Nadkar MY, Deore RA, Singh R (2008) Tumefactive demyelination. J Assoc Physicians India 56:901–903

Ng S, Butzkueven H, Kalnins R, Rowe C (2007) Prolonged interval between sentinel pseudotumoral demyelination and development of primary CNS lymphoma. J Clin Neurosci 14:1126–1129

Nilsson P, Larsson EM, Kahlon B, Nordström CH, Norrving B (2009) Tumefactive demyelinating disease treated with decompressive craniectomy. Eur J Neurol 16:639–642

Poser S, Luer W, Bruhn H, Frahm J, Bruck Y, Felgenhauer K (1992) Acute demyelinating disease. Classification and non-invasive diagnosis. Acta Neurol Scand 86:579–585

Saini J, Chatterjee S, Thomas B, Kesavadas C (2011) Conventional and advanced magnetic resonance imaging in tumefactive demyelination. Acta Radiol 52:1159–1168

Sloand EM, Rezvani K (2008) The role of the immune system in myelodysplasia: implications for therapy. Semin Hematol 45:39–48

Tan HM, Chan LL, Chuah KL, Goh NS, Tang KK (2004) Monophasic, solitary tumefactive demyelinating lesion: neuroimaging features and neuropathological diagnosis. Br J Radiol 77:153–156

Xia L, Lin S, Wang ZC, Li SW, Xu L, Wu J, Hao SY, Gao CC (2009) Tumefactive demyelinating lesions: nine cases and a review of the literature. Neurosurg Rev 32:171–179

Immunotherapies for Brain Cancer: From Preclinical Models to Human Trials

25

Thomas J. Wilson, Marianela Candolfi, Hikmat Assi, Mariela Moreno Ayala, Yohei Mineharu, Shawn L. Hervey-Jumper, Pedro R. Lowenstein, and Maria G. Castro

Contents

Abstract

Glioblastoma Multiforme (GBM) is the most common and aggressive primary brain tumor. Every year ~22,000 patients are diagnosed with GBM in the US, and less than 5% survive 5 years post-diagnosis. Thus, novel therapeutic approaches are urgently needed to improve the outcome in these patients. Immunotherapy has the potential of stimulating the immune system to eliminate GBM cells that might have spread throughout the brain. Here we will discuss the latest advances in preclinical immunotherapy for glioma, which involve the local delivery of pro-inflammatory cytokines, such as Flt3L, Type I IFNs, IL-2, IL-4, and IL-12 using gene therapy vectors and neural stem cells, or the blockade of immune-suppressive mediators such as TGF-beta, FasL and phosphorylated STAT3. Novel immunotherapeutic approaches have also been assessed in clinical trials implemented in GBM patients. These involve the systemic or local adoptive transfer of autologous immune cells activated *ex vivo* back into the patient, and the administration of dendritic cell vaccines loaded with tumor peptides or cells, which induce active immunity against GBM.

T.J. Wilson, M.D. • S.L. Hervey-Jumper, M.D.
Department of Neurosurgery, School of Medicine, University of Michigan, MSRB II, Room 4570, 1150 West Medical Center Drive, Ann Arbor, MI 48109-5689, USA

M. Candolfi, Ph.D.
Instituto de Investigaciones Biomédicas, INBIOMED, Facultad de Medicina, Universidad de Buenos Aires, Paraguay 2155, piso 10, CP 1421 Buenos Aires, Argentina

H. Assi • Y. Mineharu, M.D., Ph.D.
P.R. Lowenstein, M.D., Ph.D. • M.G. Castro, Ph.D. (✉)
Department of Neurosurgery, Department of Cell and Developmental Biology, School of Medicine, University of Michigan, MSRB II, Room 4570, 1150 West Medical Center Drive, Ann Arbor, MI 48109-5689, USA
e-mail: mariacas@med.umich.edu

M. Moreno Ayala
Instituto de Investigaciones Biomédicas, National Council of Science and Technology – University of Buenos Aires, Paraguay 2155, piso 10, University of Buenos Aires School of Medicine, Buenos Aires, Argentina

M. Hayat (ed.), *Tumors of the Central Nervous System, Volume 13*,
DOI 10.1007/978-94-007-7602-9_25, © Springer Science+Business Media Dordrecht 2014

Preclinical and clinical findings so far indicate that immunotherapy improves anti-tumor immunity in preclinical GBM models and patients, which makes it a valuable adjuvant in the treatment of GBM.

Introduction

In spite of extensive preclinical and clinical research seeking to improve the survival of glioma patients, this disease carries a dismal prognosis and remains a therapeutic challenge to neuro-oncologists and neurosurgeons. The inclusion of the chemotherapeutic agent temozolomide to the standard of care; i.e., surgical resection and radiation therapy has improved the survival of GBM patients, but still only ~5% of the patients remain alive 5 years after initial diagnosis. The low efficacy of traditional therapies has been attributed to the many challenges that this disease poses: the intrinsic resistance of GBM to radio- and chemotherapy, the unfeasibility of complete surgical resection due to the invasive nature of the tumor, the presence of blood brain barrier that limits the access of therapeutic agents into the tumor site, and the putative neurotoxicity of therapeutic agents that limit the doses that can be administered. Thus, novel therapeutic approaches are sorely needed. Immunotherapeutic strategies aim to stimulate the immune system to detect and kill GBM cells that remain after surgical resection. Researchers have aimed to improve the efficacy of immunotherapeutic approaches for the treatment of GBM over the last decade. Here we discuss the latest advances in preclinical immunotherapy for glioma and the outcomes of clinical trials implemented in GBM patients.

Standard Approaches to Malignant Gliomas in Clinical Practice

World Health Organization (WHO) grade III and IV astrocytic tumors (anaplastic astrocytoma and glioblastoma multiforme) as well as mixed oligoastrocytomas (WHO grade III) and anaplastic oligodendrogliomas (WHO grade III) constitute a class of central nervous system tumors commonly referred to as malignant gliomas. The diagnosis of a malignant astrocytoma is made by histopathologic analysis. The WHO utilizes the St. Anne-Mayo system for histologic diagnosis of these tumors and defines an anaplastic astrocytoma (grade III) as having two of the four following features: nuclear atypia, mitoses, endothelial proliferation, or necrosis. Glioblastoma multiforme (grade IV) is defined as having at least three of these features. Because the risk of systemic dissemination is low, staging and grading is done purely based on the pathology within the central nervous system and does not include a systemic survey.

The Central Brain Tumor Registry of the United States (CBTRUS) reports that between 2004 and 2008, 295,986 primary central nervous system tumors were diagnosed and reported. Of these, 16.3% were glioblastoma multiforme, making it the most common primary malignant tumor of the central nervous system. The diagnosis of glioblastoma multiforme is significantly more common in males than in females (approximately 1.5:1) and in whites compared to blacks (approximately 2:1). The age-specific incidence continues to increase throughout life and is highest in the 75–84 year-old age group. Despite the age-specific incidence increasing throughout life, the overall incidence of glioblastoma multiforme is highest in the sixth and seventh decades of life, with the median age of diagnosis being 64 years-old (Fisher et al. 2007). At this point, the only well-established risk factor for the development of a malignant glioma is ionizing radiation exposure (Fisher et al. 2007).

The prognosis for those diagnosed with a malignant glioma remains poor. From 1995 to 2008, the 1-year survival rate for glioblastoma multiforme was 35% while the 5-year survival rate was only 5%. Those diagnosed with an anaplastic astrocytoma fared slightly better with a 1-year survival rate of 61% and a 5-year survival rate of 27%. Currently, the median survival for patients diagnosed with a glioblastoma multiforme is between 12 and 15 months while the median survival for anaplastic astrocytomas is between 36 and 60 months. These patients present in a variety of ways though no presentation is pathognomonic of the disease. Common presenting symptoms include headaches, focal neurologic deficits, altered mental status, personality changes, or seizures. The headaches can be associated with nausea and vomiting and are often secondary to elevated intracranial pressure. When headaches are due to elevated intracranial pressure, they can have a characteristic temporal and positional pattern, with the headache being worse upon awakening in the morning and when in the recumbent position. At the time of diagnosis, several factors have been shown to portend a poorer prognosis including advanced age, lower Karnofsky performance status, unresectability of the tumor, and histologic features consistent with glioblastoma multiforme.

The current standard of care for patients presenting with a suspected malignant glioma is maximum operative resection followed by focal radiation therapy with concurrent and adjuvant temozolomide chemotherapy. High dose corticosteroids (often dexamethasone) can be used to rapidly decrease tumor associated edema and can improve neurologic symptoms but should be used only acutely and rapidly tapered. In addition, antiepileptic therapy should be used when the initial presentation is consistent with a seizure. One additional therapy that has been often utilized is carmustine wafer implantation into the surgical bed though this is not considered standard of care at this point.

The first step in treatment is maximal surgical resection which allows for a tissue diagnosis, cytoreduction of the tumor, and reduction in mass effect. There has been debate about the value of subtotal resection of glioblastomas versus stereotactic biopsy when extensive surgical resection is not possible. The usual limitation to surgical resection is the location of the tumor relative to eloquent areas of the brain. Recently, it was shown that resection of at least 78% of the tumor incurs a survival advantage. This study found that the factors predictive of overall survival included age, Karnofsky performance status, extent of resection, and post-operative tumor volume. The survival advantage incurred was modest but significant. The median survival was 12.5 months for those patients where a 78% tumor reduction was achieved versus 16 months when a 100% gross total resection was achieved (Sanai et al. 2011).

Temozolomide is an oral alkylating agent. Temozolomide has been standard of care since 2005 when Stupp et al. demonstrated a survival advantage utilizing radiotherapy plus concurrent and adjuvant temozolomide versus radiotherapy alone (Stupp et al. 2005). The radiotherapy dosing utilized in this trial is what continues to be the standard of care today: 60 Gy of fractionated focal radiotherapy, usually given in fractions of 1.8–2 Gy. Stupp and colleagues showed that the median survival for patients receiving radiotherapy alone was 12.1 months versus 14.6 months when concurrent and adjuvant temozolomide was given in addition to radiotherapy. Notably, the 2 year survival rate was 26.5% in the temozolomide plus radiotherapy group versus only 10.4% in the radiotherapy alone group (Stupp et al. 2005). Because of this demonstration of survival advantage, the use of concurrent and adjuvant temozolomide has become standard of care.

One molecular marker that has been shown to have prognostic significance is epigenetic silencing of the methyl-guanine methyltransferase (MGMT) gene promoter. The MGMT gene encodes a DNA-repair protein responsible for removing alkyl groups from the O^6 position of guanine. Methylation of the promoter region of this gene results in its epigenetic silencing and a decrease in this DNA repair mechanism. A survival benefit has been particularly noted in patients with a methylated promoter who receive temozolomide (Hegi et al. 2005). In patients with a methylated MGMT promoter who received temozolomide and radiotherapy, the median survival was 21.7 months versus 15.3 months for those patients receiving similar therapy but without a methylated MGMT promoter (Hegi et al. 2005).

Immunotherapeutic Strategies for Gliomas: Preclinical Efficacy

Over the last decade many immunotherapeutic approaches for the treatment of GBM have been attempted using gene therapy vectors. Most of them involve the local delivery of proinflammatory cytokines, which recruit inflammatory cells into the tumor mass (e.g., Flt3L) stimulate the host immune system (e.g., IL-12) or exert additional direct antitumor effects (e.g., IFN-α). The blockade of immunosuppressive molecules (e.g., TGF-β and FasL) has also been evaluated preclinically. Here we discuss the efficacy and safety of these preclinical immune-mediated gene therapy approaches.

Local Overexpression of Cytokines

Flt3L

Fms-like Tyrosine Kinase 3 Ligand (Flt3L) is a growth factor for hematopoietic progenitors that induces the expansion and differentiation of dendritic cells. Intracranial administration of an adenoviral vector that encodes this cytokine (Ad-Flt3L) promotes the expansion of dendritic cells and their recruitment into the brain parenchyma. Intratumor injection of Ad-Flt3L substantially increases the number of dendritic cells and other antigen presenting cells in intracranial GBMs growing in the brain of rats and mice. However, administration of Ad-Flt3L as a single therapeutic agent does not induce tumor regression in preclinical models of GBM (Candolfi et al. 2009; Curtin et al. 2009). Induction of an effective antitumor immune response requires not only an immune-stimulant but also the release of tumor antigens and intracellular inflammatory proteins from dying tumor cells. A growing body of evidence suggests that combination of Ad-Flt3L with cytotoxic agents is required in order to trigger an effective antitumor immune response that leads to tumor regression and long term survival (Candolfi et al. 2009). Dendritic cells recruited within the brain tumor mass in mice and rats treated with Ad-Flt3L seem to require activation by HMGB1—an intracellular TLR2 agonist released by dying tumor cells—in order to initiate the antitumor immune response (Candolfi et al. 2009; Curtin et al. 2009). Several proapoptotic agents induce HMGB1 release from GBM cells, including the conditionally cytotoxic thymidine kinase (TK), which in the presence of the prodrug ganciclovir (GCV) leads to apoptosis of proliferating tumor cells (Candolfi et al. 2009; Curtin et al. 2009); proapoptotic cytokines, such as FasL, TNF-α and TRAIL, which are cytotoxic in cells expressing the corresponding death receptor (Candolfi et al. 2009); radiotherapy and chemotherapeutic agents such as temozolomide (Curtin et al. 2009). Amongst these proapoptotic strategies, Ad-TK+GCV leads to the highest levels of HMGB1 release and combination of Ad-Flt3L with Ad-TK+GCV has proven the most effective in inducing tumor regression in mice and rats bearing established intracranial GBM. In rats bearing intracranial CNS-1 tumor, treatment with Ad-Flt3L+Ad-TK+GCV leads to long-term survival in over 75% of rats, while Ad-Flt3L+Ad-FasL fails to induce tumor regression. Preliminary findings of our lab show that in mice bearing intracranial GL26 or GL261 GBM or metastatic melanomas, combination of Ad-Flt3L+Ad-TK+GCV leads to tumor regression in 40–50% of the animals, while intratumor treatment with Ad-Flt3L in combination with systemic TMZ improves the median survival but fails to promote long term survival. Recent findings from our lab show that *in situ* administration of Ad-TK+Ad-Flt3L is an excellent candidate to be administered as an adjuvant with systemic DC vaccines (Fig. 25.1), which have been reported to induce antitumor immunity in GBM patients (Liau et al. 2005). We found that manipulation of the tumor microenvironment by intratumoral injection of Ad-TK+Ad-Flt3L in combination with peripheral DC vaccination led to antitumor immunity and enhanced the therapeutic efficacy when compared with each treatment alone in rats bearing large CNS-1 tumors (Mineharu et al. 2011).

An important feature of gene therapy vectors to be injected into the brain is to exhibit a safe neuropathological profile. Since gene therapy vectors designed to treat GBM patients are injected into the tumor bed after surgical resection, preclinical assessment of the neuropathology in naïve brain parenchyma is crucial in the evaluation of translational gene therapies for GBM. Intracranial injection of Ad-Flt3L and Ad-TK into the naïve brain of rats and dogs exhibits a very safe neuropathological profile, with mild acute local inflammation and without apparent signs of neurotoxicity (Candolfi et al. 2012). Nevertheless, the use of inducible promoters such as the tetracycline-responsive element (TRE), which depends

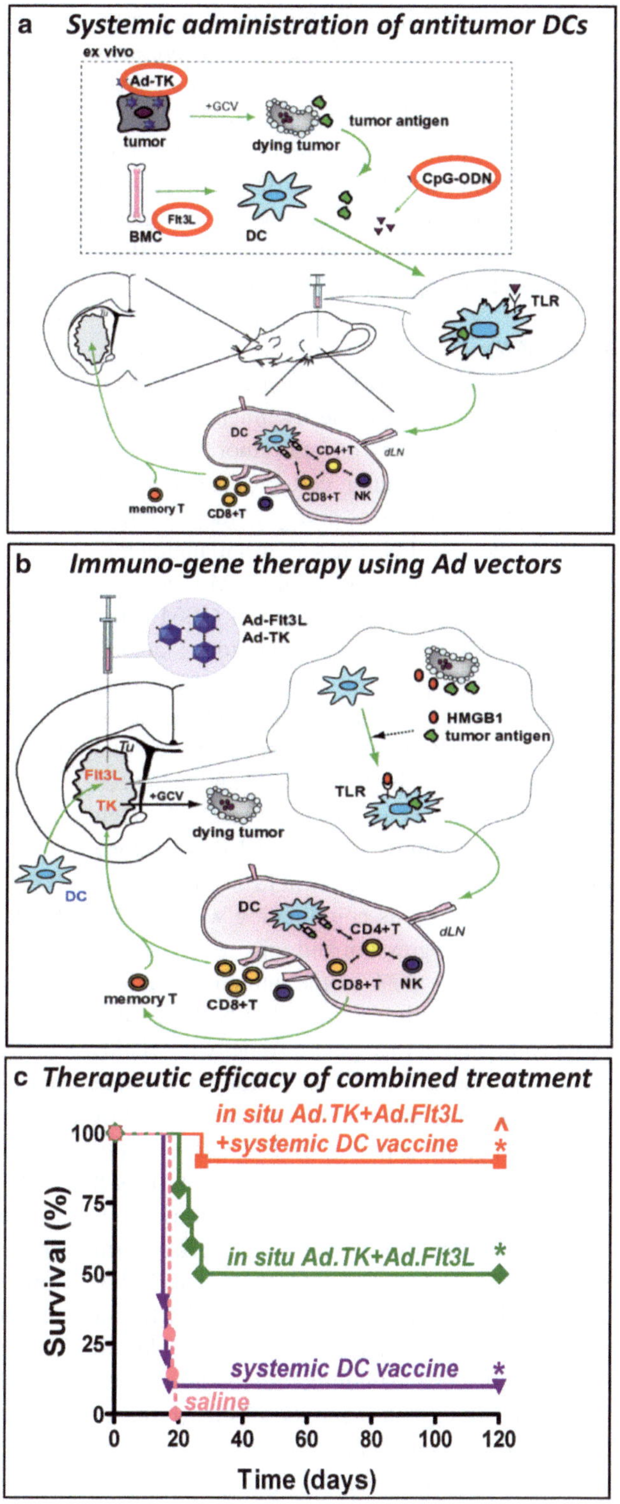

Fig. 25.1 Local administration of Ad-TK+Ad-Flt3L enhances the therapeutic efficacy of systemic antitumor DC vaccines. (**a**) DC vaccines were obtained from bone marrow precursors incubated in the presence of Flt3L, then they were loaded with Ad-TK+GCV-induced apoptotic tumor cells and activated with the TLR9 agonist CpG. DCs were administered subcutaneously in rats bearing large (day 10) intracranial CNS-1 tumors. DC vaccination was repeated 8 days later. Tumor-loaded DCs migrate to the lymph nodes and trigger an antitumor immune response that leads to tumor regression in ~10% of the animals. (**b**) Combined conditional cytotoxic/immune stimulatory gene therapy was performed by intratumoral injection of Ad-TK+Ad-Flt3L, administered in a single dose 10 days after tumor implantation. Ad-TK-induced tumor cell death leads to release of tumor antigens and intracellular inflammatory molecules, such as TLR2 agonist HMGB1, which activate DCs recruited upon Ad-Flt3L expression. Tumor-loaded DCs migrate to the draining lymph nodes and trigger an antitumor immune response that leads to tumor regression in ~50% of the animals (**c**) Kaplan–Meier survival curves of rats treated *in situ* with Ad-TK+Ad-Flt3L in combination with systemic DC vaccines. *, $p<0.05$ vs. saline, ^ $p<0.05$ vs. Ad-TK+Ad-Flt3L alone (Log rank test). Note that *in situ* immunogene therapy in combination with systemic DC vaccination significantly enhances the therapeutic efficacy of either therapy alone, leading to brain tumor regression and long-term survival in about 90% of animals

on the presence of the antibiotic tetracycline or its analogs, allows control of transgene expression by withdrawing the inducer if putative side effects develop (Curtin et al. 2008). Flt3L expression can be tightly regulated using the TRE promoter in murine, canine, and human GBM cells (Curtin et al. 2008). Expression of therapeutic transgenes under the control of inducible promoters further increases the safety of this therapeutic approach.

Type I IFNs

The delivery of IFN-α has been evaluated as a therapeutic method for the treatment of GBM. In view of its effects on antitumor immunity, angiogenesis, and tumor cell proliferation and death, IFN-α emerges as a very attractive target for GBM therapy. IFN-α acts as an antiangiogenic factor in preclinical intracranial GBM models, it directly inhibits the proliferation of GBM cells, and sensitizes them to pro-apoptotic factors. IFN-α is also a powerful stimulant of the host immune system, as it activates dendritic cells, upregulates MHC-I and –II expression, recruits immune cells into the tumor mass and enhances cellular and humoral immune responses. Although intratumoral administration of Ad-IFN-α alone does not have therapeutic benefit in intracranial rat CNS-1 tumors, its combination with Ad-TK leads to tumor regression and long term survival in approximately 50% of the animals (Candolfi et al. 2012). Administration of IFN-β has also been explored as a therapeutic strategy for GBM. Administration of IFN-β gene therapy as a single therapeutic agent did not improve the survival of mice bearing intracranial GL261 GBM (Saito et al. 2004). However, combination of IFN-β gene therapy with vaccination with tumor cell lysate-pulsed dendritic cells resulted in long term survival in 50% of treated mice (Saito et al. 2004). Sequential intratumoral delivery of Ad-IFN-α and bone marrow-derived dendritic cells led to $CD8^+$ T-cells-dependent tumor regression in 50% of GL261 tumor-bearing mice (Tsugawa et al. 2004). DCs injected intratumorally migrated to the cervical lymph nodes, where specific cytotoxic T lymphocyte (CTL) activity was detected, suggesting that this combination therapy exerts a specific antitumor immune response.

Although IFN-α was the first cytokine to be approved for cancer treatment, its considerable toxicity has limited its use. Administration of Ad-IFN-α into the naïve rat brain leads to overt local inflammation with loss of brain tissue, necrosis, and hemorrhages, as well as systemic side effects (Candolfi et al. 2012), making this vector unsuitable for intracerebral administration. However, the use of inducible or cell type-specific promoters may improve the safety profile of this approach, as they allow temporal and topographic control of transgene expression. A dual glial-specific and hypoxia-induced promoter has been recently described (Kim et al. 2011). In this dual specific vector transgene, expression is controlled by the nestin intron 2-SV40 promoter and erythropoietin enhancer for glial cell and hypoxia specific gene expression (Kim et al. 2011). Radio-inducible promoters are also useful tools to limit the toxicity of this approach. The radio-responsive early growth response gene 1 promoter has been used to drive the expression of cytotoxic molecules, such as TRAIL allowing for spatially and temporally controlled transgene expression (Tsurushima et al. 2007).

IL-2

IL-2 is a Th1 cytokine released from helper $CD4^+$ T-cells that enhances CTL-mediated antitumor immunity, which makes it an attractive target for immunotherapy of gliomas. However, its systemic administration is limited by the toxicity of this molecule, and local delivery of IL-2 requires multiple or continued intracranial administration. Gene therapy vectors are useful tools for the long term delivery of this cytokine in the brain. Delivery of IL-2 using gene therapy vectors has been combined with proapoptotic strategies in order to promote anti-glioma immunity. Plasmids and vaccinia viruses encoding IL-2 and the proapoptotic molecules p53 and/or Bax were administered intratumorally in C6 rat gliomas in nude mice (Haghighat et al. 2000). Although these treatments showed antitumor activity without apparent side effects, the animal model used to evaluate this therapy has the vital limitation of lacking an intact immune system.

An adenoviral vector encoding IL-2 for the treatment of the experimental RG2 rat GBM model in combination with Ad-TK and Ad-Flt3L has also been used (Mineharu et al. 2012). Rats bearing intracranial RG2 tumors constitute a challenging GBM model, since these tumors have been shown to be refractory to most therapeutic approaches, including radiation, chemotherapy, and immunotherapy (Mineharu et al. 2012). Overexpression of IL-2 within the RG2 tumor microenvironment stimulates the recruitment of effector T-cells and reduces the intratumor infiltration of Tregs, enhances CTL-mediated antitumor immune responses, and successfully extends the median survival of RG2 tumor-bearing rats when combined with Ad-Flt3L+Ad-TK.

Other Cytokines

Another strategy for the delivery of therapeutic transgenes into brain tumors is the use of neural stem cells (NSC), which have been reported to display similar migratory activity to that of GBM cells within the brain parenchyma (Aboody et al. 2000). NSCs have been used to deliver pro-inflammatory cytokines such as IL-4 and IL-12 into intracranial gliomas *in vivo*, leading to tumor regression in mouse and rat models of GBM (Vetter et al. 2009). NSCs exhibit tropism for disseminating glioma cells, and IL-12-secreting NSC therapy has been linked to enhanced T-cell infiltration in tumor microsatellites, promoting antitumor immunity (Vetter et al. 2009). Tumor rejection was shown to be dependent on $CD8^+$ T-cells (Vetter et al. 2009). Although neurotoxicity remains to be evaluated, local grafting of NSC emerges as a useful tool to deliver therapeutic cytokines for the treatment of GBM.

Gene therapy-mediated overexpression of proinflammatory cytokines can also be exploited in order to enhance the efficacy of antiglioma vaccines. 9L rat GBM tumor cell vaccines engineered to produce IL-4 and TK were employed to treat intracranial GBM (Okada et al. 1999). Subcutaneous immunization of rats with nonirradiated 9L-IL-4 cells or 9L-IL-4-TK cells followed by treatment with GCV protected rats from subsequent intracranial implantation of wild-type 9L tumors (Okada et al. 1999). Treatment of small but established (day 3) intracranial 9L tumors induced cellular antitumor immunity and promoted long-term survival (>100 days) in 25–45% of the rats (Okada et al. 1999), indicating that gene therapy-mediated engineering of antitumor vaccines may enhance antitumor immunity.

Blockade of Local Immunosuppressive Targets

TGF-β

One of the major immunosuppressive molecules produced by GBM cells is Transforming Growth Factor-β (TGF-β), which inhibits T and B cell proliferation and activation, reduces the secretion of pro-inflammatory molecules, and downregulates the expression of MHCII in GBM cells. In order to overcome the immune evasion of GBM, blockade of TGF-β activity has been attempted using several gene therapy approaches. Systemic delivery of TGF-β antisense oligonucleotides has been performed using polybutyl cyanoacrylate nanoparticles in rats bearing intracranial F98 GBM, in order to improve the antitumor immune response induced by peripheral immunization with tumor cells infected with Newcastle Disease Virus (Schneider et al. 2008). Although this treatment did not lead to long term survival, it improved median survival of tumor-bearing rats and led to increased levels of activated T-cells, suggesting that blockade of TGF-β may have therapeutic value as an adjuvant to immune-stimulant strategies.

Another strategy to block TGF-β activity is to overexpress soluble TGF-β receptors, which compete for the binding of the ligand to its receptor, abrogating TGF-β effects. Delivery of soluble TGF-β receptors using adenoviral vectors delays the growth of intracerebral human GBM xenografts in nude mice *in vivo* (Naumann et al. 2008). Although in this tumor model the immune effects of this strategy cannot be assessed, the direct antitumor effects of TGF-β blockade warrant further experimentation using this vector in relevant syngeneic murine models of GBM.

Blocking TGF-β activity by overexpressing decorin, a small proteoglycan that binds and inactivates TGF-β, using adenoviral vectors in the CNS-1 rat GBM model has been attempted (Biglari et al. 2004). Decorin expression was controlled by the ubiquitous hCMV promoter and the glial cell-specific GFAP promoter. Decorin overexpression slows glioma progression *in vivo* (Biglari et al. 2004). Decorin overexpression also reduces TGF-β bioactivity and expression in human cells *in vitro* and increases the local infiltration of activated T-cells *in vivo* in preclinical GBM models (Stander et al. 1998). Taken together, these findings indicate that gene therapy-mediated delivery of decorin may prove a useful adjuvant for the treatment of GBM.

FasL

Another mechanism by which GBM cells are thought to evade the immune system is the expression of FasL. Expression of FasL has been detected in human and experimental GBM and has been shown to support the growth of experimental intracranial gliomas. FasL expression has been detected in specimens of human GBM and negatively correlates with the degree of intratumoral $CD4^+$ and $CD8^+$ T-cell infiltration. FasL expression in human GBM is detected in tumor cells and endothelial cells of the tumor blood vessels, which has been proposed as a mechanism to deplete tumor infiltrating Fas^+ T-cells. Although an Ad vector encoding FasL (Ad-FasL) exerts cytotoxic effects in CNS-1 cells *in vitro* and *in vivo*, its administration in combination with the immune-stimulant Ad-Flt3L fails to induce tumor regression in rats bearing intracranial GBM (Candolfi et al. 2009).

Knock down of FasL expression using shRNA in rat GBM cells reduced their ability to induce T-cell apoptosis and increased tumor infiltration of T-cells, leading to an inhibition of intracranial tumor growth (Jansen et al. 2010). Mice bearing intracranial GL26 tumors have been shown to double the expression of FasL when compared to tumors growing in the flank (Badie et al. 2001). These findings were attributed to the expression of FasL in microglia, which is absent in subcutaneous tumors (Badie et al. 2001). Blockade of FasL activity in GL26 GBM growing in the brain altered the local immunosuppressive milieu, increasing the infiltration of leukocytes (Badie et al. 2001). These findings indicate that blockade of FasL expression may improve the response of GBM to immunotherapeutic approaches.

STAT3 Signaling Involvement in Glioma Immunotherapies

The STAT3 transcription factor has proven to be a fundamental component of tumor growth and progression. Multiple growth factors and cytokines frequently overexpressed in cancer such as EGF, FGF and IL-6 activate Janus kinase 2 (JAK2). When activated, JAK2 phosphorylates STAT3 at its Tyr705 residue. Phosphorylated STAT3 translocates to the nucleus where it activates the transcription of genes which promote the evasion of apoptosis, cell division, angiogenesis, invasion, and metastases. Histopathological analysis of brain tumors found STAT3 to be overexpressed in 53% patients with grade III anaplastic astrocytomas, grade IV GBMs, or gliosarcoma but not in patients with low grade tumors. STAT3 levels correlated with the infiltration of T-cells and median survival time of tumor bearing patients (Abou-Ghazal et al. 2008). Expression of the DNA repair enzyme MGMT was directly influenced by the presence of STAT3. Inactivation of STAT3 by small hairpin RNA leads to a marked sensitivity to the alkylating agent temozolomide (Kohsaka et al. 2012). Preclinical studies using dominant negative vectors, small molecule inhibitors, peptidomimetics, and decoy antisense oligonucleotides have convincingly demonstrated that aberrant STAT3 signaling is crucial for the survival and growth of various brain tumor types (Doucette et al. 2012).

STAT3 has also been shown to be an important mediator of immune suppression. Tumors frequently secrete factors that inhibit the maturation and activation of multiple immune lineages. The production of pro-inflammatory cytokines such as TNF-α, IL-6, RANTES, IFN-B, and IP-10 from multiple tumor lines has been shown to be controlled by STAT3 (Wang et al. 2004).

The presence of conditioned media from v-src transformed 3T3 cells in cultured dendritic cells promotes a tolerogenic profile (low MHC-II, CD80, CD86 expression, low IL-12 release) which can be reversed by disruption of STAT3 (Wang et al. 2004). STAT3 is also believed to mediate the inflammatory response in humans diagnosed with GBM. Peripheral and tumor infiltrating macrophages isolated from resected tumors produce elevated levels of pro-inflammatory cytokines when treated with a small molecule inhibitor of STAT3 lending support to the idea that STAT3 signaling is responsible for the anti-inflammatory process and associated immune-suppression (Hussain et al. 2007).

Owing to the multiple roles of STAT3 in the survival and expansion of tumors, various groups have tried to use STAT3 inhibitors to treat tumors. Kortylewski et al. (2005) have shown that deletion of STAT3 in hematopoietic cells can strengthen the immune response of several different hematopoietic lineages and restrict the growth of B16 melanoma tumors. Tumor infiltrating T-cells isolated from B16 tumor-bearing STAT3 null mice produced an abundance of IFNγ relative to wild type mice. In addition, NK cells had increased cytolytic activity. In a different tumor model, improved inhibition of tumor growth was observed after vaccinating with STAT3 deficient DCs (Iwata-Kajihara et al. 2011). In addition to an increase of IL-12, IFNγ ELISPOT assays indicated a stronger immune response in STAT3 null mice. Adoptive T-cell transfer is another way of inducing an immune response against experimental tumors. Ablating STAT3 in T-cells prior to their transfer resulted in robust antigen-specific tumor T-cell activity and growth inhibition in B16 tumors (Kujawski et al. 2010). The ability of STAT3 to conduct a diverse set of immune suppressive instructions makes it an attractive potential target as adjuvant therapy with tumor immunotherapy. Although compounds such as AZD1480 and WP1066 hold promise as potential inhibitors, they await clinical confirmation. With the use of structure-activity relationships that facilitate the screening of large libraries of compounds, the stage is set for the development of safe and efficacious STAT3 inhibitors.

Glioma Immunotherapies: Clinical Experience

Over the past century there has been speculation that the immune system plays a role in patients with glioblastoma. Anecdotal reports followed by retrospective analyses have suggested that postoperative infections lead to improved survival in glioblastoma patients (De Bonis et al. 2011). De Bonis et al. (2011) published a retrospective series of 197 patients who underwent a craniotomy for glioblastoma. Postoperative infections were noted in 5% and this set of patients was noted to have a 15-month survival benefit when compared with uninfected craniotomies for glioblastoma (De Bonis et al. 2011). It is now known that patients with glioblastoma are relatively immunosuppressed and the degree of immunosuppression inversely correlates with survival. Given these observations, strategies targeting both adoptive and active immunotherapy have led to novel therapeutics for patients with glioblastoma.

Cell Based Therapies for Malignant Gliomas

Cell based immunotherapy (also known as adoptive immunity) involves the administration of immune cells activated *ex vivo* into the patient either intravenously or via implantation within the tumor cavity. Lymphocyte activated killer cells were the first cells used in clinical trials in glioblastoma patients. These cells are collected from peripheral blood lymphocytes cultured with IL-2. This process generates cytolytic natural killer and T-cells which are capable of mounting a generalized attack, though not specifically against glioblastoma cells. Phase 1 clinical trials using this modality have been largely ineffective. Dillman et al. (2009), however, published their series of 36 patients who received autologous lymphokine activated killer cells (Dillman et al. 2009). They revealed an overall median survival of 20 months with 75% survival at 1 year. Another approach has been the use of allogenic cytotoxic T lymphocytes stimulated by autologous tumor cells as an antigen source (Tsuboi et al. 2003). One additional approach used in a clinical trial was the collection of lymphocytes from peripheral blood or lymph nodes after peripheral injection of radiation inactivated autologous tumor cells and GM-CSF. These cells are then activated *in vitro* and re-implanted into the patient (Plautz et al. 1998).

Vaccine Trials for Malignant Gliomas

Vaccine based modalities employing active immunity against glioblastoma have been used in clinical practice using either peptide or cell based therapies. Vaccination with dendritic cells is the most widely studied treatment option. To date, 20 phase I and II clinical trials involving DC vaccines for adult glioblastoma patients have been published. In the majority of clinical studies, dendritic cells are prepared using peripheral monocytes cultured with either IL4 or GM-CSF. Various studies, however, have used a blend of other agents including toll like receptor agonists, IFNγ, IL1β, and TNFα to "mature" dendritic cells. Though the antigen source varies between trials, it typically involves autologous tumor lysate, specific tumor peptides, mRNA isolated from tumor lysate, or peptides eluted from autologous tumor lysate. Vaccines are typically administered subcutaneously, intradermally, or intracranially. DC vaccines appear to be well tolerated, with the majority of toxicity being grade II effects (flu like illness, headaches, etc.). Currently there has been only one episode of peritumoral edema and stupor in the published literature (grade IV neurotoxicity) (Yamanaka et al. 2005).

DC immunotherapy appears to have a beneficial effect in some patients with glioblastoma. Of the 20 published reports, 13 have demonstrated a survival benefit compared with either a control cohort or historical controls. Yamanaka et al. (2005) published their phase I/II results of 24 patients with new and recurrent glioblastoma who received autologous tumor cells with DC plus keyhole limpet hemocyanin (Yamanaka et al. 2005). These cells were administered either intradermally alone or intradermally combined with intracranial injection. This study revealed an improved survival in patients who received both peripheral and intracranial injections, as well as patients vaccinated with mature DC. Liau et al. (2005) showed that an immunologic response in the patient (T-cell infiltration into tissues) correlated with decreased TGF-β and improved mean overall survival (23 months for vaccine cohort versus 18.3 for non-randomized controls) (Liau et al. 2005). Further studies by Wheeler et al. distinguished vaccination "responders" from "non-responders", noting a 92 week survival and greater IFNγ production compared with those who mounted an immune response after treatment (92 weeks survival for responders compared with 61 weeks survival for non-responders) (Wheeler et al. 2008).

Several tumor-associated peptides have been used in clinical application for patients with malignant gliomas. Epidermal growth factor receptor plays an important role in glioblastoma proliferation and survival. An in-frame deletion of exons 2–7 in the mRNA results in a truncated mutant, epidermal growth factor receptor variant III (EGFRviii). EGFRviii lacks a portion of the extracellular binding domain but retains the ability to constitutively activate intracellular tyrosine kinase. EGFRviii is a glioblastoma specific antigen expressed in 40% of patients and is not expressed in non-neoplastic human tissues. Sampson et al. (2009, 2010) generated a vaccine against EGFRviii specific antibodies using PEP-3 coupled with keyhole limpet hemocyanin (EGFRViii, PEPviii-KLH). They published results after a phase II clinical trial including 18 patients with EGFRviii expressing malignant gliomas. Vaccinated patients had greater overall and progression free survival rates when compared with matched control subjects (progression free survival 20% for controls and 40% for vaccinated patients at 6 months) (Sampson et al. 2010). The authors also noted that 82% of vaccinated patients lost EGFRviii expression at recurrence.

The past decade of research has shown that immune therapy for malignant glioma triggers an immune response. Many questions remain, particularly whether these therapies truly improve patient survival. Variability in approach and treatment protocol, limited numbers of patients, and absence of large clinical trials have posed barriers to answering these important questions. As our knowledge of immunotherapy expands, we hope to better understand the effects of malignant glioma on the immune system and develop novel strategies to improve therapeutic outcomes.

Discussion

Glioblastoma multiforme (GBM) remains a huge therapeutic challenge, in spite of advances in standard of care, i.e., neurosurgery, radiotherapy and chemotherapy. Harnessing the immune system of the host to specifically recognize and destroy tumor cells within the brain, without eliciting adverse effects to neighboring normal brain structures is an area of active research and development. The hallmark characteristics of GBM are its invasive nature and the fact that the tumor always recurs. Immune therapeutic approaches have the potential of finding and destroying GBM cells infiltrating the brain. Further, immunotherapy generates tumor specific memory T-cells, which have the capability of eradicating tumor recurrences. Thus, immunotherapies have tremendous potential for the effective treatment of GBM; they could be used in conjunction with current standard of care. Although several preclinical strategies have proven highly efficacious and non-toxic in preclinical animal models of GBM, these successes have not yet been translated to human clinical trials. It is expected that the data reported in this chapter will stimulate further research and the implementation of novel clinical trials based on this exciting modality.

Acknowledgements This work was supported by National Institutes of Health/National Institute of Neurological Disorders & Stroke (NIH/NINDS) Grants 1UO1-NS052465, UO1-NS052465-04S1, 1RO1-NS074387 and 1RO1-NS 057711 to M.G.C.; NIH/NINDS Grants 1R21-NS 084275 and 1RO1- NS 061107 to P.R.L. T32 NS 007222, T32 CA 009676, T32 GM 007863, UL1TR000433; and the Department of Neurosurgery, University of Michigan School of Medicine. M.C. and M.M.A. were supported by the National Council of Science and Technology (CONICET, Argentina) PIP CONICET 114-201101-00353. We are grateful to Dr. K. Murasko for her academic leadership and M. Dahlgren, D. Tomford and S. Napolitan for superb administrative support.

References

Aboody KS, Brown A, Rainov NG, Bower KA, Liu S, Yang W, Small JE, Herrlinger U, Ourednik V, Black PM, Breakefield XO, Snyder EY (2000) Neural stem cells display extensive tropism for pathology in adult brain: evidence from intracranial gliomas. Proc Natl Acad Sci U S A 97:12846–12851

Abou-Ghazal M, Yang DS, Qiao W, Reina-Ortiz C, Wei J, Kong LY, Fuller GN, Hiraoka N, Priebe W, Sawaya R, Heimberger AB (2008) The incidence, correlation with tumor-infiltrating inflammation, and prognosis of phosphorylated STAT3 expression in human gliomas. Clin Cancer Res 14:8228–8235

Badie B, Schartner J, Prabakaran S, Paul J, Vorpahl J (2001) Expression of Fas ligand by microglia: possible role in glioma immune evasion. J Neuroimmunol 120:19–24

Biglari A, Bataille D, Naumann U, Weller M, Zirger J, Castro MG, Lowenstein PR (2004) Effects of ectopic decorin in modulating intracranial glioma progression in vivo, in a rat syngeneic model. Cancer Gene Ther 11:721–732

Candolfi M, Yagiz K, Foulad D, Alzadeh GE, Tesarfreund M, Muhammad AK, Puntel M, Kroeger KM, Liu C, Lee S, Curtin JF, King GD, Lerner J, Sato K, Mineharu Y, Xiong W, Lowenstein PR, Castro MG (2009) Release of HMGB1 in response to proapoptotic glioma killing strategies: efficacy and neurotoxicity. Clin Cancer Res 15:4401–4414

Candolfi M, King G, Yagiz K, Curtin J, Mineharu Y, Muhammad AG, Foulad D, Kroeger KM, Barnett N, Josien R, Lowenstein PR, Castro MG (2012) Plasmacytoid dendritic cells in the tumor microenvironment: immune targets for glioma therapeutics. Neoplasia 14(8):757–770

Curtin JF, Candolfi M, Xiong W, Lowenstein PR, Castro MG (2008) Turning the gene tap off; implications of regulating gene expression for cancer therapeutics. Mol Cancer Ther 7:439–448

Curtin JF, Liu N, Candolfi M, Xiong W, Assi H, Yagiz K, Edwards MR, Michelsen KS, Kroeger KM, Liu C, Muhammad AK, Clark MC, Arditi M, Comin-Anduix B, Ribas A, Lowenstein PR, Castro MG (2009) HMGB1 mediates endogenous TLR2 activation and brain tumor regression. PLoS Med 6:e10

De Bonis P, Albanese A, Lofrese G, de Waure C, Mangiola A, Pettorini BL, Pompucci A, Balducci M, Fiorentino A, Lauriola L, Anile C, Maira G (2011) Postoperative infection may influence survival in patients with glioblastoma: simply a myth? Neurosurgery 69:864–868; discussion 868–869

Dillman RO, Duma CM, Ellis RA, Cornforth AN, Schiltz PM, Sharp SL, DePriest MC (2009) Intralesional lymphokine-activated killer cells as adjuvant therapy for primary glioblastoma. J Immunother 32:914–919

Doucette TA, Kong LY, Yang Y, Ferguson SD, Yang J, Wei J, Qiao W, Fuller GN, Bhat KP, Aldape K, Priebe W, Bogler O, Heimberger AB, Rao G (2012) Signal transducer and activator of transcription 3 promotes angiogenesis and drives malignant progression in glioma. Neuro Oncol 14(9):1136–1145

Fisher JL, Schwartzbaum JA, Wrensch M, Wiemels JL (2007) Epidemiology of brain tumors. Neurol Clin 25:867–890, vii

Haghighat P, Timiryasova TM, Chen B, Kajioka EH, Gridley DS, Fodor I (2000) Antitumor effect of IL-2, p53, and bax gene transfer in C6 glioma cells. Anticancer Res 20:1337–1342

Hegi ME, Diserens AC, Gorlia T, Hamou MF, de Tribolet N, Weller M, Kros JM, Hainfellner JA, Mason W, Mariani L, Bromberg JE, Hau P, Mirimanoff RO, Cairncross JG, Janzer RC, Stupp R (2005) MGMT gene silencing and benefit from temozolomide in glioblastoma. N Engl J Med 352:997–1003

Hussain SF, Kong LY, Jordan J, Conrad C, Madden T, Fokt I, Priebe W, Heimberger AB (2007) A novel small molecule inhibitor of signal transducers and activators of transcription 3 reverses immune tolerance in malignant glioma patients. Cancer Res 67:9630–9636

Iwata-Kajihara T, Sumimoto H, Kawamura N, Ueda R, Takahashi T, Mizuguchi H, Miyagishi M, Takeda K, Kawakami Y (2011) Enhanced cancer immunotherapy using STAT3-depleted dendritic cells with high Th1-inducing ability and resistance to cancer cell-derived inhibitory factors. J Immunol 187:27–36

Jansen T, Tyler B, Mankowski JL, Recinos VR, Pradilla G, Legnani F, Laterra J, Olivi A (2010) FasL gene knock-down therapy enhances the antiglioma immune response. Neuro Oncol 12:482–489

Kim HA, Park JH, Cho SH, Lee J, Lee M (2011) Glia/ischemia tissue dual specific gene expression vector for glioblastoma gene therapy. J Control Release 152(Suppl 1):e146–e148

Kohsaka S, Wang L, Yachi K, Mahabir R, Narita T, Itoh T, Tanino M, Kimura T, Nishihara H, Tanaka S (2012) STAT3 inhibition overcomes temozolomide resistance in glioblastoma by downregulating MGMT expression. Mol Cancer Ther 11:1289–1299

Kortylewski M, Kujawski M, Wang T, Wei S, Zhang S, Pilon-Thomas S, Niu G, Kay H, Mulé J, Kerr WG, Jove R, Pardoll D, Yu H (2005) Inhibiting Stat3 signaling in the hematopoietic system elicits multicomponent antitumor immunity. Nat Med 11:1314–1321

Kujawski M, Zhang C, Herrmann A, Reckamp K, Scuto A, Jensen M, Deng J, Forman S, Figlin R, Yu H (2010) Targeting STAT3 in adoptively transferred T cells promotes their in vivo expansion and antitumor effects. Cancer Res 70:9599–9610

Liau LM, Prins RM, Kiertscher SM, Odesa SK, Kremen TJ, Giovannone AJ, Lin JW, Chute DJ, Mischel PS, Cloughesy TF, Roth MD (2005) Dendritic cell vaccination in glioblastoma patients induces systemic and intracranial T-cell responses modulated by the local central nervous system tumor microenvironment. Clin Cancer Res 11:5515–5525

Mineharu Y, King GD, Muhammad AK, Bannykh S, Kroeger KM, Liu C, Lowenstein PR, Castro MG (2011) Engineering the brain tumor microenvironment enhances the efficacy of dendritic cell vaccination: implications for clinical trial design. Clin Cancer Res 17:4705–4718

Mineharu Y, Muhammad A, Yagiz K, Candolfi M, Kroeger K, Xiong W, Puntel M, Liu C, Levy E, Lugo C, Kocharian A, Allison J, Curran M, Lowenstein P, Castro M (2012) Gene therapy mediated reprogramming tumor infiltrating T cells using IL-2 and inhibiting NF-κB signaling improves the efficacy of immunotherapy in a brain cancer model. Neurotherapeutics 9(4):827–843

Naumann U, Maass P, Gleske AK, Aulwurm S, Weller M, Eisele G (2008) Glioma gene therapy with soluble transforming growth factor-beta receptors II and III. Int J Oncol 33:759–765

Okada H, Giezeman-Smits KM, Tahara H, Attanucci J, Fellows WK, Lotze MT, Chambers WH, Bozik ME (1999) Effective cytokine gene therapy against an intracranial glioma using a retrovirally transduced IL-4 plus HSVtk tumor vaccine. Gene Ther 6:219–226

Plautz GE, Barnett GH, Miller DW, Cohen BH, Prayson RA, Krauss JC, Luciano M, Kangisser DB, Shu S (1998) Systemic T cell adoptive immunotherapy of malignant gliomas. J Neurosurg 89:42–51

Saito R, Mizuno M, Nakahara N, Tsuno T, Kumabe T, Yoshimoto T, Yoshida J (2004) Vaccination with tumor cell lysate-pulsed dendritic cells augments the effect of IFN-beta gene therapy for malignant glioma in an experimental mouse intracranial glioma. Int J Cancer 111:777–782

Sampson JH, Archer GE, Mitchell DA, Heimberger AB, Herndon JE 2nd, Lally-Goss D, McGehee-Norman S, Paolino A, Reardon DA, Friedman AH, Friedman HS, Bigner DD (2009) Mol Cancer Ther 8(10): 2773–2779

Sampson JH, Heimberger AB, Archer GE, Aldape KD, Friedman AH, Friedman HS, Gilbert MR, Herndon JE 2nd, McLendon RE, Mitchell DA, Reardon DA, Sawaya R, Schmittling RJ, Shi W, Vredenburgh JJ, Bigner DD (2010) Immunologic escape after prolonged progression-free survival with epidermal growth factor receptor variant III peptide vaccination in patients with newly diagnosed glioblastoma. J Clin Oncol 28:4722–4729

Sanai N, Polley MY, McDermott MW, Parsa AT, Berger MS (2011) An extent of resection threshold for newly diagnosed glioblastomas. J Neurosurg 115:3–8

Schneider T, Becker A, Ringe K, Reinhold A, Firsching R, Sabel BA (2008) Brain tumor therapy by combined vaccination and antisense oligonucleotide delivery with nanoparticles. J Neuroimmunol 195:21–27

Stander M, Naumann U, Dumitrescu L, Heneka M, Loschmann P, Gulbins E, Dichgans J, Weller M (1998) Decorin gene transfer-mediated suppression of TGF-beta synthesis abrogates experimental malignant glioma growth in vivo. Gene Ther 5:1187–1194

Stupp R, Mason WP, van den Bent MJ, Weller M, Fisher B, Taphoorn MJ, Belanger K, Brandes AA, Marosi C, Bogdahn U, Curschmann J, Janzer RC, Ludwin SK, Gorlia T, Allgeier A, Lacombe D, Cairncross JG, Eisenhauer E, Mirimanoff RO (2005) Radiotherapy plus concomitant and adjuvant temozolomide for glioblastoma. N Engl J Med 352:987–996

Tsuboi K, Saijo K, Ishikawa E, Tsurushima H, Takano S, Morishita Y, Ohno T (2003) Effects of local injection of ex vivo expanded autologous tumor-specific T lymphocytes in cases with recurrent malignant gliomas. Clin Cancer Res 9:3294–3302

Tsugawa T, Kuwashima N, Sato H, Fellows-Mayle WK, Dusak JE, Okada K, Papworth GD, Watkins SC, Gambotto A, Yoshida J, Pollack IF, Okada H (2004) Sequential delivery of interferon-alpha gene and DCs to intracranial gliomas promotes an effective antitumor response. Gene Ther 11:1551–1558

Tsurushima H, Yuan X, Dillehay LE, Leong KW (2007) Radioresponsive tumor necrosis factor-related apoptosis-inducing ligand (TRAIL) gene therapy for malignant brain tumors. Cancer Gene Ther 14:706–716

Vetter M, Hofer MJ, Roth E, Pircher HP, Pagenstecher A (2009) Intracerebral interleukin 12 induces glioma rejection in the brain predominantly by CD8+ T cells and independently of interferon-gamma. J Neuropathol Exp Neurol 68:525–534

Wang T, Niu G, Kortylewski M, Burdelya L, Shain K, Zhang S, Bhattacharya R, Gabrilovich D, Heller R, Coppola D, Dalton W, Jove R, Pardoll D, Yu H (2004) Regulation of the innate and adaptive immune responses by Stat-3 signaling in tumor cells. Nat Med 10:48–54

Wheeler CJ, Black KL, Liu G, Mazer M, Zhang XX, Pepkowitz S, Goldfinger D, Ng H, Irvin D, Yu JS (2008) Vaccination elicits correlated immune and clinical responses in glioblastoma multiforme patients. Cancer Res 68:5955–5964

Yamanaka R, Homma J, Yajima N, Tsuchiya N, Sano M, Kobayashi T, Yoshida S, Abe T, Narita M, Takahashi M, Tanaka R (2005) Clinical evaluation of dendritic cell vaccination for patients with recurrent glioma: results of a clinical phase I/II trial. Clin Cancer Res 11:4160–4167

The Role of Hyaluronic Acid and Its Receptors in the Growth and Invasion of Brain Tumors

26

Yushan Kim and Sanjay Kumar

Contents

Abstract

Malignant gliomas induce a complex cascade of changes in the extracellular matrix of the brain during their growth and invasion. This chapter highlights those changes involving hyaluronic acid, a glycosaminoglycan that constitutes much of the brain extracellular matrix, and the biophysical and biochemical effects those changes have on glioma cells. Signaling effects of hyaluronic acid receptors will be discussed, with a focus on CD44. The implications of CD44 enrichment in cancer stem cells will be discussed. Finally, because these interactions are highly dependent on the cellular microenvironment, we will review various in vitro cell culture platforms that have been used to model glioma cell motility and invasion.

Introduction

Malignant gliomas, particularly those classified as grade III or grade IV by the World Health Organization, are among the most fatal of all cancers. Grade IV glioma, also known as glioblastoma multiforme (GBM), is highly aggressive and recalcitrant to treatment, causing a dismal prognosis of 12–15 months after survival even with aggressive multimodal therapy (Siebzehnrubl et al. 2011). While the past two decades have seen much progress in understanding the origins and mechanisms of this aggressive cancer, these advances have not translated to a significant improvement in survival time. This is due in large

Y. Kim • S. Kumar (✉)
Department of Bioengineering and California, Institute for Quantitative Biosciences, University of California, Berkeley, CA, 94720, USA
e-mail: skumar@berkeley.edu

M. Hayat (ed.), *Tumors of the Central Nervous System, Volume 13*,
DOI 10.1007/978-94-007-7602-9_26, © Springer Science+Business Media Dordrecht 2014

part to the heterogeneity and genetic instability of glioma cells, which promote evasion of anti-cancer therapies. Current first line treatments for GBM include surgical resection of the tumor, chemotherapies such as the DNA-alkylating agent temozolomide, and radiotherapy. New targeted molecular therapies against tumor angiogenesis, such as bevacizumab, a monoclonal antibody against vascular endothelial growth factor, have shown some success in reducing tumor burden but have at best modestly increased survival time (Pàez-Ribes et al. 2009). Thus, the field continues to seek out new strategies to combine with those that are already used to form the most effective anti-cancer treatment possible.

Cells receive a plethora of signals from their microenvironment, including a variety of soluble autocrine, paracrine, and endocrine factors, as well as solid-state signals such as receptors on adjacent cells (e.g. cadherins) and the extracellular matrix (ECM), which cells engage through integrins and other adhesive receptors. These signals are then integrated by the cell to regulate polarity, motility, proliferation, cell fate, and a variety of other phenotypic characteristics. In cancer, both these microenvironmental biophysical signals and the way cells sense and respond to these signals are profoundly dysfunctional. Even in macroscopically static tissues cells exert forces against one another and the ECM, and the significant influence of this mechanical "context" on the resulting signal transduction has gained appreciation. These interactions are important during tumor invasion as well, as the microstructural arrangement of both ligands and steric barriers to migration greatly impact tumor cell motility. These ECM-based cues can in turn be remodeled by resident cancer cells, further perturbing native tissue homeostasis (Kumar and Weaver 2009).

These principles apply to the growth of GBM tumors, which is in large part dependent on glioma cell invasion through brain ECM. In vitro, the spreading area, motility, and proliferation of U373-MG and U87-MG cells is dependent on matrix stiffness (Ulrich et al. 2009). Neurosurgeons frequently use the high stiffness of brain tumors relative to normal brain tissue to identify appropriate resection planes, and these stiffness differentials have also been mapped by ultrasound. Finally, topologic structures in brain are thought to guide the migration of glioma cells; white matter fiber tracts and blood vessel basement membrane can act as tissue "highways" on which glioma cells invade rapidly to remote parts of the brain. Thus, an essential component to understanding the molecular mechanisms and potential points of intervention in GBM will be to clarify the biophysical inputs that cancer cells receive from their external environment, and how they interpret these inputs. Notably, brain ECM differs significantly from many connective tissues; whereas fibronectin, collagen, and vitronectin are essential elements of connective tissues and brain vasculature, brain parenchymal tissue is distinctly poor in protein, especially fibrillar proteins, and rich in glycosaminoglycans (GAGs). This chapter will focus on one of the key components of brain ECM, the GAG hyaluronic acid (HA). In GBM, HA is upregulated in the brain matrix, and HA receptors are overexpressed. We will now examine more closely how HA and its downstream signaling effects are relevant to the promotion of tumor growth and spread.

Hyaluronic Acid and Its Receptors

Matrix Properties of Hyaluronic Acid

Hyaluronic acid (HA), a linear GAG composed of repeating disaccharide units of glucuronic acid and N-acetylglucosamine, is essential to morphogenesis, tissue homeostasis, and wound repair throughout the body. Physically, due to its high density of anionic charge, HA is very hygroscopic and promotes tissue hydration and swelling. In brain, HA is a fundamental component of the ECM, as it serves as the high molecular weight template onto which many other hyaluronic acid binding proteins anchor (Fig. 26.1a). This class of proteins (variably called hyaladherins, HA binding proteins, and link proteins) shares a highly conserved HA-binding tandem repeat domain. The most prominent of these in brain is the lectican family of chondroitin sulfate proteoglycans, including versican, aggrecan, and neurocan.

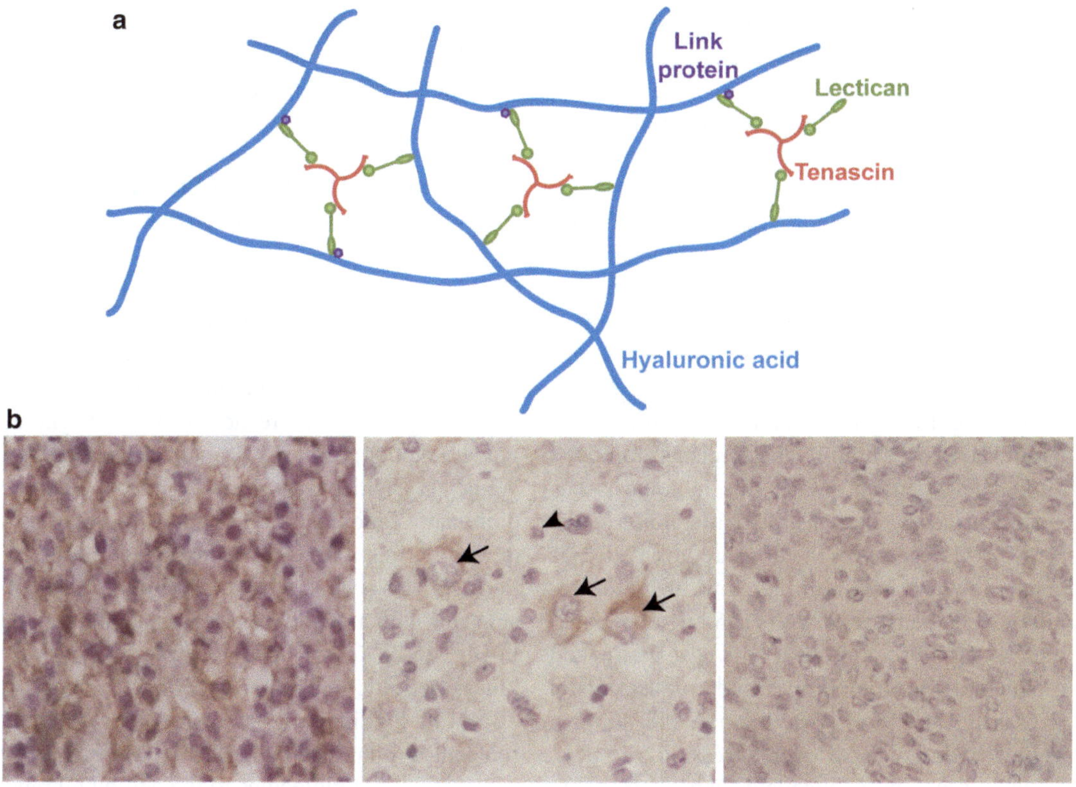

Fig. 26.1 Hyaluronic acid and CD44 are essential components of brain matrix and glioma invasion. (**a**) Schematic of HA architecture in brain. Brain matrix structure is based on high molecular weight hyaluronic acid (HA), which forms a hydrated network that is crosslinked by other biomolecules. Lecticans such as aggrecan, neurocan, versican, and brevican, are chondroitin sulfate proteoglycans that bind to HA at the N-terminus. HA-lectican binding is stabilized by link proteins. At the C-terminus, lecticans bind to the arms of tenascin-R, -C, and -W, which exist as trimeric or hexameric structures. (**b**) CD44 expression in the tumor microenvironment. Alterations in CD44 expression with tumor grade, demonstrated by immunohistochemical analysis of CD44 expression in human brain tumors by immunoperoxidase, with nuclei counterstained by eosin. An aggressive GBM tumor (*left panel*) displays intense staining of CD44. At the leading edge of the tumor (*middle panel*), neoplastic astrocytes with enlarged nuclei (*arrows*) display intense CD44 expression, while non-neoplastic cells (*arrowheads*) display little CD44 expression. In contrast, a noninvasive meningioma (*right panel*) shows weak CD44 staining. Magnification 400× (Ariza et al. 1995)

Chondroitin sulfate, the other glycosaminoglycan abundant in brain, is present as relatively minute sidearms on these proteoglycans. In contrast, HA is present as much longer chains, with molecular weight range of 100–1,000 kDa. Lecticans in turn bind to the third major class of brain matrix components, the tenascins, which are the main integrin-binding molecules in the brain. The biochemical makeup of brain stands in stark contrast to the collagen-based structures found in most tissues, so it is critical to recognize how these differences may affect cell adhesion and motility.

The exceptionally high molecular weight of HA is made possible by its unique mechanism of synthesis, which is orchestrated by the three HA synthases (HAS). Other GAGs are typically synthesized in the Golgi apparatus into short sidearms of proteoglycans, whose molecular weights are limited by vesicular transport. HA synthases, in contrast, are channel-like transmembrane proteins that sequester monosaccharides from the cytoplasm, add them onto the HA chain via glycotransferase domains, and extrude the growing linear chain out of the cell as it is synthesized. Thus, restrictions on the size of the HA molecule are not imposed by biosynthesis and transport, and a single molecule can span several microns in length.

In the developing mouse cerebellum, HA organizes into fine web-like structures, which are hypothesized to aid in the migration of interneuron precursors and oligodendroglial cell types (Baier et al. 2007). Analysis of rat brain composition as a function of age shows that HA concentration peaks shortly after birth and then drops off in adulthood. However, in brain tumors, HA secretion is again elevated (Delpech et al. 1993), suggesting that glioma cells may hijack the natural HA-based motility mechanisms employed during development. While this hypothesis has not been clearly demonstrated in glioma cells, HAS2 overexpression in fibrosarcoma cells has been shown to have a direct link to tumorigenicity; cells transfected with *HAS2* demonstrated increased proliferation in a soft agar assay, and produced larger tumors in a nude mouse model (Kosaki et al. 1999). Whether HA elevation in glioma is due to HAS overexpression *per se* or some expression-independent enhancement in HAS activity remains unclear.

In the early stages of neural crest and brain development, levels of hyaluronidases (encoded by six *hyal* genes), the primary enzymes that degrade HA, are also at their peak as matrix turnover is necessary for cell migration. Increased HAS expression is only beneficial for cell migration if hyaluronidase is concurrently secreted; increased HA adhesive contacts must also be released for productive cell movement to occur (Enegd et al. 2002). Animal studies reveal that Hyal-2 overexpression facilitates tumor angiogenesis and formation in the HA-rich brain, but not in an HA-poor subcutaneous environment (Novak et al. 1999). Upon degradation of high molecular weight HA chains by hyaluronidase secretion from tumor cells, low molecular weight HA by-products of roughly 20 or fewer disaccharides are released, which then stimulate endothelial cells in neighboring blood vessels to undergo angiogenesis (Liu et al. 1996). While this mechanism is also necessary for the initiation of wound healing and matrix remodeling, it is one of the many HA-based signals that gliomas co-opt for aberrant growth and invasion. HA-based signaling mechanisms within glioma cells themselves are discussed in the next section.

Adhesion and Signaling Effects of Hyaluronic Acid Receptors

Chief among the diverse family of HA-binding adhesion proteins is CD44, a single pass transmembrane receptor that is upregulated in a variety of solid tumors, including brain tumors, and whose expression correlates with high glioma grade and poor prognosis (Ranuncolo et al. 2002). While the field has not converged on a single, dominant "canonical" CD44 pathway, the effects of HA binding with CD44 converge on proliferation, cell survival, and anti-apoptotic fate decisions through a variety of pathways (summarized in Fig. 26.2). CD44 does not have any intrinsic kinase activity, but rather executes its signaling effects by binding to kinases and other signaling molecules via its cytoplasmic tail, thereby recruiting these molecules to the cell membrane. While some CD44-HA signaling effects can be initiated by a single binding event, the high molecular weight characteristic of HA also serves to organize an activity-rich signaling center by bringing together many receptors and their downstream signaling partners. Indeed, CD44 tends to aggregate in protein-rich caveolae or lipid rafts.

At the gene level, *CD44* transcription is suppressed by binding of p53 to a non-canonical binding sequence in the *CD44* promoter (Godar et al. 2008). Godar et al. showed that many of the oncogenic effects of p53 loss are mediated by an increase of CD44-based survival signaling. Once CD44 is translated and brought to the cell membrane, CD44 complexes with a variety of pro-tumorigenic receptor tyrosine kinases to promote their kinase activity, including EGFR and ErbB2, TGFβ receptor, and c-Met/HGFR (Jung et al. 2011).

CD44 is capable of indirectly engaging the actin cytoskeleton through its cytoplasmic tail via the adaptor proteins ankyrin, and the ERM (ezrin-ridixin-moesin) family proteins. This interaction is essential for cell motility stimulated by phorbol

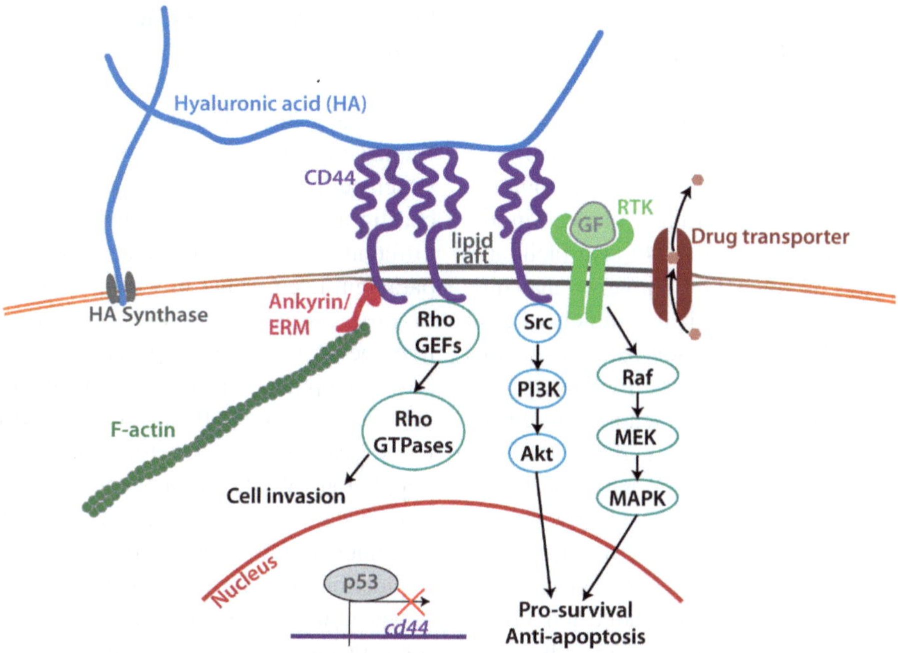

Fig. 26.2 Signal transduction mediated by hyaluronic acid (HA)-CD44 binding. HA is synthesized at the cell membrane by HA synthase, and extruded into the extracellular matrix. HA then binds to CD44, which can in turn associate with filamentous actin via ankyrin or the ezrin-ridixin-moesin (ERM) family proteins. RhoGTPases can also bind to the tail of CD44 via adaptor proteins. CD44 can associate in lipid rafts with receptor tyrosine kinases (RTKs) or multidrug efflux transporters on the cell membrane to promote their pro-survival and anti-apoptosis activity. All of these effects act in concert to promote cancer cell proliferation, motility, and chemoresistance. Transcription of *CD44* is kept in check by the tumor suppressor protein p53

ester, an analogue of diacyl glyceride (DAG). Legg et al. (2002) showed that the resulting activation of protein kinase C (PKC) triggers dephosphorylation of CD44 on Ser325 and phosphorylation on Ser291, then used fluorescence resonance energy transfer (FRET) to show that this dual change in phosphorylation reduces interaction between CD44 and ezrin. This dynamic control of ezrin association and dissociation via modulation of phosphorylation state is necessary for rapid chemotactic response to DAG gradients. Aside from anchoring to filamentous actin, CD44 also promotes cell motility by activating RhoGTPases. This occurs via recruitment of guanine nucleotide exchange factors (GEFs) to the cell membrane, facilitating interaction with their effectors, the Rho family of GTPases. Addition of soluble HA to cell culture medium quickly activates Rac-1 by recruiting the GEFs Tiam1 and Vav2, and induces lamellipodia formation within several minutes (Oliferenko et al. 2000) in both primary astrocytes and mammary epithelial cells. Similarly, in mammary epithelial cells, CD44-HA binding activates RhoA by recruiting p115 RhoGEF and myosin-mediated cell motility (Bourguignon 2008).

Other molecular mechanisms suppress the tumor-promoting effects of CD44 association with F-actin and activation of Rho GTPases. The most notable of these CD44-antagonizing molecules is merlin, the tumor suppressor protein encoded by the *NF2* gene, which blocks association with actin when bound to CD44. This also decreases CD44 affinity for HA, and thus reduces the intensity of downstream pro-tumorigenic signaling (Bai et al. 2007). In this way, CD44 acts as a molecular switch that promotes proliferation when ERM proteins are bound and quiescence when merlin is bound. Significantly, merlin is an essential antagonist of the Hippo signaling pathway, which attenuates apoptotic responses to oxidative stress and cytotoxic drugs. CD44 knockdown in glioma cells reduces merlin phosphorylation (which is required for CD44 binding) and blocks the cell survival signals downstream of Hippo, thereby inducing p53 expression and increasing survival in an animal model (Xu et al. 2010).

In addition to interacting with actin binding proteins and RhoGTPases to promote cell motility, the cytoplasmic tail of CD44 also facilitates the formation of signaling complexes that ultimately promote cell survival. Namely, CD44 recruits several Src family non-receptor kinases such as Lyn (Lin 2001), Lck, Fyn, and c-Src (Bourguignon et al. 2001), which activate PI3K-Akt signaling. Integrins are perhaps the best known and widely studied cell adhesion receptors, so it is worth noting that the cytoplasmic tail of CD44 also interacts with integrin-based focal adhesion proteins. For instance, CD44 has been shown to complex with focal adhesion kinase (FAK) in a lung cancer cell line, and cells stimulated by HA express increased activated (FAK), which in turn augments both the PI3K and MAPK pathways (Fujita et al. 2002). Thus, this extensive set of interactions allows CD44 to exert its signaling effects even in the absence of any intrinsic kinase activity.

Finally, RHAMM (receptor for hyaluronic acid mediated motility) is another HA receptor that has been studied for its pro-tumorigenic effects. The functions of RHAMM and CD44 appear to be inextricably linked; many roles of RHAMM involve supporting CD44-HA binding and downstream signaling. Additionally, in an arthritic mouse model, increased RHAMM expression has been shown to phenotypically complement CD44 gene suppression, as CD44 knockout mice develop normally (Nedvetzki et al. 2004). While the two receptors share many of the same binding partners, RHAMM does have other tumorigenic roles independent of CD44; for example, its other primary role is to facilitate the assembly of mitotic spindles required for cell division by assisting in microtubule nucleation. This role, independent of HA binding functions, is known to be especially active in breast cancer cells, where it is proposed that overexpressed RHAMM drives aberrations in mitotic spindles and thus supports genomic instability.

Significance of CD44 in Cancer Stem-Like Cells

Historically, the progression of GBM has been framed in terms of clonal evolution models, in which cellular heterogeneity arises from the stochastic accumulation of different mutations by different cells. More recently, a new prevailing paradigm is emerging in which GBM is thought to progress according to a cancer stem cell hypothesis, in which a specialized and perhaps rare subpopulation possessing the hallmark stem cell characteristics of self-renewal and multipotency give rise to a heterogeneous bulk tumor population. While components of this model remain somewhat controversial, a preponderance of evidence now supports that a stem-like subpopulation of cells exists within the heterogeneous tumor population that has much higher tumorigenic potential than the other cells. For example, only specific cell subpopulations of primary tumors are capable of histologically recapitulating human tumors when orthotopically implanted at low numbers in immunocompromised mice. These subpopulations often share key features of normal neural stem cells such as expression of neural stem cell markers (e.g. nestin), the ability to form clonal neurospheres, strong preference for laminin-based ECMs, and, most critically, the ability to self-renew and differentiate into cells positive for neural, astrocytic, and oligodendrocytic markers. As is commonly done in the field, we will refer to this stem-like population as brain cancer stem cells (BCSCs).

BCSCs can be selected from a primary tumor by culturing tumor explants in neurobasal medium supplemented with growth factors, which is also used as a neural stem cell growth medium. Some fraction of the tumor cells will form clonal neurospheres under these conditions, and these clones satisfy specific criteria including expression of specific markers, multipotency, and – critically – the ability to histologically recapitulate GBM when orthotopically implanted into immunocompromised mice. These cells are designated as BCSCs, and significantly, this population is enriched in CD44 expression. One possible functional consequence of CD44 overexpression is the promotion of multidrug resistance proteins on the cell membrane, which confer chemoresistance by pumping cytotoxic agents out of the cancer cell (Toole and Slomiany 2008). An alternative method of isolating cancer stem cell populations involves collection of "side population" cells, which are defined by low uptake of Hoechst dyes due to high expression of ABCG2 (BCRP) and ABCB1 (MDR1) drug transporters of the ABC (ATP-binding cassette) family (Hirschmann-Jax et al. 2004). This expression pattern partially accounts for the highly chemoresistant characteristics of BCSCs compared to their bulk tumor cell counterparts. BCSCs also divide, or cycle, less rapidly than bulk tumor cells, reducing the uptake and efficacy of anti-cancer drugs whose mechanism of action requires cell division.

Flow cytometry and histological analysis of human brain tumors reveals that glioma cells express a variety of these drug transporters, which are expressed more frequently in high grade tumors (Calatozzolo et al. 2005). In a malignant peripheral nerve sheath tumor cell line, CD44 forms a stable complex with ABCG2 in the plasma membrane (Slomiany et al. 2009). This complex is disrupted by adding oligomeric HA, which competes with endogenous high molecular-weight HA for CD44 binding, and leads to internalization of both CD44 and ABCG2. Finally, cells treated in this way are rendered more susceptible to apoptosis induced by the anti-cancer DNA intercalating agent doxorubicin. Since ABCG2 is highly expressed in BCSCs (Bleau et al. 2009), strategies to disrupt the association of CD44 and multidrug transporters expressed on BCSCs may prove to be effective ways of improving the effectiveness of chemotherapy treatments for malignant gliomas. The role of HA in the tumor matrix on promoting multidrug resistance in BCSCs has not yet been investigated.

While CD44 is increasingly accepted as a BCSC marker, the functional significance of CD44 enrichment in BCSCs remains unclear. Recently, Jijiwa et al. (2011) began to investigate these questions through the identification and characterization of CD44 variants in BCSCs. While CD44s is the standard isoform of the receptor, variant isoforms can be formed by alternative splicing of exons encoding the middle stalk region of the receptor; all isoforms contain the same intracellular domain and HA-binding domain. Expression of the variant isoform CD44v6 has long been known to be exclusively expressed by cancer cells of other tissues where it promotes cancer malignancy and invasiveness, but now it is known that CD44v6 is also expressed by BCSCs (Jijiwa et al. 2011). CD44v6 promotes cell survival signals by Akt phosphorylation when it binds to its secondary ligand, osteopontin. Thus, the further exploration of the functional role of CD44 expression in the BCSC specialized cell population will surely unlock greater understanding of the mechanisms of GBM progression.

In Vitro Culture Models of Brain Tumor Invasion

Any cell-level in vitro study of GBM must strike a balance between interpretability and physiological mimicry. On one extreme, cell culture paradigms employing two-dimensional tissue culture polystyrene (TCPS) surfaces are convenient and widely used, but they do not recapitulate the myriad microstructural, biophysical, and biochemical features of the in vivo brain microenvironment. When one's goal is to systematically test the regulatory role of specific, defined features of this microenvironment, it may be advantageous to use these highly defined albeit reductionist platforms. In other cases, it may be desirable to incorporate as many brain matrix characteristics as possible. We now briefly review strategies that have been used to study GBM in a cell-scale in vitro setting, with a focus on the incorporation of HA in these systems. In addition to the models described here, there are also a variety of in vivo animal models that are available for studying heterotypic cell interactions (e.g. with endothelial cells) or more clinically relevant endpoints (e.g. tumor size or animal survival).

2D Culture Models

The simplest approach for studying the effects of soluble factors such as HA is to place them in the cell culture medium itself. This format has been vitally important in identifying the many biochemical signaling effects of HA. However, from a biophysical standpoint, it does not mimic important structural aspects of large matrix molecules such as full-length adhesive proteins or full-length HA. To study the adhesive effects of these matrix macromolecules, they can be adsorbed onto glass or TCPS surfaces. However, the disadvantages to this approach are that the stiffness of these surfaces is orders of magnitude higher than most tissues including brain, that adsorption can alter the biological activity of the adsorbed ligand, and that the adsorption is nonspecific. To mitigate many of these disadvantages, polyacrylamide and other polymer hydrogel substrates have been widely used to create matrix materials with finely tunable stiffness (Wang and Pelham 1998) by varying the ratio of the monomer and crosslinker (e.g. acrylamide and bis-acrylamide). Matrix proteins such as fibronectin or collagen can be covalently attached to the hydrogel surface, often with heterobifunctional coupling agents such as sulfo-SANPAH (sulfosuccinimidyl-6-(40-azido-20-nitrophenylamino) hexanoate) which may be conjugated to the gel surface with UV irradiation and to ECM proteins via NHS-ester chemistry. Aside from the selective covalent attachment of desired proteins using this chemistry, serum proteins do not adsorb onto polyacrylamide, allowing for a highly controlled biochemical and biomechanical surface.

More recently, soft lithographic techniques have emerged as a powerful method to selectively pattern precise 2D geometries of adhesive proteins set on an adhesion-resistant background (Chen et al. 1997). Typically, a molded stamp of polydimethylsiloxane (PDMS) is coated with adhesive protein, and stamped onto a bioinert layer. The effects of cell shape, cell size, and cell-cell interactions have been well explored with this technique, as well as "1D" studies of cell migration on thin confined paths.

Cell motility is another important characteristic of cancer cells that can be measured in vitro. Time-lapse microscopy can be used to track the paths of migrating cells over time, yielding metrics such as cell speed and persistence length. In trans-well (Boyden chamber) assays, which enable the assessment of invasive cell motility, cells are seeded on top of polymer membranes with pore diameters on the same length scale of the cell. Cells are then challenged with a chemoattractant placed in the bottom compartment, thus inducing migration through the pores. The number of cells that successfully traverse a pore and arrive in the bottom chamber in a given time may then be used as a metric of invasive potential. A wide range of pore sizes is available, and the filters can be coated with specific ECM proteins to assess the effect of these proteins on migration.

While 2D models may be considered reasonable representations of certain native tissue geometries such as epithelial and endothelial monolayers, many other cell types, including glioma cells, are fully ensconced by ECM and/or cells in all three dimensions. Moreover, multiple lines of experimental evidence have revealed that glioma cell behavior often depends strongly on matrix dimensionality, i.e. 2D vs. 3D (Beadle et al. 2008). One way to create a 3D environment directly from a 2D surface is to use "sandwich" cultures, in which cells are first cultured on a 2D surface, and then another matrix layer is placed on the apical side of the cells. This method is widely used for the culture of mammary epithelial cells between layers of Matrigel and has been shown to reproduce defining phenotypic features of 3D matrices (Lee et al. 2007). In addition to its experimental convenience, this geometry recapitulates key features of anatomical interfaces along which glioma cells invade, such as the basement membrane of blood vessels and the glia limitans externa.

3D Culture Models

Many tissue architectures, including that of the brain parenchyma, are more accurately recapitulated by a "true" 3D structure in which cells are embedded within, rather than sandwiched between, ECM. This has significant implications for a number of cell behaviors relevant to glioma invasion, particularly cell motility. In 2D, lamellipodium formation at the leading edge is not geometrically constrained, and productive forward movement of the cell body is restricted by rupture of adhesions at the trailing edge. However in 3D, cells must either exert force to extrude through voids in the matrix, or enzymatically degrade the matrix to clear a path for productive movement. However, the use of 3D matrices presents unique challenges not encountered in 2D culture, including limited throughput, potentially uneven exposure to oxygen and soluble factors, reduced suitability for high-resolution imaging, and difficulty harvesting cells at high numbers for post-hoc analysis.

As described above, the task of imaging individual cells using light microscopy can be problematic in 3D cultures; images contain optical contributions from material above and below the focal plane, and cells may migrate into and out of focus. To overcome these and other challenges, multicellular tumor spheroid models are a convenient and widely-used method of tracking invasive behavior. Spheroids are formed by culturing cells in hanging droplets, in which the cells aggregate into dense spherical masses. The spheroids are then implanted during, or less commonly after, gel solidification. Local nutrient and oxygen gradients are automatically imposed by the geometry of the spheroids, and this induces cells on the periphery of the spheroid to invade outward if the matrix is permissive to migration (analogous to the familiar scratch wound assay in 2D). These built-in gradients are a major advantage of this method, which also mimics some aspects of multicellular tumor growth. Although single cells cannot be tracked for long periods of time, migration of the overall population can be monitored simply by tracking the invasive radius or projected area. In a variation on this paradigm, cells can also be cultured on the surfaces of microcarrier beads and then implanted, which has the benefit of standardizing aggregate size and limiting necrosis but may not recapitulate important radial cell-cell interactions found in tumors.

In terms of design of matrix materials, the most basic 3D matrices are biologically-derived protein gels such as collagen, fibrin, and Matrigel. While all of these model systems continue to serve as important discovery platforms for GBM, they have several limitations. First, matrix properties cannot be tuned over a wide range of stiffnesses, as these materials are intrinsically quite soft and cannot easily be stiffened or softened independent of other critical properties. For example, manipulation of matrix stiffness by increasing protein density also alters ligand density and the pore size of the matrix, which may sterically hinder migrating cells and alter solute diffusion. To overcome the first problem, our group has developed a collagen-agarose hybrid system, in which the addition of inert agarose stiffens the matrix without simultaneously adding more ligand density (Fig. 26.3b) (Ulrich et al. 2011). However, in this system, agarose also introduces steric barriers and restricts cell-directed remodeling of the collagen fibers.

Biologically-derived materials that do not form a gel on their own can (and in some cases must) also be covalently modified and crosslinked in order to form a solid hydrogel. For instance, HA does not spontaneously assemble into gels but can be made to do so if chemically modified and covalently crosslinked (Ananthanarayanan et al. 2011; Burdick and Prestwich 2011). By adding reactive sites on an otherwise unreactive biopolymer, stiffness and density can be controlled by extent of functionalization, crosslinking, and the molecular weight of the starting material (Fig. 26.3c). HA can also be used as a scaffold to embed matrix proteins (e.g. fibronectin, laminin) that do not easily assemble into gels on their own.

An alternative strategy to native ECM formulations is the use of synthetic polymers as a "blank slate" on which to build modular bioactive components. Both the polymerization and crosslinking chemistries of polyacrylamide, the most commonly used 2D synthetic hydrogel, are highly cytotoxic, rendering this material inappropriate for hydrogel assembly around cells. Instead, pre-polymerized materials such as polyethylene glycol (PEG) or polyvinyl alcohol are used in conjunction with more biocompatible crosslinking chemistries. Biocompatible photoinitiators and Michael Additions are broadly used for these crosslinking reactions. To add cell-adhesion functionality, full-length proteins can

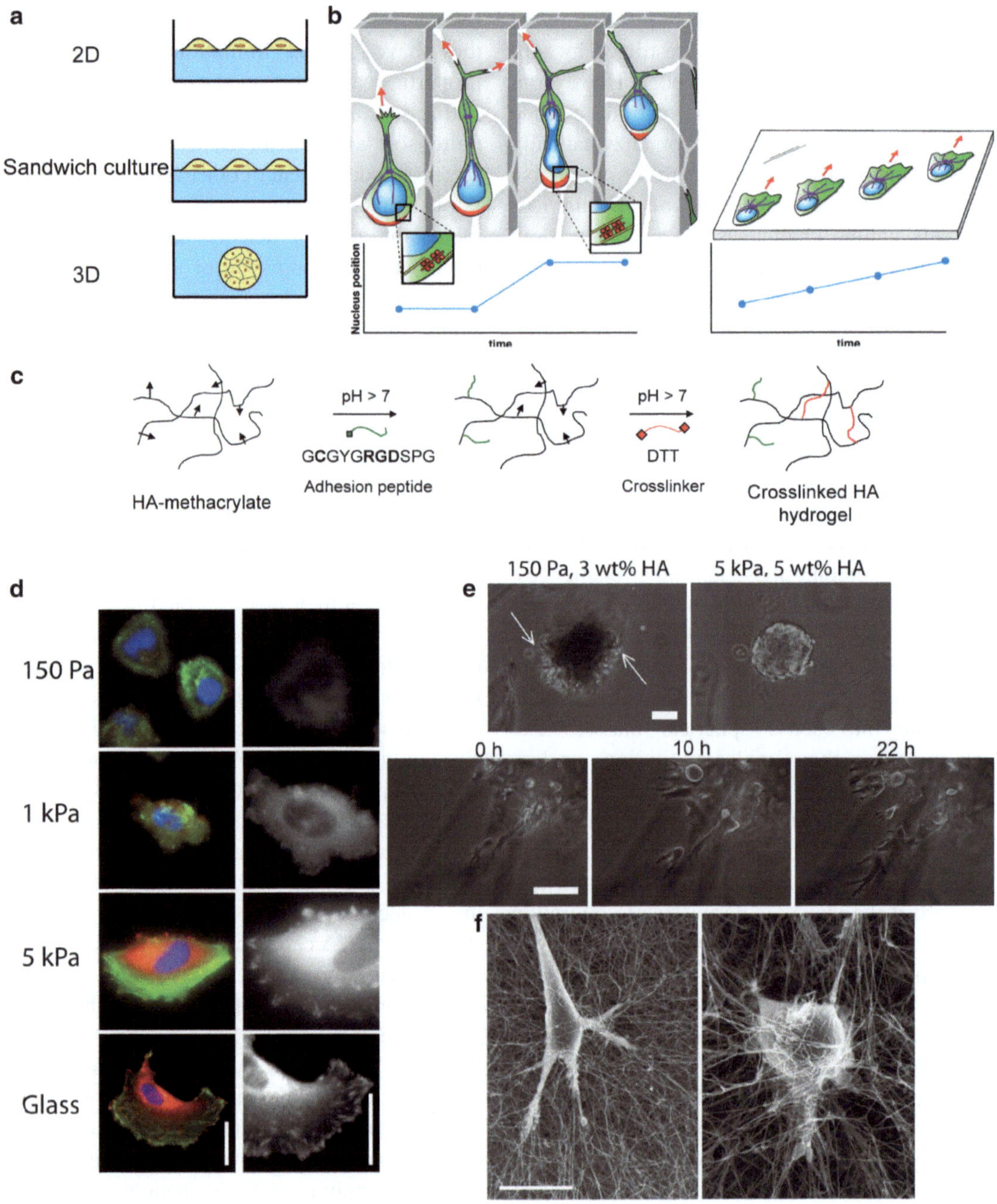

Fig. 26.3 In vitro cell culture models of GBM. (**a**) Schematic of various in vitro geometries. In 2D, cells lie on a flat monolayer, either on glass, tissue culture polystyrene, or an engineered or reconstituted matrix. In sandwich culture, they lie at an interface between two layers of matrix. In 3D, cells are surrounded on all sides by matrix either as single cells or in a spheroid as diagrammed. (**b**) Schematic of differing cell migration in 3D environment of brain (*left*) versus 2D (*right*). In 3D, glioma cells extend highly dynamic and branched leading protrusions to extrude cell bodies and nuclei through small voids between neighboring cells in order to migrate. In 2D, cell migration is not spatially hindered, and lamellipodial protrusion occurs freely, resulting in smooth gliding movement (Beadle et al. 2008). (**c**–**e**) Synthetically modified matrix based on hyaluronic acid (Ananthanarayanan

be embedded in the matrix prior to crosslinking, or the polymer can be covalently functionalized with bioadhesive peptide sequences. Finally, enzymatically degradable peptide sequences can also be incorporated into the crosslinks to allow for matrix degradation as cells proliferate and/or migrate.

More recently, some investigators have adapted a third, distinct approach based on the use of decellularized matrices (Fig. 26.3f), in which native tissue is denuded of cells using chemical detergents, which in the case of brain leaves behind matrix proteins, GAGs, and other brain-specific components (Crapo et al. 2012). This material can also be liquefied by pepsin digestion, but re-assembles after injection in vivo. This approach can have great advantages over building matrix scaffolds from the ground up, but faces unique challenges such as immune response, residual DNA, and sample-to-sample variability, though these are of greater concern for regenerative medicine applications than in vitro studies or in vivo implantation into immunocompromised animals. These materials have not been extensively explored in the context of glioma invasion, but may hold great promise as a sort of middle ground between reconstituted matrix preparations and in vivo paradigms. Given the history of reconstituted and synthetic ECM scaffolds, it is reasonable to expect that future efforts will focus in part on manipulation of properties such as porosity and matrix stiffness in these decellularized systems.

Discussion

CD44 was first discovered in the 1980s by the immunology community, which quickly found that the cell surface molecule is involved in a variety of diverse functions such as leukocyte homing and T-cell activation. By the early 1990s, cancer researchers began to find yet another CD44 function: that it is differentially expressed in many cancers, and plays an important role in cancer invasion and metastasis. Today, the quest to fully understand the processes of this complex molecule is still not complete; much remains to be discovered about not only the exact signaling pathways in which CD44 is involved, but to also understand how these numerous and at times conflicting signals are interpreted or exploited by cancer cells.

In GBM, aberrant CD44-HA mediated pathways play a critical role in aiding, and perhaps even driving, the highly motile, malignant, and chemoresistant properties of gliomas that make them so fatal. In this review, we have discussed how elevation of HA secretion in tumor matrix and high CD44 expression in glioma cells combine to promote cell invasion and survival in GBM. In the specialized BCSC population, we raised the prospect that the high expression level of CD44 may be more than simply a correlative marker and may play key functional roles in maintaining proliferation and chemoresistance patterns that are hallmarks of malignant gliomas.

Fig. 26.3 (continued) et al. 2011). (**c**) Chemical modification scheme, in which synthetically methacrylated HA can react with free thiol groups, such as those on cysteine-containing resides or the bifunctional crosslinker dithiothreitol (DTT) (Ananthanarayanan et al. 2011). (**d**) U373-MG cells on 2D HA functionalized with RGD show a stiffness-dependent response to spreading, actin (*green*) stress fiber formation, and vinculin-based focal adhesions (*red*). (**e**) Spheroids of U373-MG cells embedded in RGD-functionalized HA demonstrate dependence on stiffness and density of hydrogel (*top panel*). In the 150 Pa 3 wt% gel, cells migrate outward in a highly branched and dynamic manner (Ananthanarayanan et al. 2011). (**f**) Collagen-agarose hybrid matrices. The collagen-agarose system is another example of a biologically derived matrix. Scanning electron micrographs of pure collagen matrix (*left panel*) show that a U373-MG glioma cell can exert forces needed to bundle collagen fibrils and spread. In contrast, the addition of agarose prior to collagen gelation intercepts this bundling process, and the cell remains rounded (Ulrich et al. 2011)

Finally, we discussed the rational design of in vitro GBM studies, ranging from highly simplified models to complex ones that incorporate important structural, biochemical, and biomechanical aspects of native tissue.

This body of knowledge leaves many open questions unanswered. Among the most critical of these is whether CD44 expression is a requirement for GBM malignancy, or whether it is a downstream effector that happens to promote further cancer invasion. Thus far, several clues suggest that CD44 may contribute in significant ways to glioma progression, such as its previously discussed role in executing many of the cancer-promoting effects of *p53* loss. However, the extent to which CD44 alone is intrinsically responsible for triggering tumorigenesis is not yet clear. This question intersects with our emerging understanding of the tumor-initiating role of BCSCs, which often happen to overexpress CD44. While CD44 supports chemoresistance in these cells, it remains unknown whether CD44 overexpression in this subpopulation causally drives this process. To decipher these puzzles, combined approaches must be brought to bear from the fields of cell-ECM biology, bioengineering, and cancer biology. The list of questions these multidisciplinary approaches could help address is a long one. For example, what are the individual roles of CD44 isoform expression on GBM progression? How do CD44-matrix linkages and their downstream signaling crosstalk with other important pathways in brain, including aberrant growth factor and integrin signaling? To what extent does the CD44 receptor transduce force-based signals, as integrins have long been recognized to do? Finally, as animal models of GBM become increasingly sophisticated, researchers will have more powerful tools to study the impact of these questions in an in vivo system, and ultimately the genetic and molecular origins of GBM.

While animal models are critical to cancer research, in vitro studies are also a powerful platform to study cell-ECM interactions in a controlled and manipulatable manner. CD44 and other cell surface molecules act as the interface through which the cell communicates with its surroundings, so brain ECM composition and structure, and cancer-induced changes in these properties, compose an essential piece of the puzzle. Therefore, another challenge in this field is the incorporation of brain-specific features into ECM models used to study GBM growth and invasion, including those having to do with ligand type and presentation and scaffold mechanics. With our exploration of various in vitro platforms for studying glioma cell behavior, we reviewed some methods developed to isolate and study certain aspects of the distinct microstructural and biophysical properties of brain. A central limitation of the current generation of HA-based scaffolds is the lack of fine microstructural and topological control, with the creation of multicomponent, hybrid matrices with the same molecular arrangements as those found in the brain remaining a particular challenge. Finally, there remains a great need for methods to incorporate larger anatomic structures in brain, such as blood vessels and white matter fiber tracts. In this regard, decellularized matrices may offer an important and relatively unique opportunity to perform "top down" matrix engineering, in which post hoc synthetic modifications on the matrix could be made to study their effects on glioma cell motility in an environment that retains all of the native ECM of brain.

Ultimately, these future findings and methods can be leveraged to design new therapies that make anti-cancer regimens more effective. As we have discussed, several studies have begun to explore the HA-CD44 interaction as a potential "druggable target," with promising results. Examples include addition of oligomeric HA to competitively bind and reduce the receptor clustering effect of high molecular weight HA, or addition of hyaluronidase to enzymatically cleave it. As discussed earlier, these strategies should be systematically tested in increasingly physiomimetic models of human GBM, using not only the endpoints of cell proliferation and apoptosis which can be screened in vitro, but also heterotypic cell-cell interactions, angiogenesis, and mean survival time which can only be tested in vivo.

Perhaps an even greater challenge than understanding how cancer cells synthesize and respond to signals through one receptor such as CD44 is to understand how they fit in with the greater picture, composed of an entire slew of signals from other adhesion and growth factor receptors, and biophysically-imposed constraints. In particular, the great focus of the field on the dominance of integrin-based adhesion and tumorigenic signaling calls into question how integrin-ligand and CD44-HA receptor-ligand pairs might interact, and what relative role they have in GBM progression. The field's attempts to answer these questions have the potential to not only inform greater therapeutics to treat GBM, but also to understand and treat other cancers throughout the body.

References

Ananthanarayanan B, Kim Y, Kumar S (2011) Elucidating the mechanobiology of malignant brain tumors using a brain matrix-mimetic hyaluronic acid hydrogel platform. Biomaterials 32:7913–7923

Ariza A, López D, Mate JL, Isamat M, Musulén E, Pujol M, Ley A, Navas-Palacios JJ (1995) Role of cd44 in the invasiveness of glioblastoma multiforme and the noninvasiveness of meningioma: an immunohistochemistry study. Hum Pathol 26:1144–1147

Bai Y, Liu Y-J, Wang H, Xu Y, Stamenkovic I, Yu Q (2007) Inhibition of the hyaluronan-cd44 interaction by merlin contributes to the tumor-suppressor activity of merlin. Oncogene 26:836–850

Baier C, Baader SL, Jankowski J, Gieselmann V, Schilling K, Rauch U, Kappler J (2007) Hyaluronan is organized into fiber-like structures along migratory pathways in the developing mouse cerebellum. Mat Biol 26:348–358

Beadle C, Assanah MC, Monzo P, Vallee R, Rosenfeld SS, Canoll P (2008) The role of myosin ii in glioma invasion of the brain. Mol Bio Cell 19:3357–3368

Bleau A-M, Hambardzumyan D, Ozawa T, Fomchenko EI, Huse JT, Brennan CW, Holland EC (2009) Pten/pi3k/akt pathway regulates the side population phenotype and abcg2 activity in glioma tumor stem-like cells. Cell Stem Cell 4:226–235

Bourguignon LYW (2008) Hyaluronan-mediated cd44 activation of RhoGTPase signaling and cytoskeleton function promotes tumor progression. Semin Cancer Biol 18:251–259

Bourguignon LYW, Zhu H, Shao L, Chen YW (2001) Cd44 interaction with c-src kinase promotes cortactin-mediated cytoskeleton function and hyaluronic acid-dependent ovarian tumor cell migration. J Biol Chem 276:7327–7336

Burdick JA, Prestwich GD (2011) Hyaluronic acid hydrogels for biomedical applications. Adv Mater 23:H41–H56

Calatozzolo C, Gelati M, Ciusani E, Sciacca F, Pollo B, Cajola L, Marras C, Silvani A, Vitellaro-Zuccarello L, Croci D, Boiardi A, Salmaggi A (2005) Expression of drug resistance proteins pgp, mrp1, mrp3, mrp5 and gst-π in human glioma. J Neurooncol 74:113–121

Chen CS, Mrksich M, Huang S, Whitesides GM, Ingber DE (1997) Geometric control of cell life and death. Science 276:1425–1428

Crapo PM, Medberry CJ, Reing JE, Tottey S, van der Merwe Y, Jones KE, Badylak SF (2012) Biologic scaffolds composed of central nervous system extracellular matrix. Biomaterials 33:3539–3547

Delpech B, Maingonnat C, Girard N, Chauzy C, Maunoury R, Olivier A, Tayot J, Creissard P (1993) Hyaluronan and hyaluronectin in the extracellular matrix of human brain tumour stroma. Eur J Cancer 29A:1012–1017

Enegd B, King JAJ, Stylli S, Paradiso L, Kaye AH, Novak U (2002) Overexpression of hyaluronan synthase-2 reduces the tumorigenic potential of glioma cells lacking hyaluronidase activity. Neurosurgery 50: 1311–1318

Fujita Y, Kitagawa M, Nakamura S, Azuma K, Ishii G, Higashi M, Kishi H, Hiwasa T, Koda K, Nakajima N, Harigaya K (2002) Cd44 signaling through focal adhesion kinase and its anti-apoptotic effect. FEBS Lett 528:101–108

Godar S, Ince TA, Bell GW, Feldser D, Donaher JL, Bergh J, Liu A, Miu K, Watnick RS, Reinhardt F, McAllister SS, Jacks T, Weinberg RA (2008) Growth-inhibitory and tumor-suppressive functions of p53 depend on its repression of cd44 expression. Cell 134:62–73

Hirschmann-Jax C, Foster AE, Wulf GG, Nuchtern JG, Jax TW, Gobel U, Goodell MA, Brenner MK (2004) A distinct "side population" of cells with high drug efflux capacity in human tumor cells. Proc Natl Acad Sci U S A 101:14228–14233

Jijiwa M, Demir H, Gupta S, Leung C, Joshi K, Orozco N, Huang T, Yildiz VO, Shibahara I, de Jesus JA, Yong WH, Mischel PS, Fernandez S, Kornblum HI, Nakano I (2011) Cd44v6 regulates growth of brain tumor stem cells partially through the akt-mediated pathway. PLoS ONE 6:e24217

Jung T, Gross W, Zoller M (2011) Cd44v6 coordinates tumor matrix-triggered motility and apoptosis resistance. J Biol Chem 286:15862–15874

Kosaki R, Watanabe K, Yamaguchi Y (1999) Overproduction of hyaluronan by expression of the hyaluronan synthase has2 enhances anchorage-independent growth and tumorigenicity. Cancer Res 59:1141–1145

Kumar S, Weaver V (2009) Mechanics, malignancy, and metastasis: the force journey of a tumor cell. Cancer Metastasis Rev 28:113–127

Lee GY, Kenny PA, Lee EH, Bissell MJ (2007) Three-dimensional culture models of normal and malignant breast epithelial cells. Nat Method 4:359–365

Legg JW, Lewis CA, Parsons M, Ng T, Isacke CM (2002) A novel pkc-regulated mechanism controls cd44 ezrin association and directional cell motility. Nat Cell Biol 4:399–407

Lin YH (2001) The osteopontin-cd44 survival signal involves activation of the phosphatidylinositol 3-kinase/akt signaling pathway. J Biol Chem 276:46024–46030

Liu D, Pearlman E, Diaconu E, Guo K, Mori H, Haqqi T, Markowitz S, Willson J, Sy MS (1996) Expression of hyaluronidase by tumor cells induces angiogenesis in vivo. Proc Natl Acad Sci U S A 93:7832–7837

Nedvetzki S, Gonen E, Assayag N, Reich R, Williams RO, Thurmond RL, Huang JF, Neudecker BA, Wang FS, Turley EA (2004) Rhamm, a receptor for hyaluronan-mediated motility, compensates for cd44 in inflamed cd44-knockout mice: a different interpretation of redundancy. Proc Natl Acad Sci U S A 101:18081

Novak U, Stylli SS, Kaye AH, Lepperdingerm G (1999) Hyaluronidase-2 overexpression accelerates intracerebral but not subcutaneous tumor formation of murine astrocytoma cells. Cancer Res 59:6246–6250

Oliferenko S, Kaverina I, Small JV, Huber LA (2000) Hyaluronic acid (ha) binding to cd44 activates rac1 and induces lamellipodia outgrowth. J Cell Biol 148:1159–1164

Pàez-Ribes M, Allen E, Hudock J, Takeda T, Okuyama H, Viñals F, Inoue M, Bergers G, Hanahan D, Casanovas O (2009) Antiangiogenic therapy elicits malignant progression of tumors to increased local invasion and distant metastasis. Cancer Cell 15:220–231

Ranuncolo SM, Ladeda V, Specterman S, Varela M, Lastiri J, Morandi A, Matos E, Joffe EBDK, Puricelli L, Pallotta MG (2002) Cd44 expression in human gliomas. J Surg Oncol 79:30–36

Siebzehnrubl FA, Reynolds BA, Vescovi A, Steindler DA, Deleyrolle LP (2011) The origins of glioma: E pluribus unum? Glia 59:1135–1147

Slomiany MG, Dai L, Bomar PA, Knackstedt TJ, Kranc DA, Tolliver L, Maria BL, Toole BP (2009) Abrogating drug resistance in malignant peripheral nerve sheath tumors by disrupting hyaluronan-cd44 interactions with small hyaluronan oligosaccharides. Cancer Res 69:4992–4998

Toole BP, Slomiany MG (2008) Hyaluronan: a constitutive regulator of chemoresistance and malignancy in cancer cells. Semin Cancer Biol 18:244–250

Ulrich TA, de Juan Pard EM, Kumar S (2009) The mechanical rigidity of the extracellular matrix regulates the structure, motility, and proliferation of glioma cells. Cancer Res 69:4167–4174

Ulrich TA, Lee TG, Shon HK, Moon DW, Kumar S (2011) Microscale mechanisms of agarose-induced disruption of collagen remodeling. Biomaterials 32: 5633–5642

Wang Y-L, Pelham RJ Jr (1998) Preparation of a flexible, porous polyacrylamide substrate for mechanical studies of cultured cells. Methods Enzymol 298:489–496

Xu Y, Stamenkovic I, Yu Q (2010) Cd44 attenuates activation of the hippo signaling pathway and is a prime therapeutic target for glioblastoma. Cancer Res 70: 2455–2464

Neonatal Hypoxic-Ischemic Brain Damage: Human Umbilical Cord Blood Mononuclear Cells Transplantation

27

Pedro M. Pimentel-Coelho and Rosalia Mendez-Otero

Contents

Abstract

In the last decades, a great effort has been made to understand the physiopathology of neonatal hypoxic-ischemic encephalopathy (HIE), the main cause of long-term neurological impairments in term neonates. This effort was recently marked by a landmark advance in the treatment of HIE, represented by the significant neuroprotective effect of therapeutic hypothermia, in an example of successful translation of preclinical research findings to the bedside. However, at least 40% of the cooled infants will still die or have moderate/severe neurological disability, indicating that newer therapies are absolutely necessary. In this regard, umbilical cord blood mononuclear cells (UCBC), which are readily available for transplantation in the first hours after birth, have been shown to improve the neurological function when transplanted in several models of brain injury, including HIE. In this chapter we give a concise overview of recent studies evaluating the potential therapeutic role of UCBC transplantation in animal models of HIE. We also discuss the potential mechanisms underlying the action of these cells in the newborn brain and the current effort to translate these observations to patients in several ongoing clinical trials.

P.M. Pimentel-Coelho (✉) • R. Mendez-Otero
Instituto de Biofísica Carlos Chagas Filho, Universidade Federal do Rio de Janeiro, Rio de Janeiro, Brazil
e-mail: pedrompc@biof.ufrj.br

M. Hayat (ed.), *Tumors of the Central Nervous System, Volume 13*,
DOI 10.1007/978-94-007-7602-9_27,

Introduction

Neonatal hypoxic-ischemic encephalopathy (HIE) is a clinical syndrome characterized by the presence of neurological symptoms in the first days of life of term newborns. Although neonatal encephalopathy can be caused by several different etiologies, including genetic, metabolic and endocrine disorders, the term HIE is used for the cases that meet the criteria for perinatal asphyxia. One of the most important criteria for the diagnosis of perinatal asphyxia is the presence of metabolic acidosis in umbilical artery blood (pH < 7.0 and base deficit > 12 mmol/L), which has an incidence of 3.7 of 1,000 term live births. In a recent meta-analysis, it was shown that 23% of the neonates with this degree of acidosis had neonatal neurologic morbidity or mortality (Graham et al. 2008). Other criteria are also used to suggest the perinatal timing of asphyxia, such as abrupt changes in fetal heart rate, a low Apgar score, neuroimaging evidences, or the presence of a sentinel event (i.e., uterine rupture, placental abruption or umbilical cord prolapse).

HIE has an incidence of 2.5 of 1,000 term live births in developed countries (Graham et al. 2008). In the last years, the results of six large clinical trials have shown that therapeutic hypothermia for 48–72 h, initiated within 6 h after birth, can reduce death or neurological disability in children with HIE. Since then, therapeutic hypothermia has become the standard treatment for HIE. In spite of that, 40–50% of the treated children will not benefit from this therapy and will still die or have significant long-term neurological impairments (Higgins et al. 2011). Therefore, new therapies that could improve the survival and prevent the neurodevelopmental deficits in these children are urgently need. In recent years, a large body of evidence has suggested that cell therapies hold promising potential for the treatment of brain disorders. In particular, the umbilical cord blood (UCB) represents a rich source of cells with therapeutic potential, which are readily available for transplantation after birth.

Umbilical Cord Blood

UCB units can be collected *in utero* or from the delivered placenta and stored in public or private cord blood banks. Since the first UCB hematopoietic stem cell transplantation in a patient with Fanconi anemia by Dr. Eliane Gluckman's group in 1988, more than 20,000 UCB transplants have been performed in children and adults. Cryopreserved UCBC from a HLA-matched unrelated donor can be found in public cord blood banks and are readily available for the treatment of malignant and non-malignant disorders, such as acute and chronic myeloid and lymphoid leukemias, myelodysplastic syndromes, hemoglobinopathies, primary immunodeficiencies and inborn errors of metabolism. Besides the rapid accessibility, one important advantage of using UCBC is the fact that it is not strictly necessary to have a complete HLA match, given that UCB elicits less acute and chronic graft vs. host disease (GVHD) than the transplantation of peripheral blood or bone marrow cells. On the other hand, disadvantages of UCB transplantation include the lower number of collected cells, especially for the treatment of adult patients, and the delayed engraftment of neutrophils and platelets (Broxmeyer 2011).

When stored in private banks, UCBC can be used for the treatment of biological siblings or for autologous transplantation in experimental clinical trials. In the pediatric field, these trials are testing the safety and the efficacy of transplanting UCB mononuclear cells from the own child in the first days after HIE or several months/years after the antenatal/neonatal brain damage in children with cerebral palsy, as will be discussed further on. The UCB mononuclear cell fraction contains numerous cell types, including hematopoietic stem/progenitor cells (HSPC), monocytes, lymphocytes, endothelial progenitor cells and a small number of mesenchymal stem cells (MSC).

One interesting study has also identified a common progenitor ($CD34^{+}$ $CD45^{+}$ $CD133^{+}$ $CD38^{+}$) for endothelial, lymphoid and myeloid precursors in the UCB. These progenitors improved the perfusion in the hindlimb of mice subjected to femoral artery ligation, although the donor cells rarely differentiated into endothelial cells, suggesting that a paracrine mechanism was involved in the effect. On the other hand, when these UCB progenitors were differentiated into endothelial cells prior to transplantation, the transplanted cells incorporated into blood vessels in the ischemic hindlimb, indicating their potential for revascularization (Ramos et al. 2010).

Monocytes are important circulating immune effector cells, which can differentiate into macrophages, dendritic cells and microglial cells under inflammatory conditions. Both monocytes subsets ($CD14^{high}$ $CD16^{-}$ and $CD14^{+}$ $CD16^{+}$) are present in the UCB, in a frequency similar to that found in the adult peripheral blood. However, UCB monocytes may have some functional differences when compared to adult peripheral blood monocytes, such as an increased production of tumor necrosis factor-alpha (TNF-α) and IL-12p70 upon stimulation with the bacterial component peptidoglycan (Sohlberg et al. 2011). The therapeutic potential of monocytes was recently suggested by a recent study showing that monocyte-derived macrophages play an important role in the protection of retinal ganglion cell and in the proliferation of retinal progenitor cells after a retinal injury (London et al. 2011). However, the role of monocytes after cerebral ischemia remains elusive.

Among the subtypes of T lymphocytes present in the UCB, regulatory T cells (Treg) are those with the highest therapeutic potential. $CD4^{+}$ $CD25^{+}$ Treg can be easily obtained from the UCB and present a potent immunosuppressive function after isolation and culture (Godfrey et al. 2005). In addition, Treg have been shown to be neuroprotective in animal models of amyotrophic lateral sclerosis, Parkinson's disease and stroke. This effect is mediated by the potent immunomodulatory capacity of Treg and probably involves a shift of microglial phenotype towards a neuroprotective microglial response (Gendelman and Appel 2011).

Finally, MSC is an adherent fibroblast-like cell population that can be expanded in culture and induced to differentiate into specialized mesenchymal cells, such as osteoblasts, chondrocytes and adipocytes. MSC reside in perivascular niches and can be obtained from most of the tissues in the body. The number of circulating MSC in the UCB seems to be low, given that MSC cultures cannot be obtained from all UCB units. However, a recent study has observed that other factors, such as the volume of blood and the time between UCB collection and cell processing, are critical for the successful isolation of MSC. In this study, MSC could be isolated from 90% of the UCB units with at least 90 ml of blood, collected less than 2 h before the cell processing (Zhang et al. 2011).

Intracerebral transplantation of human UCB-derived MSC, 3 days after HIE, improved the neurological function of the treated animals. Although the mechanisms involved in this effect were not evaluated, the transplanted cells have not differentiated into neurons (Xia et al. 2010). Similar results have been demonstrated by previous studies transplanting bone marrow-derived MSC in animal models of HIE. For instance, one of these studies reported that intracerebral MSC transplantation promotes a better functional recovery through a combination of neuroprotection, stimulation of endogenous neurogenesis and oligodendrogenesis and a reduction in microglial proliferation (van Velthoven et al. 2010). Given the low numbers of MSC in the UCB, these cells are probably not involved in the therapeutic effects observed after the transplantation of freshly isolated UCB mononuclear cells in HIE. Nevertheless, MSC obtained from the UCB or from the Wharton's jelly, a soft connective tissue of the umbilical cord, could be a promising option for the treatment of HIE. One important limitation of this approach is the time required to expand these cells in culture, making unfeasible the transplantation of autologous MSC in the acute/subacute phase of the injury. Alternatively, heterologous MSC transplantation could be performed, considering the low immunogenicity and the immunoregulatory capacity of MSC. However, a recent study observed that the crosstalk between transplanted MSC and recipient T lymphocytes could inhibit MSC-mediated tissue regeneration (Liu et al. 2011). Therefore, the safety and the therapeutic potential of heterologous MSC transplantation still needs to be carefully investigated in animal models of HIE.

Umbilical Cord Blood Mononuclear Cells Transplantation in Animal Models of Hypoxic-Ischemic Encephalopathy

The first study evaluating the effects of UCB mononuclear cell transplantation in the Rice-Vannuci model of HIE was published by Meier et al. (2006). This model consists of permanent unilateral common carotid artery ligation followed by systemic hypoxia (8% oxygen-balance nitrogen) for several minutes/hours in post-natal day 7 rats. The damage is restricted to the hemisphere ipsilateral to the common carotid artery occlusion, affecting the cerebral cortex, thalamus, striatum, hippocampus and subcortical white matter (Johnston et al. 2005).

Injection of 1×10^7 cells, 24 h after the hypoxic-ischemic injury, decreased the locomotor deficits, as assessed by the footprint analysis 20 days after the treatment (Meier et al. 2006). Recently, they have also shown that the same treatment protocol results in long-term improvements in the sensorimotor outcome (by using the foot print analysis and the cylinder test), 7 weeks after the injury (Geissler et al. 2011). Moreover, these effects in neurological function were accompanied by the restoration of somatosensory cortical processing. For instance, electrophysiological mapping revealed that UCBC transplantation prevented the decrease in the size of the cortical hind paw representation, as well as restored the receptive field size of neurons in the damaged hemisphere. However, the volume of the brain lesion was not affected by the treatment (Geissler et al. 2011). In addition, other studies from three independent research groups have evaluated the therapeutic potential of UCBC transplantation in the Rice-Vannuci animal model of HIE.

Our group has injected 2×10^6 mononuclear cells intraperitoneally, 3 h after the hypoxic-ischemic insult. The timing of the injections was chosen based on the findings of preclinical studies on hypothermia, in which a neuroprotective effect could be obtained when the treatment was initiated within the first 5.5 h after the injury (Higgins et al. 2011). Our results showed that UCBC transplantation preserved two primitive neurological reflexes that are affected by HIE: negative geotaxis and cliff aversion reflexes. This effect was mediated by a reduction of caspase-3 activation and neuronal death in the striatum, 24 h after the injury, suggesting a neuroprotective role for these cells. We have also observed a lower number of activated microglial cells in the cerebral cortex of the treated animals, 1 week after the insult, indicating an immunomodulatory role of UCBC in the hypoxic-ischemic brain (Pimentel-Coelho et al. 2010).

One interesting study has also demonstrated that the intravenous injection of a lower number of cells (1.5×10^4 cells), 7 days after the injury, can reduce the deficits in motor symmetry and motor coordination in the elevated body swing test and in the rotarod performance test. Furthermore, the treatment increased the levels of the neurotrophic factors nerve growth factor (NGF), glial cell line-derived neurotrophic factor (GDNF) and brain-derived neurotrophic factor (BDNF) in the brain, as well as enhanced synaptic plasticity in the hippocampus. The administration of mannitol, a drug that permeabilizes the blood brain barrier, potentiated the therapeutic effects of UCBC, further improving the motor function. However, given that mannitol has not increased the number of UCBC reaching the brain, it is still unknown how it contributed to enhance the effect of these cells in the damaged brain (Yasuhara et al. 2010). In contrast, one study has reported that UCBC transplantation (1×10^7 cells, intravenously injected 24 h after HIE) has no effect on motor and cognitive function, as assessed by the Morris water maze and by four motor tests. In addition, the treatment did not change the volume of the brain lesion (de Paula et al. 2009). Taken together, most of the studies observed an improvement of the sensorimotor function of the animals that received UCBC infusions. However, the cell dose, the route of cell deliver and the timing of transplantation were not uniform between the studies (Table 27.1). Therefore, future studies should compare different cell doses and routes of cell administration, determining the time window and the optimal conditions for UCBC transplantation.

Table 27.1 Main results of studies evaluating the potential therapeutic role of human umbilical cord blood mononuclear cells in the Rice-Vanucci animal model of HIE

Model (animal, duration of hypoxia)	Dose; route; timing	Engraftment (days after transplantation)	Functional outcome	Cellular/ molecular effects	References
P7 Wistar rats, 80 min	1×10^7 cells, IP; 24 h after HI	Large number of cells in the ischemic hemisphere (20 days)	Improved motor outcome (footprint analysis)	N/A	Meier et al. (2006)
P7 Wistar rats, 120 min	1×10^7 cells, IV; 24 h after HI	Few cells in the brain (24 h, 1 week and 3 weeks)	Spatial memory deficits persisted (Morris water maze)	Infarct size was not changed	de Paula et al. (2009)
P7 Sprague-Dawley rats, 150 min	$1,5 \times 10^4$ cells, IV; 7 days after HI	Few cells in the ischemic hippocampus (14 days)	Improved motor outcome (rotarod and elevated body swing test)	Increased levels of GDNF, BDNF and NGF in the brain Increased hippocampal CA1 dendritic density	Yasuhara et al. (2010)
P7 Lister-Hooded rats, 90 min	2×10^6 cells, IP; 3 h after HI	Few cells in the ischemic cortex and striatum (2 days)	Prevented deficits in neonatal reflexes (cliff aversion and negative geotaxis)	Decreased neuronal death in the striatum Decreased microglial activation in the cortex	Pimentel-Coelho et al. (2010)
P7 Wistar rats, 80 min	1×10^7 cells, IP; 24 h after HI	Large number of cells in the ischemic hemisphere (48 days)	Decreased sensorimotor deficits (cylinder test)	Infarct size was not changed Reduced neural processing impairments in the primary somatosensory cortex	Geissler et al. (2011)

BDNF brain-derived neurotrophic factor, *GDNF* glial cell line-derived neurotrophic factor, *HI* hypoxia-ischemia, *NGF* nerve growth factor, *IP* intraperitoneal, *IV* intravenous, *P7* postnatal day 7

Biodistribution of the Transplanted Umbilical Cord Blood Mononuclear Cells

Meier et al. (2006) have found numerous donor UCBCs in the brain of hypoxic-ischemic rats, up to 41 days after the intraperitoneal transplantation (performed 24 h after the injury). The cells were localized only in the damaged brain regions, especially in areas with an increased expression of the chemokine stromal-derived factor-1 (SDF-1). Furthermore, they have demonstrated that SDF-1 expression is upregulated in reactive astrocytes after the injury and that the migration of UCBC to the ischemic brain could be blocked by the administration of SDF-1 neutralizing antibodies. Interestingly, there were no evidences of neuronal or glial differentiation of the transplanted cells (Meier et al. 2006; Rosenkranz et al. 2010; Geissler et al. 2011). In contrast, two other research groups have failed to find a large number of donor cells in the hypoxic-ischemic brain, when UCBC were intravenously transplanted 7 days after the injury or intraperitoneally injected 3 h after the injury, despite the functional benefits of the treatment (Yasuhara et al. 2010; Pimentel-Coelho et al. 2010). These apparently divergent observations could be attributed to the different timing of transplantation or to the different cell dose used in each study. Nevertheless, they indicate that the therapeutic effects of UCBC transplantation do not depend on the migration of the transplanted cells to the brain or on the differentiation of the donor cells into neural cells. Therefore, these studies suggest that neuronal replacement is not the mechanism of action of transplanted UCB mononuclear cells in HIE.

This is an important observation, given that some studies have reported that UCBC could differentiate into neurons under specific culture conditions. One of these studies has demonstrated that the UCB has a population of unrestricted somatic stem cells that grow as adherent cell colonies in the presence of dexamethasone. These cells can be induced to differentiate into neuron-like cells, although only a minority (11%) of the differentiated cells possesses functional voltage-gated sodium channels (Greschat et al. 2008). However, when UCB-derived unrestricted somatic stem cells were transplanted in an animal model of acute spinal cord trauma, there was no evidence of neural differentiation of the graft cells. Nevertheless, the cell therapy reduced the lesion size and promoted axonal regrowth in the injured spinal cord, which in turn resulted in an improved locomotor recovery of the treated animals. In this study, the authors have also identified hepatocyte growth factor as an important chemoattractant for UCB unrestricted somatic stem cells and they have demonstrated that the secreted factors present in the conditioned medium of these cells can stimulate neurite outgrowth in vitro (Schira et al. 2012). Taken together, these studies indicate that UCBC transplantation is not a cell replacement therapy. Multiple mechanisms seem to be involved in the therapeutic effects of UCBC, which are probably mediated by a paracrine effect.

Cellular and Molecular Mechanisms of Action of Umbilical Cord Blood Mononuclear Cells

The neuroprotective effects observed after UCB mononuclear cell transplantation in animal models of cerebral ischemia, such as in HIE and stroke (Pimentel-Coelho et al. 2010; Vendrame et al. 2006), were also replicated in several in vitro models of neuronal death. These studies are contributing to clarify the molecular and signaling pathways involved in UCBC-mediated neuroprotection and to identify which cell populations are involved in this effect.

For instance, it was observed that UCBC increase the survival of neuronal cells in an in vitro model of glutamate-induced toxicity, through the activation of the Akt pro-survival pathway (Dasari et al. 2008). Similarly, UCBC protect oligodendrocytes against oxygen/glucose deprivation in vitro, as well as in an animal model of stroke, through activation of Akt and upregulation of the antioxidant enzyme peroxiredoxin 4 (Rowe et al. 2012). However, it still needs to be investigated whether UCB would protect oligodendrocyte progenitors, which are highly susceptible to HIE.

Regarding the UCB cell types contributing to neuroprotection, it has been demonstrated that the $CD133^+$ population, which encompass both HSPC and endothelial progenitor cells, decreased apoptosis and prevented the deleterious effects of hypoxia on axonal growth in organ co-cultures of the cerebral cortex and spinal cord from 3-day-old neonatal rats (Tanaka et al. 2010). However, one study has observed that UCB $CD133^+$ cells and $CD133^-$ cells (the UCB mononuclear cell fraction depleted of $CD133^+$ cells) provided a similar degree of neuroprotection in an in vitro model of neuronal hypoxia (Reich et al. 2008). Furthermore, UCB mononuclear cell transplantation provides an increased protection after stroke, than the transplantation of UCB $CD34^+$ cells or than the mononuclear fraction depleted of $CD34^+$ cells ($CD34^-$ cells), indicating that multiple cell types are involved in this effect. Both $CD34^+$ and $CD34^-$ cells decreased the infarct volume and improved the motor function of the treated animals to a similar extent, although less efficiently than the complete mononuclear cell fraction. However, while the mononuclear cells and the CD34-enriched fraction protected hippocampal neurons in a model of oxygen and glucose deprivation, the CD34-depleted fraction had no effects on neuronal viability (Boltze et al. 2012). Taken together, these observations suggest that multiple cell types may contribute to the therapeutic effect of the UCB mononuclear cell fraction and indicate a role of $CD34^+$ cells, which includes a population of HSPC and endothelial progenitor cells, in neuroprotection.

Interestingly, neuroprotection can be achieved by either direct or indirect co-cultivation of UCBC with post-hypoxic neuronal cells (Reich et al. 2008), suggesting that secreted soluble factors are largely responsible for the neuroprotective effects of UCBC. Accordingly, freshly isolated UCBC express higher levels of the mRNA of several neurotrophic factors, including BDNF, GDNF, NGF, neurotrophin-3 (NT-3) and NT-5, when compared to peripheral blood mononuclear cells (Fan et al. 2005). In addition, BDNF, vascular endothelial growth factor (VEGF), NT-4, NT-5 and several cytokines and chemokines can be found in the conditioned medium of UCBC cultures (Fan et al. 2005; Newman et al. 2006). These evidences, coupled to the absence of neural differentiation of the transplanted cells, suggest that UCBC might exert their therapeutic actions through a paracrine effect in the hypoxic-ischemic brain.

Besides neuroprotection, other mechanisms may underlie the functional benefits observed after UCBC transplantation in animal models of cerebral ischemia. For instance, Taguchi et al. (2004) have shown that UCBC stimulate angiogenesis and neurogenesis after stroke. Moreover, the effect of UCBC treatment on neurogenesis was abolished by the administration of endostatin, an antiangiogenic agent. Given that neurogenesis and angiogenesis are coupled and are regulated by the same growth factors (i.e., VEGF and angiopoietins), it is possible to speculate that some of the effects of UCBC might be related to the secretion of angiogenic factors.

Another line of evidence suggests that UCBC may have an anti-inflammatory effect after cerebral ischemia. UCBC treatment decreased the number of activated microglial cells in the cortex in a model of HIE (Pimentel-Coelho et al. 2010). In addition, intravenous UCBC transplantation reduced the infiltration of B lymphocytes and decreased the expression of pro-inflammatory cytokines in the brain of rats subjected to middle cerebral artery occlusion (MCAO; Vendrame et al. 2005). Although this anti-inflammatory effect is not yet completely understood, it has been suggested that UCBC can modulate the response of the spleen after stroke in rats, suggesting a systemic immunomodulatory role of UCBC treatment. In this regard, UCBC transplantation prevented the stroke-induced alterations in spleen size, cellular composition and function. This effect was accompanied by an increased expression of the mRNA of interferon-gamma and of the anti-inflammatory cytokine IL-10 in the spleen (Vendrame et al. 2006).

Therefore, UCBC exert their beneficial effects in the ischemic brain through a combination of multiple mechanisms. Future efforts should be made to understand which trophic/growth factors are involved in these mechanisms. Genetic manipulation of the cells before transplantation could increase the expression and the secretion of potentially therapeutic neurotrophic/growth factors by UCBC. In this regard, a recent study has reported a non-viral lipofection technique to obtain UCB $CD34^+$ cells stably expressing the GDNF gene (Yu et al. 2010). It is also necessary to identify which cell populations contribute to the therapeutic effects and whether the depletion or the enrichment of one (or more) cell population(s) could further improve the benefits on neurological function.

Clinical Translation

Currently, two clinical trials are assessing the safety and the feasibility of autologous UCBC transplantation in HIE. In the first study, conducted at Duke University, newborns with HIE will be treated with up to four infusions of 5×10^7 cells/kg (depending on the number of available cells), within the first 14 postnatal days. Participants will be assessed by neurodevelopmental and neuroimaging examination for 1 year (http://clinicaltrials.gov; Identifier: NCT00593242). The number of cells transplanted in this clinical trial (even considering that the percentage of mononuclear cells will depend on the procedure used for reducing the volume of cord blood) is not so different from the dose used in the preclinical studies discussed above. In these studies, the cell dose ranged from 1.5×10^4 to 1×10^7 mononuclear UCBC (approximately 1.1×10^6–7.5×10^8 cells/kg of body weight). Nevertheless, future dose-response studies should address what is the optimal number of cells required for a long-term therapeutic effect in HIE.

The second trial has started recruiting participants in January 2012, in Mexico. In this study, autologous non-cryopreserved UCB $CD34^+$ cells will be intravenously transplanted in term newborns with HIE, within the first 48 h after birth. Participants will be followed by clinical examination for 1 year (http://clinicaltrials.gov; Identifier: NCT01506258). In both studies, the results will be compared to historic controls.

In addition, there are two ongoing clinical trials evaluating the safety and effectiveness of autologous UCBC transplantation in children with cerebral palsy (CP), a group of non-progressive syndromes of motor and posture impairments caused by a wide range of acquired brain disorders during the antenatal or perinatal period. HIE is the cause of cerebral palsy in at least 14% of the cases, especially (but not exclusively) causing dyskinetic or spastic tetraplegic cerebral palsy (Graham et al. 2008). The trial at Georgia Health Sciences University, is recruiting children with a non-progressive motor disability, from 1 to 12 years, who were unable to sit independently by 12 months of age or unable to walk independently by 18 months of age, while the study at Duke University is treating children with spastic CP with diplegia, hemiplegia, or quadriplegia, ranging from 1 to 6 years of age. Both studies are randomized, crossover trials, in which both treatment groups (i.e., placebo first, or UCBC treatment first) will receive UCB mononuclear cells, but at different points in the study (http://clinicaltrials.gov; Identifiers NCT01072370 and NCT01147653, respectively).

In contrast to the clinical studies that are treating neonates in the acute/subacute phase of HIE, clinical trials in children with cerebral palsy are not supported by the findings of preclinical studies. The main reason for this difference is the lack of good animal models of cerebral palsy. Only one model, consisting of intrauterine asphyxia in near-term rabbits, leads to persistent hypertonic motor deficits that resemble cerebral palsy (Johnston et al. 2005). Still, this model may mimic just one of the many possible (and not yet completely understood) causes of CP.

Furthermore, the benefits of UCBC transplantation were observed when the cells were administered within the first 7 days after HIE in rats. Therefore, it is still unknown if UCBC would provide any benefits once the lesion is already established, several months or years after the antenatal/perinatal brain insult. Future studies should address whether UCB mononuclear cells could stimulate endogenous mechanisms of brain regeneration, such as angiogenesis and neurogenesis, or increase brain plasticity, when administered in the chronic phase of the injury, in animal models of CP and HIE.

Finally, autologous intravenous UCB transplantation is being evaluated in a phase I/II clinical trial for the treatment of children with traumatic brain injury (http://clinicaltrials.gov; Identifiers NCT01251003) and. an ongoing clinical trial conducted at Ain Shams University, in Egypt, is evaluating the feasibility of autologous UCB mononuclear cell transplantation in low birth weight preterm neonates, as a possible therapy to prevent the neurodevelopmental delay associated with preterm birth (http://clinicaltrials.gov; Identifier: NCT01121328).

The rationale behind these studies (summarized in Table 27.2) is the fact that autologous intravenous UCBC transplantation is considered to be safe. Accordingly, a recent study has reported that intravenous infusion of autologous UCBC is safe and feasible in young children with acquired neurological injuries. The median age at infusion was 27 months and most of the children had CP (76%) or congenital hydrocephalus (12%). Only three of the 184 patients experienced anaphylactic reactions (wheezing with or without urticaria), 2–10 min after the infusion was initiated, which were resolved after discontinuation of the infusion and treatment with a first-generation antihistamine and bronchodilators (Sun et al. 2010).

Table 27.2 Current phase I/II clinical trials evaluating the safety and feasibility of autologous umbilical cord blood cell transplantation in newborns and children (registered on http://clinicaltrials.gov)

Disease	Cell dose	Age	Local	References
HIE	1–4 infusions of 5×10^7 cells/kg (volume reduced UCB cells)	Newborns (up to 14 days after birth)	Duke University, USA	NCT00593242
HIE	N/A (freshly isolated $CD34^+$ cells)	Newborns (up to 48 h after birth)	Hospital Universitario Dr. Jose E. Gonzalez, Mexico	NCT01506258
Developmental delay in low birth weight preterm neonates	N/A (UCB mononuclear cells)	Newborns (up to 30 days after birth)	Ain Shams University, Egypt	NCT01121328
Cerebral palsy	$\geq 1 \times 10^7$ nucleated cells/kg	1–12 years	Georgia Health Sciences University, USA	NCT01072370
Cerebral palsy	$\geq 1 \times 10^7$ nucleated cells/kg	1–6 years	Duke University, USA	NCT01147653
Pediatric traumatic brain injury	Maximum dose of 10×10^9 cells/kg	18 months to 17 years (6–18 months after the injury)	The University of Texas Health Science Center, USA	NCT01251003

UCB umbilical cord blood, *HIE* neonatal hypoxic-ischemic encephalopathy, *N/A* information not available

On the other hand, the transplantation of allogeneic cells could offer some risks to the children, such as the risk of graft-versus-host disease. Furthermore, the potential need for immunosuppression could bring additional risks. However, one clinical trial at Sung Kwang Medical Foundation in Republic of Korea has evaluated the safety and efficacy of allogeneic UCBC transplantation, in combination with erythropoietin injections, in children with CP (10-month-old to 10 year-old children; http://clinicaltrials.gov; Identifier: NCT01193660).

Future Direction

Given that therapeutic hypothermia is currently the standard treatment for HIE, future preclinical and clinical studies should evaluate the effects of UCB mononuclear cell transplantation in combination with hypothermia. It is also necessary to evaluate the effects of UCBC treatment in models of HIE in large animals with a gyrencephalic brain and increased cerebral white matter. For instance, models of HIE in piglets and sheep were fundamental for the evaluation of the efficacy of therapeutic hypothermia for HIE (Johnston et al. 2005).

Finally, freshly isolated and cryopreserved human UCB $CD133^+$ and $CD34^+$ cells can be reprogrammed to pluripotency, generating induced pluripotent stem cells (iPS; Giorgetti et al. 2009; Ye et al. 2009). UCBC-derived iPS formed teratomas when transplanted in immunocompromised mice, as well as generated embryoid bodies with high efficiency, which could differentiate into cells of the three embryonic germ layers in vitro. Under specific culture conditions, UCBC-derived iPS could also give rise to dopaminergic neurons. Interestingly, while at least three transcription factors (OCT4, SOX2 and KLF-4) are required to generate iPS from keratinocytes and fibroblasts, UCBC could be reprogrammed by the transduction of only two transcription factors (OCT4 and SOX2), indicating an increased reprogramming susceptibility of these cells. In addition, while the generation of iPS from keratinocytes and fibroblasts requires the establishment of primary cultures for up to a few weeks, UCBC can be reprogrammed after just 1–4 days in culture. In addition, UCBC are young cells with less accumulated genetic mutations, readily available for use, which may also represent an advantage for future cell replacement therapies using iPS-derived neural cells (Giorgetti et al. 2009; Ye et al. 2009).

In conclusion, UCB mononuclear cell transplantation improved the sensorimotor outcome in most of the preclinical studies using the Rice-Vannuci animal model of HIE. Several mechanisms contributed to this effect, including neuroprotection, induction of synaptic plasticity and immunomodulation. All these mechanisms seem to result from a paracrine effect, rather than by the replacement of lost cells. Future studies should address important questions regarding the optimal conditions for transplantation. Furthermore, it will be important to understand which cells and molecules are involved in the beneficial effects, in order to improve the therapy, by means of genetic manipulation and enrichment/depletion of specific cell populations. Finally, autologous UCBC transplantation is safe in children and the results of the first clinical trials in newborns with HIE and in children with CP are expected for the next years.

References

Boltze J, Reich DM, Hau S, Reymann KG, Strassburger M, Lobsien D, Wagner DC, Kamprad M, Stahl T (2012) Assessment of neuroprotective effects of human umbilical cord blood mononuclear cell subpopulations in vitro and in vivo. Cell Transpl 21(4):723–737

Broxmeyer HE (2011) Insights into the biology of cord blood stem/progenitor cells. Cell Prolif 44(Suppl 1):55–59

Dasari VR, Veeravalli KK, Saving KL, Gujrati M, Fassett D, Klopfenstein JD, Dinh DH, Rao JS (2008) Neuroprotection by cord blood stem cells against glutamate-induced apoptosis is mediated by Akt pathway. Neurobiol Dis 32:486–498

de Paula S, Vitola AS, Greggio S, de Paula D, Mello PB, Lubianca JM, Xavier LL, Fiori HH, Dacosta JC (2009) Hemispheric brain injury and behavioral deficits induced by severe neonatal hypoxia-ischemia in rats are not attenuated by intravenous administration of human umbilical cord blood cells. Pediatr Res 65:631–635

Fan CG, Zhang QJ, Tang FW, Han ZB, Wang GS, Han ZC (2005) Human umbilical cord blood cells express neurotrophic factors. Neurosci Lett 380:322–325

Geissler M, Dinse HR, Neuhoff S, Kreikemeier K, Meier C (2011) Human umbilical cord blood cells restore brain damage induced changes in rat somatosensory cortex. PLoS One 6:e20194

Gendelman HE, Appel SH (2011) Neuroprotective activities of regulatory T cells. Trends Mol Med 17:687–688

Giorgetti A, Montserrat N, Aasen T, Gonzalez F, Rodriguez-Piza I, Vassena R, Raya A, Boue S, Barrero MJ, Corbella BA, Torrabadella M, Veiga A, Izpisua Belmonte JC (2009) Generation of induced pluripotent stem cells from human cord blood using OCT4 and SOX2. Cell Stem Cell 5:353–357

Godfrey WR, Spoden DJ, Ge YG, Baker SR, Liu B, Levine BL, June CH, Blazar BR, Porter SB (2005) Cord blood CD4(+)CD25(+)-derived T regulatory cell lines express FoxP3 protein and manifest potent suppressor function. Blood 105:750–758

Graham EM, Ruis KA, Hartman AL, Northington FJ, Fox HE (2008) A systematic review of the role of intrapartum hypoxia-ischemia in the causation of neonatal encephalopathy. Am J Obstet Gynecol 199:587–595

Greschat S, Schira J, Kury P, Rosenbaum C, de Souza Silva MA, Kogler G, Wernet P, Muller HW (2008) Unrestricted somatic stem cells from human umbilical cord blood can be differentiated into neurons with a dopaminergic phenotype. Stem Cells Dev 17:221–232

Higgins RD, Raju T, Edwards AD, Azzopardi DV, Bose CL, Clark RH, Ferriero DM, Guillet R, Gunn AJ, Hagberg H, Hirtz D, Inder TE, Jacobs SE, Jenkins D, Juul S, Laptook AR, Lucey JF, Maze M, Palmer C, Papile L, Pfister RH, Robertson NJ, Rutherford M, Shankaran S, Silverstein FS, Soll RF, Thoresen M, Walsh WF (2011) Hypothermia and other treatment options for neonatal encephalopathy: an executive summary of the Eunice Kennedy Shriver NICHD workshop. J Pediatr 159:851 e1–858 e1

Johnston MV, Ferriero DM, Vannucci SJ, Hagberg H (2005) Models of cerebral palsy: which ones are best? J Child Neurol 20:984–987

Liu Y, Wang L, Kikuiri T, Akiyama K, Chen C, Xu X, Yang R, Chen W, Wang S, Shi S (2011) Mesenchymal stem cell-based tissue regeneration is governed by recipient T lymphocytes via IFN-gamma and TNF-alpha. Nat Med 17:1594–1601

London A, Itskovich E, Benhar I, Kalchenko V, Mack M, Jung S, Schwartz M (2011) Neuroprotection and progenitor cell renewal in the injured adult murine retina requires healing monocyte-derived macrophages. J Exp Med 208:23–39

Meier C, Middelanis J, Wasielewski B, Neuhoff S, Roth-Haerer A, Gantert M, Dinse HR, Dermietzel R, Jensen A (2006) Spastic paresis after perinatal brain damage in rats is reduced by human cord blood mononuclear cells. Pediatr Res 59:244–249

Newman MB, Willing AE, Manresa JJ, Sanberg CD, Sanberg PR (2006) Cytokines produced by cultured human umbilical cord blood (HUCB) cells: implications for brain repair. Exp Neurol 199:201–208

Pimentel-Coelho PM, Magalhaes ES, Lopes LM, deAzevedo LC, Santiago MF, Mendez-Otero R (2010) Human cord blood transplantation in a neonatal rat model of hypoxic-ischemic brain damage: functional outcome related to neuroprotection in the striatum. Stem Cells Dev 19:351–358

Ramos AL, Darabi R, Akbarloo N, Borges L, Catanese J, Dineen SP, Brekken RA, Perlingeiro RC (2010) Clonal analysis reveals a common progenitor for endothelial, myeloid, and lymphoid precursors in umbilical cord blood. Circ Res 107:1460–1469

Reich DM, Hau S, Stahl T, Scholz M, Naumann W, Emmrich F, Boltze J, Kamprad M (2008) Neuronal hypoxia in vitro: investigation of therapeutic principles of HUCB-MNC and CD133+ stem cells. BMC Neurosci 9:91

Rosenkranz K, Kumbruch S, Lebermann K, Marschner K, Jensen A, Dermietzel R, Meier C (2010) The chemokine SDF-1/CXCL12 contributes to the 'homing' of umbilical cord blood cells to a hypoxic-ischemic lesion in the rat brain. J Neurosci Res 88:1223–1233

Rowe DD, Leonardo CC, Recio JA, Collier LA, Willing AE, Pennypacker KR (2012) Human umbilical cord blood cells protect oligodendrocytes from brain ischemia through Akt signal transduction. J Biol Chem 287(6):4177–4187

Schira J, Gasis M, Estrada V, Hendricks M, Schmitz C, Trapp T, Kruse F, Kogler G, Wernet P, Hartung HP, Muller HW (2012) Significant clinical, neuropathological and behavioural recovery from acute spinal cord trauma by transplantation of a well-defined somatic stem cell from human umbilical cord blood. Brain 135(Pt 2):431–446

Sohlberg E, Saghafian-Hedengren S, Bremme K, Sverremark-Ekstrom E (2011) Cord blood monocyte subsets are similar to adult and show potent peptidoglycan-stimulated cytokine responses. Immunology 133:41–50

Sun J, Allison J, McLaughlin C, Sledge L, Waters-Pick B, Wease S, Kurtzberg J (2010) Differences in quality between privately and publicly banked umbilical cord blood units: a pilot study of autologous cord blood infusion in children with acquired neurologic disorders. Transfusion 50:1980–1987

Taguchi A, Soma T, Tanaka H, Kanda T, Nishimura H, Yoshikawa H, Tsukamoto Y, Iso H, Fujimori Y, Stern DM, Naritomi H, Matsuyama T (2004) Administration of CD34+ cells after stroke enhances neurogenesis via angiogenesis in a mouse model. J Clin Invest 114:330–338

Tanaka N, Kamei N, Nakamae T, Yamamoto R, Ishikawa M, Fujiwara H, Miyoshi H, Asahara T, Ochi M, Kudo Y (2010) CD133+ cells from human umbilical cord blood reduce cortical damage and promote axonal growth in neonatal rat organ co-cultures exposed to hypoxia. Int J Dev Neurosci 28:581–587

van Velthoven CT, Kavelaars A, van Bel F, Heijnen CJ (2010) Repeated mesenchymal stem cell treatment after neonatal hypoxia-ischemia has distinct effects on formation and maturation of new neurons and oligodendrocytes leading to restoration of damage, corticospinal motor tract activity, and sensorimotor function. J Neurosci 30:9603–96011

Vendrame M, Gemma C, de Mesquita D, Collier L, Bickford PC, Sanberg CD, Sanberg PR, Pennypacker KR, Willing AE (2005) Anti-inflammatory effects of human cord blood cells in a rat model of stroke. Stem Cells Dev 14:595–604

Vendrame M, Gemma C, Pennypacker KR, Bickford PC, Davis Sanberg C, Sanberg PR, Willing AE (2006) Cord blood rescues stroke-induced changes in splenocyte phenotype and function. Exp Neurol 199:191–200

Xia G, Hong X, Chen X, Lan F, Zhang G, Liao L (2010) Intracerebral transplantation of mesenchymal stem cells derived from human umbilical cord blood alleviates hypoxic ischemic brain injury in rat neonates. J Perinat Med 38:215–221

Yasuhara T, Hara K, Maki M, Xu L, Yu G, Ali MM, Masuda T, Yu SJ, Bae EK, Hayashi T, Matsukawa N, Kaneko Y, Kuzmin-Nichols N, Ellovitch S, Cruz EL, Klasko SK, Sanberg CD, Sanberg PR, Borlongan CV (2010) Mannitol facilitates neurotrophic factor upregulation and behavioural recovery in neonatal hypoxic-ischaemic rats with human umbilical cord blood grafts. J Cell Mol Med 14:914–921

Ye Z, Zhan H, Mali P, Dowey S, Williams DM, Jang YY, Dang CV, Spivak JL, Moliterno AR, Cheng L (2009) Human-induced pluripotent stem cells from blood cells of healthy donors and patients with acquired blood disorders. Blood 114:5473–5480

Yu G, Borlongan CV, Ou Y, Stahl CE, Yu S, Bae E, Kaneko Y, Yang T, Yuan C, Fang L (2010) In vitro non-viral lipofectamine delivery of the gene for glial cell line-derived neurotrophic factor to human umbilical cord blood CD34+ cells. Brain Res 1325:147–154

Zhang X, Hirai M, Cantero S, Ciubotariu R, Dobrila L, Hirsh A, Igura K, Satoh H, Yokomi I, Nishimura T, Yamaguchi S, Yoshimura K, Rubinstein P, Takahashi TA (2011) Isolation and characterization of mesenchymal stem cells from human umbilical cord blood: reevaluation of critical factors for successful isolation and high ability to proliferate and differentiate to chondrocytes as compared to mesenchymal stem cells from bone marrow and adipose tissue. J Cell Biochem 112:1206–1218

Pathological Angiogenesis: An Overview

28

Jennifer Roth, Rajiv D. Desai, Robert Friesel, and Peter C. Brooks

Contents

Abstract

The concept that tumor growth depends in large part on angiogenesis or the development of new blood vessels from pre-existing vessels has stimulated great interest in developing novel anti-angiogenic reagents for the treatment of malignant tumors. While encouraging results from human clinical trials have lead to the approval of new anti-angiogenic drugs, these novel therapeutics have thus far exhibited relatively limited clinical impact. Furthermore resistance to the anti-angiogenic effects of this new class of compounds has been observed. With the growing realization that the molecular and cellular mechanisms that regulate pathological angiogenesis may in part be different from that of physiological angiogenesis, new efforts are being placed on understanding the pathological angiogenic cascade as it occurs within tissue specific microenvironments. In this regard, we will provide a brief overview of the molecular and cellular mechanisms thought to control pathological angiogenesis. In addition, we will briefly discuss the general importance of understanding how angiogenesis may be governed in part by an organ microenvironment with particular attention focused on pathological angiogenesis within the central nervous system.

J. Roth • R.D. Desai • R. Friesel • P.C. Brooks (✉)
Center for Molecular Medicine, Maine Medical Center Research Institute, 81, Research Drive, Scarborough, ME 04074, USA
e-mail: brookpl@mmc.org

M. Hayat (ed.), *Tumors of the Central Nervous System, Volume 13*,
DOI 10.1007/978-94-007-7602-9_28,

Introduction

To efficiently exchange oxygen and nutrients needed for tissue expansion and to maintain homeostasis, multicellular organisms have evolved highly integrated strategies to develop elaborate circulatory networks of blood and lymphatic vessels. The formation of these circulatory loops is controlled by cues derived from both the non-cellular and cellular compartments within a particular tissue microenvironment. Given the complexity and tissue specific mechanisms that may contribute to the development of these two different circulatory systems, we will focus the majority of our attention on a general overview of the mechanisms by which new blood vessels form from pre-existing vessels (angiogenesis) and in particular, we will focus on the integrated network of cellular and molecular processes thought to govern pathological angiogenesis.

It is clear from many prior studies that the formation of blood vessels during normal physiological events, such as embryonic development, shares a number of common cellular and molecular regulators with that of pathological vessel formation. However, while similarities exist, it is also becoming evident that there are a number of differences between physiological and pathological angiogenesis. Gaining an in depth molecular understanding of these differences are necessary for the development of more effective antiangiogenic strategies that specifically target pathological angiogenesis within a particular tissue microenvironment. While providing a general overview of pathological angiogenesis, we will also highlight specific examples of the unique features of pathological vessel formation as it occurs within the Central Nervous System (CNS) and how these differences may help guide future development of new and optimized strategies to treat pathological neovascularization.

Development of Functional Blood Vessels

The formation of functional blood vessels can occur by at least 3 processes including, vasculogenesis, arteriogenesis and angiogenesis (Fig. 28.1). Vasculogenesis is the de novo formation of blood vessels from precursor cells such as the hemangioblast (Bohnsack and Hirscchi 2004). Hemangioblasts in turn can give rise to both hematopoietic progenitors as well as angioblasts, the later of which are thought to ultimately give rise to endothelial cells. Endothelial progenitor cells (EPCs) are thought to contribute to the formation of blood vessels during embryonic development or vasculogenesis (Bohnsack and Hirscchi 2004). Interestingly, emerging evidence now suggests that these EPCs as well as bone marrow derived myeloid cells may also contribute to postnatal angiogenesis during wound healing and tumor growth (Tilki et al. 2009). However, the extent of the contribution of these EPCs and other bone marrow derived cells to postnatal blood vessel formation is not completely understood and may depend in part, on the particular tissue microenvironment associated with the active angiogenic process in question. This important concept has recently been highlighted in relationship to the contribution of postnatal vasculogenesis to the formation of recurrent Glioblastoma Multiforme (GBM) and its associated blood vessels within the CNS (Kioi et al. 2010).

In contrast to vasculogenesis, arteriogenesis can be described as the formation of functional blood vessels by the activation and expansion of pre-existing small non-functional collateral vessels. Angiogenesis, which will be the major focus of this overview, is the formation of new blood vessels from pre-existing blood vessels. Angiogenesis can be divided into intussusceptive and sprouting angiogenesis. During intussusception, vessels are functionally split by the formation of intra-luminal tissue walls, which result in the formation of higher order branching structures (Nico et al. 2010). In general, this form of vascular remodeling does not require extensive endothelial cell proliferation or extracellular matrix (ECM) remodeling and may occur more often during embryonic development than during postnatal vessel formation. However, examples of intussusceptive angiogenesis have been described in some tumor types such as GBM (Nico et al. 2010). Interestingly, evidence now suggests that recovery of some tumor blood vessels following anti-angiogenic treatment may occur by

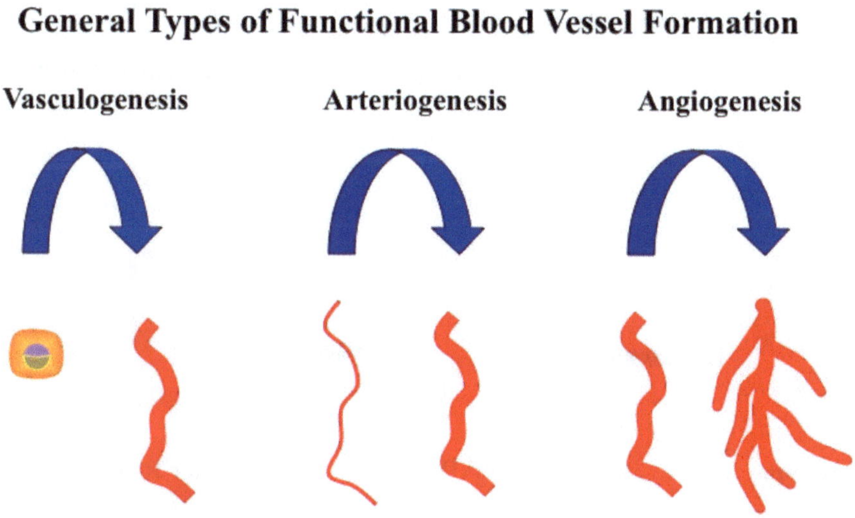

Fig. 28.1 General mechanisms of blood vessel formation. The formation of functional blood vessels can occur by several mechanisms. Three major mechanisms by which blood vessels may form include, Vasculogenesis, Arteriogenesis and Angiogenesis. Vasculogenesis is the process by which new blood vessels form from precursor cells such as hemangioblasts. Arteriogenesis is the process by which functional blood vessels form from small non-functional collateral vessels. Angiogenesis is the process by which new blood vessels form from pre-existing vessels

intussusceptive rather than sprouting angiogenesis (Nico et al. 2010). Given that intussusception does not require significant cellular proliferation, treatments with conventional anti-angiogenic drugs might have little impact on tumor perfusion under these circumstances.

Sprouting angiogenesis on the other hand is distinct from that of intussusception, as specialized endothelial tip cells are activated by a gradient of angiogenic stimuli to induce remodeling of the local ECM and directional invasion and migration into the surrounding tissue. Interestingly, the formation of the activated endothelial tip cells has been suggested to be controlled at least in part, by Notch/Dll4 signaling (Hellstrom et al. 2007). Studies have shown that Notch/Dll4 signaling may act to regulate the formation of endothelial tip cells and reduce sprouting in vivo. These activated endothelial cells along with their associated endothelial stalk cells are thought to organize into a solid cord of cells, undergo a complex process of lumen formation or canularization and finally, recruit supporting accessory cells to stabilize the newly forming blood vessel. This continuum of events within the angiogenic cascade can be organized into three interconnected phases (Fig. 28.2) including an initiation phase, an invasive phase and a maturation phase. The complex cellular and molecular mechanisms involved in the formation of functional blood vessels require fine-tuning and cooperation among numerous cell types and a wide variety of molecular regulators. For example, while Vascular Endothelial Cell Growth Factor (VEGF) is well known to stimulate vessel formation, signaling mechanisms have evolved to limit excessive vascular growth in order to optimize vessel function to provide adequate tissue perfusion. When these finely tuned mechanisms go out of balance, it can lead to either uncontrolled vessel formation and/or the formation of non-productive vessels. To this end, studies have shown that in addition to stimulating angiogenesis, VEGF can induce expression of regulators that act to limit new blood vessel formation. An example of this negative feed back system involves the VEGF induced Notch/Dll4 signaling loop which may limit the number of endothelial tip cells formed during angiogenesis (Hellstrom et al. 2007). Moreover, VEGF may also limit functional blood vessel formation by altering PDGFβ signaling in smooth muscle cells and the subsequent recruitment of pericytes to stabilize the developing vessel (Greenberg et al. 2008).

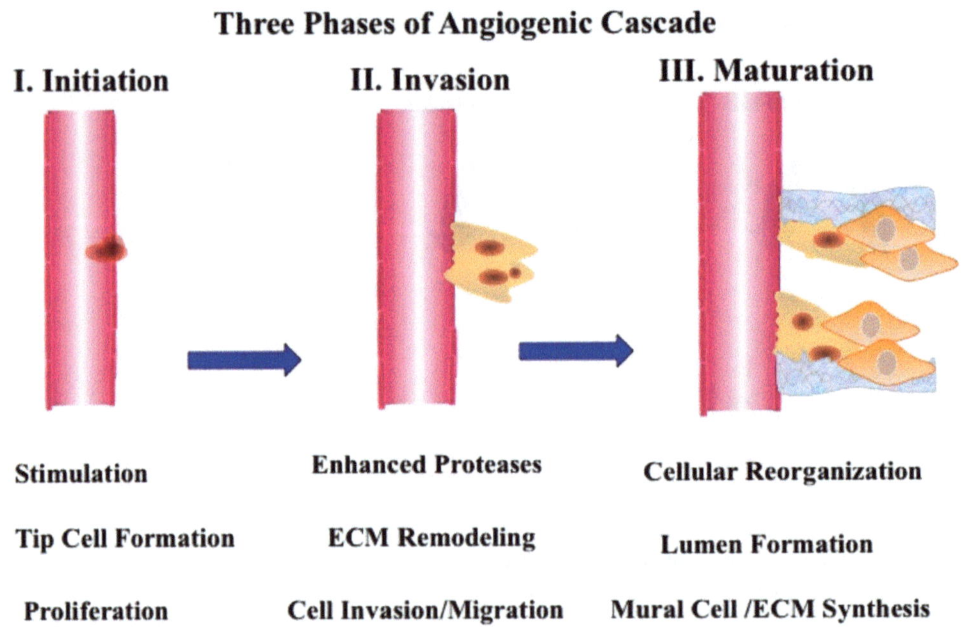

Fig. 28.2 **The angiogenic cascade.** Angiogenesis or the process by which new blood vessels form from pre-existing vessels requires a highly integrated set of molecular and cellular events. The interconnected cascade of events that govern angiogenesis can be organized into three stages. The initiation phase can be characterized by growth factor stimulated formation of tip cells and endothelial cell proliferation. The invasive phase of angiogenesis can be characterized by enhanced expression of proteolytic enzymes, localized extracellular matrix (ECM) remodeling and cellular invasion and migration. Cellular re-organization, lumen formation, recruitment of mural cells and synthesis of new ECM components to provide support for the newly forming vessel characterize the maturation phase

Thus, as with most signaling pathways that govern complex biological processes, the tissue specific repertoire of regulatory factors and the extent of the activation of negative feed back loops are critical for productive vessel formation.

Additional layers of complexity involved with angiogenesis may be observed when considering tissue specific angiogenesis. For example, distinct organ microenvironments may require unique combinations of the various classes of regulatory components needed for successful completion of the angiogenic cascade. New evidence is accumulating that tissue specific alteration in the expression of small noncoding RNAs called miRNAs may play an important role in fine-tuning the expression of crucial angiogenesis regulatory molecules. In fact, investigators have identified several miRNAs, which regulate angiogenesis. Interestingly, the miRNA-296 was increased in association with brain tumor endothelial cells as compared to normal brain endothelial cells (Wurdinger et al. 2008). Moreover, miRNA-132 has also been shown to be upregulated in tumor endothelial cells as compared to normal endothelial cells and has been suggested to play a critical role in promoting the angiogenic switch (Anand et al. 2010). Given these observations, it would not be surprising that pathological angiogenesis occurring within one particular organ may not be the same as that occurring in a different organ system. This emerging idea needs to be considered as more emphasis is being placed on the broad use of anti-angiogenic drugs for the treatment of malignant tumors. Here, we will focus on the cellular and molecular mechanisms that regulate pathological angiogenesis with an emphasis on vessel formation within the unique microenvironment of the central nervous system.

Cellular and Molecular Regulators of the Angiogenic Cascade

While it is likely that tissue specific mechanisms exist to control angiogenesis within a given tissue microenvironment, the overall angiogenic cascade can be organized into a continuum of three interconnected steps (Fig. 28.2). We will briefly discuss several well-characterized examples of the major cellular and molecular regulators of angiogenesis as they relate to the functional steps that occur during new blood vessel formation. While it has long been appreciated that endothelial and mural cells such as vascular smooth muscle cells and pericytes play active roles in physiological and pathological angiogenesis, new evidence is providing critical insight into the functional roles of numerous additional cell types in the angiogenic cascade. Studies have provided compelling evidence that cell types as diverse as fibroblast, tumor cells, bone-marrow derived progenitor cells as well as inflammatory cells such as macrophages and mast cells all contribute to the formation of new blood vessels (Contois et al. 2009). These important findings are crucial to our overall mechanistic understanding of angiogenesis, as these diverse cell types are known to represent many of the major tissue specific sources of miRNAs that may ultimately contribute to controlling the relative expression of pro and anti-angiogenic factors. Thus, it is interesting to note that these accessory cells may well represent important new therapeutic targets to help control pathological angiogenesis.

While tumor cells are well known to express many soluble pro-angiogenic growth factors, cytokines and chemokines such as VEGF, PDGF, FGF, IL-8, and SDF-1, a number of inflammatory cells such as mast cell and macrophages also express many similar pro-angiogenic molecules. In addition, distinct cell types may also contribute to the generation of endogenous angiogenesis inhibitors formed as fragments of larger molecules such as Angiostatin, PEX, Endostatin, and several NC1 domains of collagen type-IV (Contois et al. 2009). While the molecular mechanisms by which these inhibitors function in vivo is not completely understood, the relative levels of these factors in relationship to overall levels of pro-angiogenic molecules helps to dictate whether or not the angiogenic switch is activated and whether new blood vessels begin to form. Therefore, as it can be appreciated from this brief summary, many diverse cell types beyond endothelial cells play critical roles in angiogenesis.

In addition to the expression of growth factor, cytokines, chemokines and their respective cell surface receptors, other classes of regulatory molecules also play crucial roles in angiogenesis including several different types of adhesion molecules. For example, during the initiation phase, pro-angiogenic stimulation by growth factors can lead to alterations in endothelial cell-cell and cell-ECM interactions. Studies have suggested that stimulation by growth factors such as VEGF can alter the functional localization and activity of Vascular Endothelial Cadherin (VE-Cadherin), a well-characterized cell-cell adhesion molecule known to play a role in angiogenesis (Dejana et al. 2008). This altered localization and functional activity can lead to disruption of endothelial cell-cell contacts thereby promoting vascular leakage of serum proteins that contributes to the formation of a modified ECM and enhance cellular invasion and migration. This modified ECM may also provide selective homing signals to facilitate localization of additional angiogenic accessory cells. Growth factor stimulation can also enhance the expression and activation of a class of transmembrane cell adhesion receptors termed integrins. An expanding group of integrins including $\alpha 1\beta 1$, $\alpha 2\beta 1$, $\alpha 4\beta 1$, $\alpha 5\beta 1$, $\alpha 6\beta 4$ $\alpha v\beta 3$, $\alpha v\beta 5$ and $\alpha v\beta 8$ are all though to play roles in regulating angiogenesis (Contois et al. 2009). It is important to point out that these alterations in adhesion are not restricted solely to endothelial cells but may occur in other cells known to play roles in neovascularization. This altered adhesive activity can in turn contribute to changes in expression of proteolytic enzymes, another critical group of angiogenic regulatory molecules.

Importantly, integrin mediated signaling within a variety of cell types such as endothelial cells; fibroblasts and tumor cells can result in enhanced expression and activity of proteolytic enzymes such as matrix metalloproteinases (MMPs) and serine proteases (Contois et al. 2009). Release of various chemoattractants can act to recruit additional cell types such as macrophages, mast cells and bone marrow derived myeloid cells, which collectively contribute to overall levels of proteolytic enzymes at sites of new vessel growth. This localized increase in proteolytic activity can in turn further modify the vascular ECM. This newly remodeled vascular ECM has been shown to play an active role in facilitating angiogenesis by mechanisms as diverse as enhancing the release of matrix bound growth factors such as VEGF, to exposing unique matrix immobilized cryptic ECM epitopes (MICEE) to facilitate localized integrin signaling events that contribute to adhesion, migration and proliferation (Contois et al. 2009).

While growth factors, cell adhesion molecules, ECM proteins, proteolytic enzymes and many other factors all contribute to new blood vessel development, they do not function in isolation, but rather work cooperatively within an integrated network to allow productive angiogenesis to occur. During the initiation phase of angiogenesis, the balance between pro and anti-angiogenic regulatory molecules become shifted in favor of angiogenic stimulation, resulting in a significant increase in the relative expression of a number of soluble pro-angiogenic factors such as VEGF, FGF2, IGF-1, IL-8 and SDF-1. The relative levels of these angiogenic factors may be fine tuned by miRNAs. Conversely, the expression of certain miRNAs can be governed by angiogenic growth factors. For example, VEGF and FGF2 have recently been shown to mediate induction of miRNA-132 in endothelial cells (Anand et al. 2010). As mentioned above, this increased expression of pro angiogenic growth factors can originate from a variety of cell types within the microenvironment including tumor cells, inflammatory infiltrates, cancer associated fibroblasts as well as numerous bone marrow derived myeloid cells. Secreted pro-angiogenic molecules can bind their respective cell surface receptors on endothelial cells to activate a cascade of down stream signaling networks such as P13K/Akt pathways that ultimately lead to enhanced cellular proliferation, altered expression and activation of cell adhesion molecules and enhanced expression of proteolytic enzymes (Jiang and Lui 2009). Interestingly, recent evidence suggests that miRNA-126 may specifically regulate down stream signaling associated with both MAP/Erk and PI3k/Akt signaling pathways during new vessel formation (Wang et al. 2008). In fact, the ability of miRNA-126 to modulate angiogenesis was shown to be associated in part, with its ability to repress expression of the sprouty-related protein SPRED1, which is thought to modulate MAP/Erk signaling (Wang et al. 2008). Interestingly, members of the Sprouty family, which are thought to negatively regulate MAP/Erk signaling, may also regulate angiogenesis. For example, recent studies indicate that overexpression of Sprouty-4 may inhibit embryonic vascular sprouting, while Sprouty-1 may inhibit endothelial cells proliferation in part by altering levels of the cyclin dependent kinase inhibitors P21 and P27 (Lee et al. 2001, 2010). In addition to impacting the MAP/Erk pathway, miRNA-126 has also been shown to modulate the PI3k/Akt pathway by altering the expression of the PI3kinase regulatory subunit p85β (Wang et al. 2008). Intriguing new evidence suggests that growth factor-mediated upregulation of miRNA-132 in endothelial cells may repress expression of p120RASGAP, which results in increased RAS activity and enhanced endothelial cell proliferation and angiogenesis in vivo (Anand et al. 2010). In fact, antagomers directed to block miRNA-132 were shown to inhibit angiogenesis and tumor growth in multiple models (Anand et al. 2010). Finally, pro-angiogenic stimulation can be further enhanced by growth factors stimulation of additional accessory cells, which collectively add to the overall proliferative loop within the local microenvironment.

The initiation phase of angiogenesis is followed closely by the invasive phase. Proteolytic remodeling of the vascular basement membrane as well as the interstitial matrix can result in structural changes within the ECM. These changes may contribute to the release of matrix bound growth factors to enhance cellular proliferation and migration. In addition, biomechanical alterations in the structural integrity of ECM molecules such as collagen and laminin may specifically expose cryptic regulatory sites that are recognized by cell adhesion receptors such as integrins (Xu et al. 2001; Akula et al. 2007). These new receptor ligand interactions may allow initiation of distinct signaling cascades that are largely restricted to sites of matrix remodeling. Recent evidence has indicated that cellular interactions with these newly exposed cryptic sites within the matrix may regulate proliferation, survival and facilitate invasion and migration during angiogenesis and other invasive cellular processes (Xu et al. 2001; Akula et al. 2007). In fact, antibodies directed to these cryptic sites have been shown to inhibit angiogenesis, tumor growth and metastasis in vivo (Xu et al. 2001; Akula et al. 2007). These studies have recently lead to the translation of this new approach into a human clinical trial. In particular, a Phase I clinical trial using a novel Mab termed TRC093 directed to a cryptic collagen epitope has been completed (Robert et al. 2010). This novel approach of targeting highly selective cryptic sites within the ECM, rather than direct targeting of individual vascular cells, resulted in little if any toxicity and importantly, evidence of antitumor activity along with reduced levels of circulating VEGF was observed (Robert et al. 2010).

In contrast to promoting angiogenesis, proteolytic enzymes may also lead to limiting new blood vessel formation. For example, protease mediated degradation of ECM proteins may result in release of inhibitory fragments of collagen such as endostatin and several NC1 domains of collagen type-IV (Kessenbrock et al. 2010). It is interesting to point out that certain MMPs such as MMP-8 may be associated with reduced tumor progression, which might be due to multiple mechanisms such as to excessive proteolysis leading to inactivation of cell adhesion receptors, degradation of angiogenic growth factors and elevated release of endogenous ECM-derived angiogenesis inhibitors (Kessenbrock et al. 2010). Collectively, these studies again illustrate the fine balance that is needed between pro and anti-angiogenic signaling events for productive angiogenesis to occur.

During the maturation phase, endothelial cells begin to reorganize into cords of cells. These reorganized endothelial cells exhibit altered phenotypic characteristics by establishing new cell-cell interactions mediated by cell adhesion molecules such as cadherins. New signaling cascades facilitated by both cell-cell as well as integrin mediated interaction with the surrounding ECM contributes to the formation of a functional lumen by a process termed canularization. While relatively little is known concerning the molecular mechanisms regulating canularization, recent studies have implicated diverse molecules such integrin $\alpha 2\beta 1$, MMPs, and VE-Cadherin in lumen formation (Davis et al. 2007). In addition to lumen formation during vessel maturation, the measured release of growth factors and chemokines are thought to help recruit mural cells, pericytes and potentially, bone marrow derived progenitor cells to support the newly forming vessel. The newly sequestered supporting cells also contribute to the expression of ECM molecules which help construct new vascular basement membranes, which when taken together with the accessory cells, provides mechanical support for the maturing vessel. As can be appreciated from this brief overview of the general steps in the angiogenic cascade, it would not be surprising that numerous differences exist between physiological and pathological angiogenesis.

Physiological vs. Pathological Angiogenesis

It has been known for decades that distinct differences exist between normal blood vessels and blood vessels that form during pathological processes such as tumor growth. One of the most obvious differences involves a striking alteration in morphology. In contrast to quiescent blood vessels, tumor blood vessel are often characterized as being highly proliferative, leaky, tortuous vessels with altered basement membranes and reduced numbers of supporting cells. Studies have shown that tumor endothelial cells may exhibit altered expression of a variety of regulatory molecules that contribute to the unique phenotype of tumor blood vessels and this may provide important molecular insight into pathological angiogenesis. For example, studies have shown elevated expression of efflux pump proteins such as ABCG2 in GBM blood vessels as well as increased levels of apoptosis regulators such survivin (Zhang et al. 2003). Moreover, following laser capture micro-dissection, isolated GBM endothelial cells exhibited enhanced resistance to cytotoxic drugs, elevated levels of growth factors and cytokines such VEGF, IL-8 and CXCL12/SDF1, and altered expression of miRNAs that regulate angiogenesis such as miRNA-296 (Wurdinger et al. 2008). Interestingly, endothelial cells derived from multiple tumor types were shown to express elevated levels of miRNA-132 as compared to quiescent endothelial cells (Anand et al. 2010). Finally, GBM endothelial cells exhibited increased migratory capacity as compared to their normal counterparts. All of these factors and many others have been shown to impact new blood vessel development and thus may contribute to angiogenesis in a tissue specific manner.

In addition to the specific differences between normal and tumor endothelial cells described above, differences have also been documented within the local non-cellular microenvironment. For example, RGD containing provisional matrix proteins such as fibrin, vitronectin and fibronectin are often found elevated in association with tumor blood vessels as compared to quiescent vessels. These observations may not be surprising as soluble forms of these ECM proteins exist in the circulation and tumor vessels are often leaky. In particular, vitronectin, which is normally a relatively minor component of the interstitial matrix, was highly expressed in association with brain tumors and associated blood vessels (Gladson and Cheresh 1991). Moreover, alternatively spliced forms of fibronectin, which contain EIIIA and EIIIB domains are rarely found in association with normal blood vessels but have been readily detected in tumor vessels (Santimaria et al. 2003). Structurally altered forms of ECM proteins such as collagen and laminin were also shown to be expressed in association within tumor blood vessels (Xu et al. 2001; Akula et al. 2007). These observations are of considerable significance as cellular interactions with these ECM proteins may modify the response of vascular cells to soluble pro and anti- angiogenic mediators.

Impact of the Non-cellular Microenvironment on Angiogenesis

As discussed above, the numerous cell types that produce and respond to soluble pro and anti-angiogenic components have to be coordinated to allow angiogenesis to occur in a productive manner (Fig. 28.3). This coordination is regulated in large part by interactions with the distinct ECM microenvironments associated with the particular tissue within which angiogenesis is occurring. For purposes of this discussion, the local ECM can be organized into two general compartments including the vascular basement membrane and interstitial matrix. It would be well beyond the scope of this review to describe all the various components of each of the ECM compartments, thus we will only briefly highlight a few well-studied examples to illustrate critical points. The vascular basement membranes, which underlay blood vessels, are composed of numerous ECM proteins. Major components of the vascular basement membrane that help form a thin sheet-like structure include distinct isoforms of collagen type-IV and laminin. Molecules such as nidogen/enactin and perlecan connect these sheet-like networks of ECM proteins. Additional genetically distinct forms of collagen as well as different fibronectin isoforms can also be found in association with the basement membranes depending on the particular organ microenvironment and developmental stage. Beyond the sheet-like vascular basement membrane is the loosely arranged interstitial matrix composed largely of fibrillar collagen such as collagen type-I, II and III as well as fibronectin. In addition to these major interstitial glycoproteins an interconnected network of other matricellular ECM protein and proteoglycans can be found. These include, Thrombospondin and several heparin sulfate proteoglycans. The relative ratios and distribution of these individual ECM components can vary substantially depending on the particular tissue microenvironment.

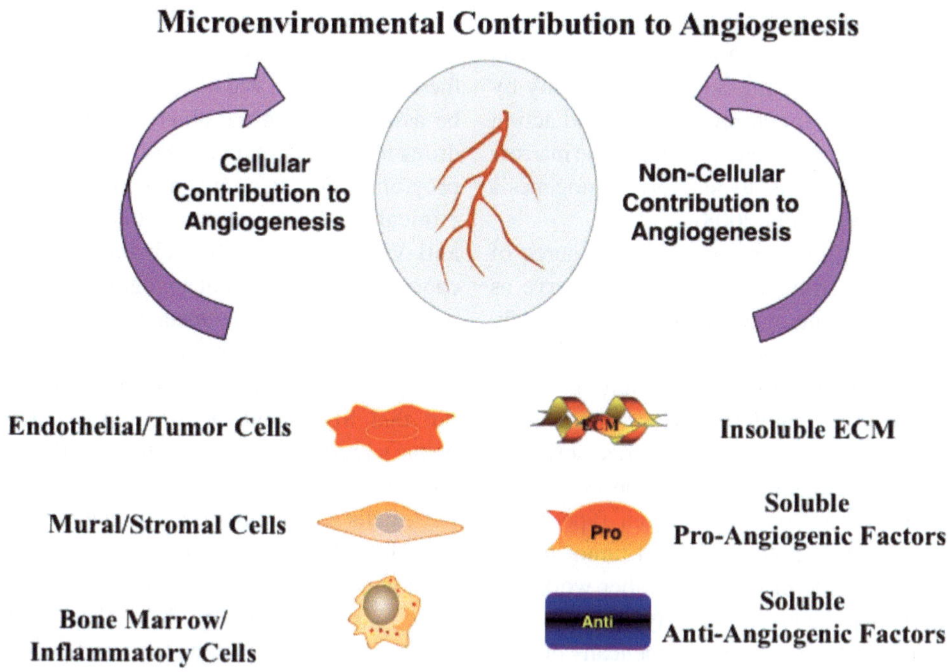

Fig. 28.3 Microenvironment contribution to angiogenesis. The formation of blood vessels can be regulated by a variety of components within a particular tissue microenvironment. An integrated group of distinct cell types as well as soluble and insoluble angiogenic regulatory factors function cooperatively to regulate the angiogenic switch within a given tissue microenvironment. Shifts in the relative abundance of these regulatory components help provide mechanisms to control tissue specific angiogenesis

The complex interwoven networks of ECM proteins provide unique and often tissue specific sources of angiogenesis regulators. For example, beyond providing mechanical support for blood vessels, the ECM is known to act as a reservoir of numerous growth factors such as VEGF and FGF2. These growth factors can be proteolytically released as sources for soluble growth factor gradients to facilitate directional migration (Chen et al. 2010). In addition, matrix immobilization of growth factors may alter their functional activity. In this regard, some studies suggest that these immobilized growth factors may be recognized in their ECM-bound forms and specifically stimulate receptor-mediated angiogenic signaling pathways. For example, new studies have indicated that matrix bound forms of VEGF may initiate sustained VEGFR2 activation leading to differential tyrosine phosphorylation, ultimately promoting sustained activation of P38 MAPkinase (Chen et al. 2010). This unique signaling pathway depends on cellular interactions with ECM bound VEGF and the formation of a β1 integrin complex with VEGFR2.

In addition to functioning as a source of growth factors, the tissue specific ECM composition serves as distinct ligands for integrin receptors. Integrin-mediated interactions with the ECM is well known to regulate diverse cellular processes ranging from altering the local expression of growth factors and their receptors and proteolytic enzymes, to modulating cell cycle progression, proliferation, migration and cell survival (Contois et al. 2009). Proteolytic remodeling of the local ECM has also been shown to release a variety of functional fragments from proteins such as collagen, laminin, fibrinogen, fibronectin and elastin. These bioactive ECM fragments termed Matrikines have been shown to play many roles in angiogenesis and tumor progression (Adair-Kirk and Senior 2008). Studies have documented that some of these proteolytic ECM fragments have pro-angiogenic activity by stimulating vascular cell proliferation and acting to recruit inflammatory cells as well as bone marrow derived myeloid cells to support angiogenesis (Adair-Kirk and Senior 2008).

Just as the ECM can be a specific source of pro-angiogenic stimulation, it can also serve as reservoir of anti-angiogenic factors. As we briefly mentioned above, proteolytic enzymes are known to remodel ECM components resulting in the release and/or accumulation of anti-angiogenic bioactive fragments. A well-characterized example of these anti-angiogenic ECM fragments includes endostatin, a fragment of collagen type-XVIII that was shown to potently inhibit tumor angiogenesis in multiple models. In further work, our laboratory and others previously demonstrated that several NC1 domains from genetically distinct forms of basement membrane collagen type-IV could bind specific integrin receptors and inhibit angiogenesis in vivo (Petitclerc et al. 2000). It is interesting to note that a common phenomenon is rapidly emerging in that many of the anti-angiogenic ECM fragments interact with integrin receptors and that this interaction plays a significant role in their biological activity. Taken together, these studies help illustrate the critical importance of localized reciprocal communication between cells and the ECM in regulating neovascularization.

The Unique Microenvironment of the Central Nervous System and Its Implications for Controlling Pathological Angiogenesis

The well-accepted role for angiogenesis in contributing to tumor growth has lead to numerous studies focused at developing novel strategies to inhibit angiogenesis for the treatment of cancer. It was originally thought that targeting relatively stable endothelial cells would have the added advantage of escaping drug resistance commonly seen with compounds aimed at unstable malignant tumor cells. However, new experimental findings from both basic research and clinical trials demonstrate that resistance to anti-angiogenic therapy can occur, and in some circumstance, may be associated with the ability of anti-angiogenic drugs to actively induce anti-angiogenesis rescue programs. For example, new observation from clinical studies evaluating the effects of the anti-VEGF antibody Bevacizumab in colorectal cancer noted significant upregulation of several molecules thought to enhance angiogenesis and tumor progression such as SDF-1, CXCR4 and CXCL6 (Xu et al. 2009). It is interesting to point out that SDF-1 interactions with CXCR4 has been suggested to promote mobilization of CD11b+bone marrow derived myeloid cells and these cells are thought to play a functional role in vasculogenic recovery of blood vessels in glioblastoma tumors following radiation therapy in mice (Kioi et al. 2010). These observations and many others are beginning to provide new insight that may help explain in part, the lack of sustained therapeutic benefits observed with certain anti-angiogenic strategies such as those targeting VEGF signaling. For example, the development of recurrent glioblastoma following an initial inhibition with anti-angiogenic drugs was associated with the development of new vessels resulting largely from vasculogenesis rather than angiogenesis. This vasculogenic response was associated with infiltration of CD11b+bone marrow cells, which are known to secrete a variety of pro angiogenic factors (Kioi et al. 2010). Interestingly, evidence from both animal models and human clinical trials suggest that while blocking VEGF can be associated with transient inhibition of vessel growth and some improvement of symptoms such as edema, these responses may be temporary and can be followed by enhanced tumor cell invasion (Xu et al. 2009). This increased tumor cell invasion may be associated with the anti-angiogenic drugs stimulating hypoxic recovery programs that facilitate influx of bone marrow progenitor cells, which secrete a number of pro-invasive factors such as proteolytic enzymes and additional mitogenic growth factors. Alternatively, brain tumors may co-opt existing blood vessels to provide at least some of their perfusion requirements. Given that vascular cooption does not require active sprouting, drugs designed to target sprouting angiogenesis would be expected to have minimal effects on tumor growth under these circumstances. Finally, intriguing new work suggests that within certain tissue microenvironments, pre-existing vessels may be physically translocated to areas of hypoxia by biomechanical tissue contraction (Kilarski et al. 2009). This relocalization of existing blood vessels may not respond to conventional anti-angiogenic drugs providing additional ways by which blood vessels may escape anti-angiogenic therapy. Given the critical importance of the non-cellular ECM in modulating the response of cells to both pro and anti-angiogenic factors, the possibility exists that these shifts in mechanisms by which blood vessels recover following anti-angiogenic treatments may be controlled in part by specific organ microenvironments. In this regard, the CNS represents a unique tissue microenvironment composed of a variety of tissue specific stromal cells and ECM components. The ECM composition of the CNS is remarkably different from that of other organ systems in that there is relatively little if any expression of fibrilar collagens, yet the CNS is associated with significant levels of distinct ECM components such as hyuronic acid and several heparin sulphate proteoglycans within the interstitial compartment. Moreover, the vascular basement membranes of the brain blood vessels exhibit distinct ECM composition and structure as compared to other vascular beds. For example, some CNS basement membranes can be composed predominately of 411 and 511 laminin isoforms, collagen type –IV and perlecan, while brain parenchyma basement membranes have been shown to express elevated levels of 111 and 211 laminin isoforms, collagen type-IV and agrin (Engelhardt and Sorokin 2009). These observations are important since integrins and other cell adhesion receptors used by vascular cells to sense their immediate microenvironment differentially bind to these ECM proteins and thereby may alter the cells response to angiogenic factors.

During embryonic neurovascular development, a tight functional association exists between vascular cells and nerve cells. Studies have suggested that blood vessel growth within the CNS occurs in specific patterns controlled in part by endothelial cells following radial glial cell tracts (Adams and Eichmann 2010). The formation of functional CNS blood vessels involve numerous regulatory molecules, which are known to contribute to the development of the nervous system, including neuropilin, a VEGF co-receptor, and several others involved in neuronal guidance such as semaphorins, Eph/Ephrins, and the Slit/Robo signaling systems (Adams and Eichmann 2010). These and other molecules contribute to the formation of the neurovascular unit, a unique structural feature of CNS blood vessels not commonly observed in other vascular beds. These structures regulate the selective permeability of the Blood Brain Barrier (BBB). In general, these neurovascular units are composed of brain endothelial cells forming lateral tight junctions composed of proteins such as claudins, occludins and other adhesion junction proteins. The endothelial cell layers can be associated in some circumstances with two different basement membranes, one associated with the endothelial cells and one in association with the closely apposed astroglial cells. These basement membranes can be composed of distinct isotypes of laminin, collagen type-IV, and perlecan, some of which are different from that found in other vascular beds. Additional cell types that help ensheath the endothelial cell layer include astrocytes, pericytes, and leptomeningial cells. Collectively, this unique structural composition provides the highly selective BBB, which specifically limits permeability and molecular access into the brain parenchyma based on physiochemical properties such as molecular size, shape, charge and aqueous and lipid solubility.

Given the unique microenvironment of the CNS, it would not be surprising that different mechanisms function to regulate pathological angiogenesis in the brain as compared to other tissues. To this end, a considerable body of evidence has implicated integrin $\alpha v\beta 3$ in regulating pathological angiogenesis as antagonists of $\alpha v\beta 3$ have been shown to inhibit tumor angiogenesis in multiple animal models and Cilengitide, an RGD containing antagonist that inhibits $\alpha v\beta 3$ and $\alpha v\beta 5$ is currently being evaluated in late stage human clinical trials for the treatment of glioblastoma with encouraging results (Reardon et al. 2008). Despite the considerable evidence demonstrating a functional role of αvβ3 in angiogenesis, αvβ3 null mice exhibited relatively few defects in embryonic vascular beds outside of the CNS (Bader et al. 1998). Interestingly however, vascular defects were noted within the brains of these mutant mice (Bader et al. 1998). Moreover, similar cerebral vascular defects were observed in β8 null mice, suggesting the possibility of disruption of brain vascular specific integrin interactions with ECM components (Proctor et al. 2005). Given these observations, it is interesting to point out that the α5 chain of laminin expressed within basement membranes of the brain vasculature is associated with a freely accessible RGD site that can be recognized by αvβ3 integrin (Sasaki and Timpl 2001). Other RGD containing ECM proteins such as Tenascin-C and Vitronectin can be highly expressed during brain tumor angiogenesis. Interestingly, a short isotype of VEGF, termed $VEGF^{121}$ was observed to be elevated within the microenvironment of gliomas growing intracranially, but not within similar tumors implanted subcutaneously (Guo et al. 2001). In other studies, high levels of activated αvβ3 and elevated levels of VEGF was detected within metastatic breast tumor cells growing intracranially, but remarkably this was not observed when similar tumors were growing in the mammary fat pad (Lorger et al. 2009). Collectively, these interesting experimental findings provide further support for the concept that specific tissue microenvironments may differentially govern complex biological processes such as angiogenesis by distinct mechanisms.

Conclusions

With the rapid expansion of basic and clinical research in angiogenesis, surprising new molecular insight into the complex mechanisms that govern new blood vessel formation is emerging. Exciting new experimental findings concerning the functional roles of endothelial progenitor cells and other bone marrow derived myeloid cells to pathological angiogenesis as well as important new molecular insight into the mechanisms that may contribute to anti-angiogenesis resistance, provides a strong foundation for the development of more efficacious anti-angiogenic clinical strategies. In addition to the more in-depth molecular understanding of pathological angiogenesis, the emerging concept that pathological angiogenesis may be controlled in part by tissue specific mechanisms, might ultimately lead to the development of more effective therapeutic approaches designed to inhibit angiogenesis in an organ specific manner. Taken together with the recent clinical validation of inhibiting angiogenesis as a useful strategy to help manage malignant human tumors, studying angiogenesis from a perspective that integrates both the cellular and non-cellular compartments of a tissue might lead to uncovering previously unappreciated ways by which pathological angiogenesis may be controlled.

Acknowledgements This work was supported in part by grant ROI CA91645 to PCB and NCRR grant P20RR15555 and RO1 H165301 to Robert Friesel and subproject to PCB. We would like to apologize to all those investigators whose important work was not discussed do to space limitations.

References

Adair-Kirk TL, Senior RM (2008) Fragments of extracellular matrix mediators of inflammation. Int J Biochem Cell Biol 40:1101–1110

Adams RH, Eichmann A (2010) Axon guidance molecules in vascular patterning. Cold Spring Harb Perspect Biol 2:a001875

Akula A, Roth JM, Cuant M, Policarpio D, Liebes L, Brooks PC (2007) Inhibition of angiogenesis and tumor metastasis by targeting a matrix immobilized cryptic extracellular matrix epitope in laminin. Cancer Res 9:4353–4363

Anand S, Majeti BK, Acevedo LM, Murphy EA, Mukthavaram R, Scheppke L, Huang M, Shields DJ, Lindquist JN, Lapinski PE, King PD, Weis SM, Cheresh DA (2010) MicroRNA-132-mediated loss of p120RasGAP activates the endothelium to facilitate pathological angiogenesis. Nat Med 16:909–916

Bader BL, Rayburn H, Crowley D, Hynes RO (1998) Extensive vasculogenesis, angiogenesis, and organogenesis precede lethality in mice lacking all alpha v integrins. Cell 4:507–519

Bohnsack B, Hirscchi KK (2004) Red light, green light: signals that control endothelial cell proliferation during embryonic vascular development. Cell Cycle 3:1506–1511

Chen TT, Luque A, Lee S, Anderson SM, Segura T, Iruela-Arispe ML (2010) Anchorage of VEGF to the extracellular matrix conveys differential signaling responses to endothelial cells. J Cell Biol 188:595–609

Contois L, Akula A, Brooks PC (2009) Integrins as "functional hubs" in the regulation of pathological angiogenesis. Semin Cancer Biol 19:318–328

Davis GE, Koh W, Straman AN (2007) Mechanisms controlling human endothelial lumen formation and tube assembly in three-dimensional extracellular matrices. Birth Defects Res Embryo Today 4:270–285

Dejana E, Orsenigo F, Lampugnani MG (2008) The role of adherens junctions and VE-cadherin in the control of vascular permeability. J Cell Sci 13:2115–2122

Engelhardt B, Sorokin L (2009) The blood–brain and the blood-cerebrospinal fluid barriers: function and dysfunction. Semin Immunopathol 31:497–511

Gladson CL, Cheresh DA (1991) Glioblastoma expression of vitronectin and the alpha v beta 3 integrin. Adhesion mechanism for transformed glial cells. J Clin Invest 6:1924–1932

Greenberg JI, Shields DJ, Barillas SG, Acevedo LM, Murphy E, Huang J, Scheppke L, Stockmann C, Johnson RS, Angle N, Cheresh DA (2008) A role for VEGF as a negative regulator of pericyte function and vessel maturation. Nature 7223:809–813

Guo P, Xu L, Pan S, Brekken RA, Yang ST, Whitaker GB, Nagane M, Thorpe PE, Rosenbaum JS, Su Huang HJ, Cavenee WK, Cheng SK (2001) Vascular endothelial growth factor isoforms display distinct activities in promoting tumor angiogenesis at different anatomic sites. Cancer Res 23:8569–8577

Hellstrom M, Phng L, Hofmann JJ, Wallgard E, Coultas L, Lindblom P, Alva J, Nilsson A, Karlsson L, Gaiano N, Yoon K, Rossant J, Iruela-Arispe ML, Gerhardt H, Betsholtz C (2007) Dll4 signaling through notch regulates formation of tip cells during angiogenesis. Nature 445:776–780

Jiang BH, Lui LZ (2009) PI3K/PTEN signaling in angiogenesis and tumorigenesis. Adv Cancer Res 102:19–65

Kessenbrock K, Plaks V, Werb Z (2010) Matrix metalloproteinases: regulators of the tumor microenvironment. Cell 141:52–67

Kilarski WW, Samolov B, Petersson L, Kvanta A, Gerwins P (2009) Biochemical regulation of blood vessel growth during tissue vascularization. Nat Med 6:657–664

Kioi M, Vogel H, Schultz G, Hoffman RM, Harsh G, Brown JM (2010) Inhibition of vasculogenesis, but not angiogenesis, prevents the recurrence of glioblastoma after irradiation in mice. J Clin Invest 120:694–705

Lee SH, Schloss DJ, Jarvis L, Krasnow MA, Swain JL (2001) Inhibition of angiogenesis by mouse sprouty protein. J Biol Chem 276:4128–4133

Lee S, Tri M, Nguyen B, Kovalenko D, Adhikari N, Grindle S, Polster SP, Friesel R, Ramakrishnan S, Hall J (2010) Sprouty1 inhibits angiogenesis in association with up-regulation of p21 and p27. Mol Cell Biochem 338:255–261

Lorger M, Kruger JS, O'Neal M, Stalin K, Felding-Habermann B (2009) Activation of tumor cell integrin $\alpha v\beta 3$ controls angiogenesis and metastatic growth in the brain. PNAS 106:10666–10671

Nico B, Crivellato E, Guidolin D, Annese T, Longo V, Finato N, Vacca A, Ribatti D (2010) Intussusceptive microvascular growth in human glioma. Clin Exp Med 10:93–98

Petitclerc E, Boutaud A, Prestayko A, Xu J, Sado Y, Ninomiya Y, Sarras MP Jr, Hudson BG, Brooks PC (2000) New functions for non-collagenous domains of human collagen type IV. Novel integrin ligands inhibiting angiogenesis and tumor growth in vivo. J Biol Chem 275:8050–8061

Proctor JM, Zang K, Wang D, Wang R, Reichardt LF (2005) Vascular development of the brain requires beta8 integrin expression in the neuroepithelium. J Neurosci 25:9940–9948

Reardon DA, Nabors LB, Stupp R, Mikkelsen T (2008) Cilengitide: an integrin-targeting arginine-glycine-aspartic acid peptide with promising activity for glioblastoma multiforme. Expert Opin Investig Drugs 17:1225–1265

Robert F, Gordon MS, Rosen LS, Mendelson DS, Mulay M, Adams BJ, Alvarez D, Theuer CP, Leigh BR (2010) Final results from a phase 1 study of TRC093 (humanized anti-cleaved collagen antibody) in patients with solid cancer. ASCO annual meeting abst # 3038. San Diego, CA

Santimaria M, Moscatelli G, Viale GL, Giovannoni L, Neri G, Viti F, Leprini A, Borsi L, Castellani P, Zardi L, Neri D, Riva P (2003) Immunoscintigraphic detection of the ED-B domain of fibronectin, a marker of angiogenesis, in patients with cancer. Clin Cancer Res 9:571–579

Sasaki T, Timpl R (2001) Domain IVa of laminin alpha5 chain is cell-adhesive and binds beta1 and alphaV-beta3 integrins through Arg-Gly-Asp. FEBS Lett 509:181–185

Tilki D, Hohn HP, Ergun B, Rafii S, Ergun S (2009) Emerging biology of vascular wall progenitor cells in health and disease. Trend Mol Med 15:501–509

Wang S, Aurora AB, Johnson BA, Qi X, McAnally J, Hill JA, Richardson JA, Bassel-Duby R, Olson EN (2008) The endothelial-specific micrRNA miR-126 governs vascular integrity and angiogenesis. Dev Cell 15:261–271

Wurdinger T, Tannous B, Saydam O, Skog J, Grau S, Soutscheck J, Weissleder R, Breakfield XO, Krichevsky AM (2008) miR-296 regulates growth factor receptor overexpression in angiogenic endothelial cells. Cancer Cell 14:382–393

Xu J, Rodriguez D, Petitclerc E, Kim JJ, Hangai M, Moon YS, Davis GE, Brooks PC (2001) Proteolytic exposure of a cryptic site within collagen type IV is required for angiogenesis and tumor growth in vivo. J Cell Biol 154:1069–1079

Xu L, Duda DG, di Tomaso E, Ancukiewicz M, Chung DC, Lauwers GY, Samuel R, Czito BG, Lin PC, Poleski M, Bentley R, Clark JW, Willet CG, Jain RK (2009) Direct evidence that Bevacizumab, an anti-VEGF antibody, up-regulates SDF1α, CXCL4, CXCL6 and neuropilin1 in tumors from patients with rectal cancer. Cancer Res 69:7905–7910

Zhang W, Mojsilovic-Petrovic J, Andrade MF, Zhang H, Ball M, Stanimirovic DB (2003) Expression and functional characterization of ABCG2 in brain endothelial cells and vessels. FASEB J 17:2085–2087

Use of Mobile Phones and Brain Cancer Risk in Children?

29

Martin Röösli, Maria Feychting, and Joachim Schüz

Contents

Abstract

Use of mobile phones among children and adolescents has steeply increased over the last decade and this has raised concerns about a potential increased brain tumor risk in this age group from radiofrequency electromagnetic waves emitted by mobile phones. Until now, only one multi-center case-control study and various incidence rate time trend analyses have addressed the issue. Overall these data do not suggest an increased brain tumor risk from using mobile phones. However, some uncertainties remain with respect to heavy mobile phone use, with early life exposure, with rare histological subtypes or with longer latencies and subsequent development of brain tumors in adulthood. In this special situation, with a steep increase in exposure prevalence (mobile phone use), the availability of virtually complete cancer registry data in various countries, and the limited number of known other environmental co-risk factors, result of incidence time trends analyses are considered more informative than in many other situations. Thus, further monitoring of childhood brain tumor incidence rate time trends is warranted given the dramatic public health consequences of even a small individual risk increase.

M. Röösli (✉)
Department of Epidemiology and Public Health, Swiss Tropical and Public Health Institute, Socinstrasse 57, PO Box, 4002, Basel, Switzerland
e-mail: martin.roosli@unibas.ch

M. Feychting
Institute of Environmental Medicine, Karolinska Institutet, Stockholm, Sweden

J. Schüz
Section of Environment and Radiation, International Agency for Research on Cancer (IARC), Lyon, France

M. Hayat (ed.), *Tumors of the Central Nervous System, Volume 13*,
DOI 10.1007/978-94-007-7602-9_29, © Springer Science+Business Media Dordrecht 2014

Introduction

The role of environmental factors in the etiology of childhood brain tumors is not well understood. Various environmental exposures have been postulated to cause pediatric brain tumors. High doses of ionizing radiation, as applied for childhood leukemia treatment, have been found to considerably increase the risk of developing a brain tumor (Pollack and Jakacki 2011). Contrary to the clear association with high dose of ionizing radiation, investigations of low dose of ionizing radiation and other environmental risk factors have produced mixed results. N-nitroso compounds are found in foods that contain nitrite (e.g., meat, fish and cheese) and have been shown to be carcinogenic in various animal types (Baldwin and Preston-Martin 2004). However, epidemiological studies reported inconsistent associations between maternal exposure to N-nitroso compounds during pregnancy and the risk of brain tumors in her offspring. Similarly, numerous epidemiological studies have attempted to evaluate whether pesticide exposure of the mother or the child increases the risk for pediatric brain tumors, but have not found compelling evidence for such an association (McKinney 2005).

Recently, the role of infectious agents in carcinogenesis has obtained increasing attention. Studies that used number of siblings or daycare attendance as proxies for exposure to infectious agents have produced inconsistent results (Connelly and Malkin 2007). However, a few studies suggest a link between viral infection of the mother during pregnancy and development of a malignant nervous system tumor (Baldwin and Preston-Martin 2004). The infections included rubella, mumps, varicella, herpes zoster and influenza. Whether allergic conditions are inversely correlated with glioma risk in children, as has been observed in adults, has not been widely investigated (McKinney 2005).

Head injuries and birth trauma have been suspected to cause brain tumors but seem unlikely to be relevant etiologic factors (Baldwin and Preston-Martin 2004). In a meta-analysis of 12 studies, little support was found for the hypothesis that smoking during pregnancy increases the brain tumor risk of the offspring (Huncharek et al. 2002). Interestingly, various studies reported an association for preconceptional smoking status of father and/or mothers and development of a brain tumor (Baldwin and Preston-Martin 2004); preconceptional DNA damage due to smoking or due to other toxic substances may be relevant for carcinogenesis.

Quite a few studies addressed the question whether exposure to extremely low frequency magnetic fields (ELF-MF) increases the risk for brain tumors in children. A pooled analysis of ten case-control studies did not reveal consistent evidence for such an association (Kheifets et al. 2010). Regarding preconceptional occupational exposure of the parents to ELF-MF, a German case-control study found no indication for an adverse effect (Hug et al. 2010).

In summary, apart from high doses of ionizing radiation, no other environmental exposure has been identified to play a major role in the etiology of childhood brain tumors, although, due to sparse data, many possible risk factors cannot be ruled out either.

Use of Mobile Phones Among Children

Use of mobile phones among children and adolescents has been steeply increasing during the last decade. In the 2007 the German Health Interview and Examination Survey for Children and Adolescents (KiGGS) showed that more than 50% of the 11–17 year olds reported to use their mobile phones for at least 30 min/day, for various purposes including text messaging, calls or data download (Lampert et al. 2007). Given the ongoing dispersion of wireless communication technologies, the amount of use among children is expected to increase even more, and regular use may also shift to younger age groups.

The increase in mobile phone use among children and adolescents has raised concerns about a potential brain tumor risk in this age group. It has been speculated that children may be more vulnerable to emissions from mobile

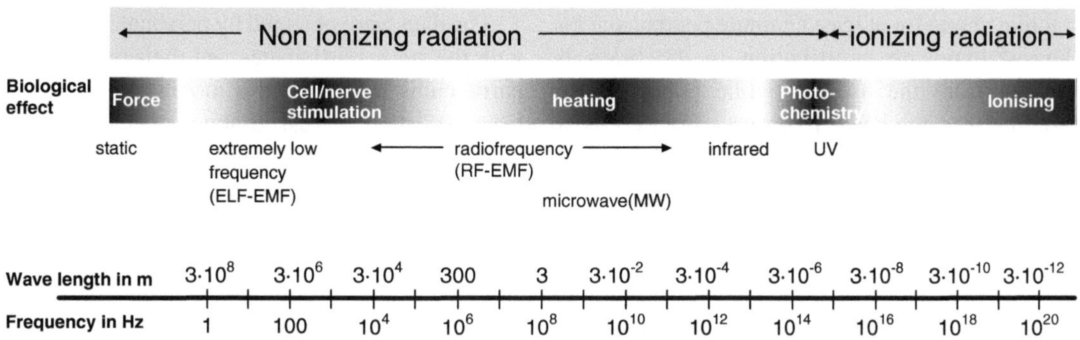

Fig. 29.1 Overview about the electromagnetic spectrum

phones because they have a developing nervous system (Kheifets et al. 2005). Their brain tissue is more conductive than that of adults because of its higher water content and ion concentration. Moreover, mobile phone radiation penetrate into regions that are deeper in their brains because of the smaller head circumference compared to adults (Wiart et al. 2008).

Exposures from Mobile Phones

Mobile phones emit electromagnetic waves in the frequency range of 800–2,300 MHz, which belongs to the radiofrequency spectrum, sometimes also called microwave radiation. Figure 29.1 shows an overview of the electromagnetic spectrum. Radiofrequency electromagnetic fields (RF-EMF) from mobile phone belong to the non-ionizing part of the spectrum. Thus, the photon energy is too weak to ionize molecules (Challis 2005) and a direct DNA damage due to RF-EMF exposure is impossible. Absorption of RF-EMF is known to heat biological tissue due its electrical conductivity. The absorbed energy in the tissue can be modeled based on the characteristics of the emitter and the tissues, or it can be measured using head phantoms. To prevent harmful heating of the body, the ICNIRP (International Commission on Non-Ionizing Radiation Protection) has suggested protection guidelines. According to these basic restrictions the average specific energy absorption rate (SAR) for localized sources like mobile phone handset emissions in any contiguous 10 g tissue region has to be less than 10 W/kg for occupational exposures or 2 W/kg for the general public (ICNIRP and (International Commission on Non-Ionizing Radiation Protection) 1998). Apart from the thermal interaction, no other mechanism has been consistently identified how RF-EMF interacts with biological tissue in a way that could be relevant for carcinogenesis.

In addition to mobile phones many other wireless communication devices emit RF-EMF, such as W-LANs (wireless local area networks), cordless phones or broadcast transmitters. Thus, the question arises how relevant is brain exposure from mobile phones compared to exposures from other sources. The absorbed energy of the body depends mainly on the source strengths and the distance to the source. Mobile phones are relatively strong transmitters since they have to reach longer distances than other daily RF-EMF sources (e.g., W-LAN, cordless phones). Obviously, fixed site transmitters used for broadcasting or mobile communication have a stronger output power than mobile phones. However, distance to the receiver is considerable larger than for mobile phone handsets which operates close to the body. In general, the SAR decreases with the square of the distance to the source, although the situation can be much more complex close to the source. For example, if the SAR value for a mobile phone is 1.0 W/kg at a distance of 0.5 cm from the head, this value would decrease by approximately a factor of 64 (=0.016 W/kg) at a distance of 4 cm. As a consequence, for a person who is using a mobile phone for at least a minute per day, the far largest exposure contribution to the brain is arising from the use of mobile phones when holding it to the head. Exposures from cordless phones are also relevant, because these devices are held to the head as well; the emitted power was lower compared to the second generation of mobile phones (Global System for Mobile Communications), but may be equally important in terms of exposure assessment compared to the UMTS (Universal Mobile Telecommunications System) mobile phone technology (Gati et al. 2009). Currently, the fourth generation (LTE: Long Term Evolution) is implemented. Preliminary measurements suggest that exposure levels of handsets will be comparable with the second generation of mobile phones. The exposure contributions from all other sources in the everyday environment are negligible for the brain (Frei et al. 2009). And there is neither any other part of the body which is exposed to any other sources on such a large population scale like the brain from the use of wireless phones. A few workers such as RF heater sealer or broadcast technicians may experience high exposure levels as well, however, this concerns only a relatively small population and is not relevant for children. For this reason epidemiologic research on the carcinogenicity of RF-EMF or microwave radiation has mainly used wireless phone use as a proxy for exposure.

Case-Control Studies

So far, results on the association between wireless phone use and childhood brain tumor have been published from one multicenter case-control study conducted in Denmark, Norway, Sweden and Switzerland (CEFALO study) (Aydin et al. 2011c). An additional multicenter case-control study is ongoing in 13 countries (http://www.mbkds.net/), however, no results are available up to now.

In CEFALO, all children and adolescents aged 7–19 years and diagnosed with a brain tumor between 2004 and 2008 in the participating countries were eligible for the study. For each patient, two controls of the same age, gender and region of residence were randomly selected from population registries.

The data were obtained by personal interviews with the study participants and their parents. All participants having used a mobile phone for at least 20 calls were asked about their usage patterns prior to diagnosis. The frequency and duration of mobile phone use was inquired for various time periods as well as the preferred side of the head and the use of hands-free kits. Furthermore, connection data from mobile phone operators were obtained whenever possible.

Data on other possible risk factors for brain tumors such as diagnostic x-ray radiation, infectious diseases and head injuries during childhood were also collected and considered in the analysis.

The association between mobile phone use and brain tumor risk was evaluated by comparing the duration and intensity of mobile phone use between cases and controls in conditional logistic regression analyses. In addition, several sensitivity analyses were conducted. For instance, the brain tumor incidence rates in Swedish children and adolescents between 1990 and 2008 as registered in the nationwide cancer registry were compared with the number of mobile phone users assuming various scenarios for hypothetical increased risks related to mobile phone use.

Overall, 352 patients and 646 controls participated in the study. The participation rates were 83% for cases and 71% for controls. Fifty-five percent of patients and 51% of controls reported regular mobile phone use (at least one call per week during at least 6 months). In the primary analysis, brain tumor risk was not significantly associated with regular mobile phone use (Odds ratio [OR] = 1.36, 95% Confidence interval [CI]: 0.92–2.02). Other exposure metrics, such as time since first mobile phone use or cumulative number and duration of calls, were also not significantly associated with brain tumors and no consistent exposure-response relationship was observed (Fig. 29.2). The risk of tumors in the brain regions most highly exposed by mobile phones (temporal lobe, frontal lobe and cerebellum) was not associated with regular mobile phone use. Regarding preferred side of the head for using the mobile phone, tumors did not occur more often on the ipsilateral compared to the contralateral side. Cordless phone use and use of

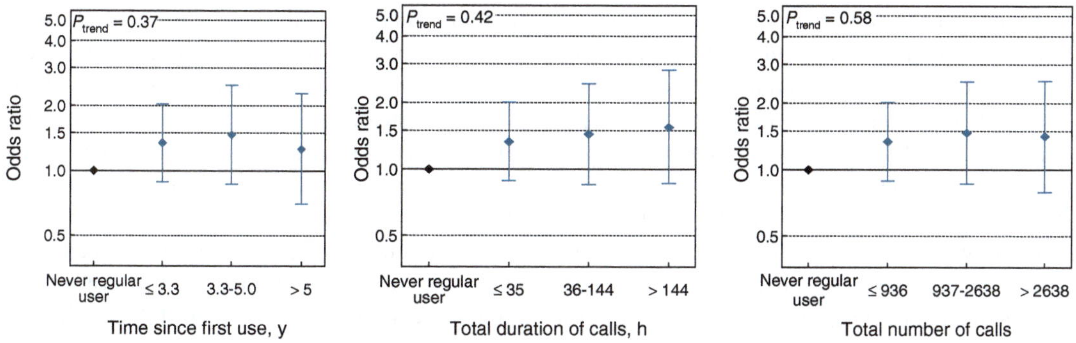

Fig. 29.2 Odds ratio (OR) and 95% confidence intervals from the multicenter CAFALO case-control study on brain tumors associated with times since first subscription (in years), cumulative duration of mobile phone use (in hours) and cumulative number of calls (Aydin et al. 2011c)

baby monitors in early childhood was also not related to an increased brain tumor risk. Objective operator data were available for only a third of the study participants who reported to have a mobile phone subscription. In this small, not randomly selected, subset, brain tumor risk was elevated for participants with the longest period since first subscription (>2.8 years) (OR = 2.15 [95%CI: 1.07–4.29]). Calculations demonstrated that such a risk, if true, would have resulted in an increase of brain tumor incidence by approximately 50% in the last few years. Such an increase was not observed in Swedish children and adolescents (Aydin et al. 2011c).

Overall, the pattern of the risk results does not suggest a causal association. First, in most analyses there was no consistent exposure-response association observed. Second, the brain tumor risk was not elevated in brain regions that are most exposed when using a mobile phone. Third, the brain tumor incidence in children and adolescents in Sweden, where the most recent data were available, has rather decreased than increased between 2000 and 2008. Thorough sensitivity analyses found no indication that the results were biased due to selective participation of controls or more pronounced overestimation of mobile phone use among patients compared to controls (Aydin et al. 2011a).

The most striking result of CEFALO is a statistically significant association between the duration of mobile phone use and brain tumor risk in the small subset of the study sample with operator data. Objective data are presumed to be more reliable than self-reported data (Aydin et al. 2011b). However, the absent increase of the brain tumor incidence based on high quality registry data strongly contradicts the observed associations. One alternative explanation might be that more patients than controls succeeded in making operator data for more distant time periods available, by having reported changes in subscriptions more completely than controls. Also, cases may have changed subscriptions or phone numbers less often than controls. Thus, mobile phone operator data would cover more distant time periods in cases than controls. This would lead to an erroneous notion that cases started to use mobile phones earlier than controls. Another possibility is the presence of prodromal symptoms before diagnosis in some case patients. To provide frail children with better protection, their parents may have given them a mobile phone subscription for use in case of emergencies.

Ecological Studies

Because mobile phone use among children and adolescents has steeply increased since the mid-1990s, a brain tumor risk after a few years of mobile phone use is to be expected to affect the brain cancer incidence rate in this particular age group. Thus, various researchers have evaluated whether brain tumor incidence time trends have shown increases in the last few years.

Using data collected by the Surveillance, Epidemiology and End Results (SEER) Program for the United States, brain tumor incidence time trends of the age group 0–19 years were separately estimated for 1977–1991 (introduction of computerized tomography, CT, and magnetic resonance imaging, MRI) and for 1992–2006, when mobile phones became more prevalent (Inskip et al. 2010). An analysis of the gender specific incidence trends revealed stable rates for both periods and genders. Another analysis of the SEER data for the 5–19 years old persons revealed also stable incidence time trends between 1990 and 2007 (Boice and Tarone 2011).

A study from the Nordic countries found stable incidence rates of malignant and benign childhood central nervous system neoplasms in children aged 0–10 years between 1985 and 2006, and concluded that major changes in environmental risk factors are unlikely (Schmidt et al. 2011). For the children aged 10–14 years a statistically significant increase in incidence of 1.02% per year was observed for the full period. An analysis of the Nordic country incidence data for the age group 5–19 years did not reveal increasing incidence rates between 1990 and 2008 (Söderqvist et al. 2011).

In England, no time trends in newly diagnosed brain cancer cases in England was observed between 1998 and 2007 among adolescents aged 10–20 years (de Vocht et al. 2011), and the same pattern was observed for malignant brain tumor incidence rates in children and adolescents 0–19 years old in Australia 2000–2008 (Dobes et al. 2011).

In conclusion, incidence rate data from high quality cancer registries with virtually complete registration rates do not indicate an increase of brain cancers among children and adolescents up to 2008.

Discussion and Conclusion

Use of mobile phones among children and adolescents is still a relatively novel phenomenon. So far only little epidemiological research has been conducted to elucidate the role of mobile phone use in the etiology of childhood brain tumors. There is only one case-control study available and a few analyses from pediatric brain tumor incidence rates from the United States, England, Australia, and the Nordic Countries. Altogether these studies do not indicate that use of mobile phone is a major risk factor for childhood brain tumor.

So far, many more studies have been conducted on adult brain tumors and use of mobile phones. Most of these studies did not find an increased risk and recent reviews concluded that within about 10–15 years after first use of mobile phones it is highly unlikely that there is a material increase in the risk of brain tumors in adults (Repacholi et al. 2012; Swerdlow et al. 2011). Data for longer time periods are lacking. The IARC (International Agency for Research on Cancer) has classified RF-EMF as "possibly carcinogenic based on limited evidence in humans and in experimental animals" (Baan et al. 2011).

Regarding childhood brain tumor studies, there are several inherent limitations in the available literature. First, the CEFALO study indicates that amount and duration of mobile phone use was until recently relatively low among children and adolescents. In addition, regular use at young age (e.g., <10 years) seems also to be a relatively rare and recent phenomenon. Thus, any risk associated with heavy mobile phone use, with exposure early in life is likely not to be captured with the available data. The available data are also insufficient to identify a potential risk associated with a rare histological subtype. It is also conceivable that exposure in childhood may increase the risk of brain tumor in adulthood. One Swedish case-control study has actually conducted such an analysis and reported that the risk to develop astrocytoma in adulthood is about five times higher for persons who started to use mobile phones before the age of 20 years and about four times higher for cordless phone use before 20 (Hardell and Carlberg 2009). However, recent studies demonstrated that risk of this order of magnitude would result in a strong increase in the incidence time trends if they were true (Deltour et al. 2012; Little et al. 2012). Such an increase was not observed, suggesting that the observed association in the questionnaire-based case-control study is an artifact. Other case-control studies have not found increased risks in the youngest age-group (Lahkola et al. 2007).

The strength of case-control studies compared to ecological analyses of incidence time trends is the availability of individual exposure data. Thus, confounding factors can be considered in the analyses as was done in the CEFALO study. Confounding factors cannot be considered in temporal incidence rate analyses. Time trends may thus be affected by many other factors which are changing over time. It may be conceivable that an increase of brain tumors due to increased mobile phone use among children is compensated by a decrease of cases due to some other factors. However, it is difficult to imagine what kind of factor this could be given the limited known impact of environmental exposures in the etiology of childhood brain tumors.

Another advantage of the individual data in a case-control study is the possibility to conduct more sophisticated exposure-response analyses than in ecological studies. Thus, ecological fallacy can be prevented. However, in case-control studies exposure data has to be collected retrospectively, either by personal interviews or by obtaining records from the mobile phone operators. Both approaches were conducted in the CEFALO study and both approaches have severe limitations. To recall past mobile phone use is a challenge and substantial uncertainty cannot be avoided (Aydin et al. 2011b). There is concern that cases may overestimate exposure compared to controls, although there was no indication for such kind of recall bias in CEFALO (Aydin et al. 2011a). Objectively recorded data on mobile phone use by the operator is presumed to be more reliable. However, retrospectively collected operator records are as well prone to error and cannot be regarded as gold standard for several reasons. Not all operators store traffic data long enough or data are not traceable anymore. Moreover, children and adolescent tend to switch operators often according to their cost plan and thus may not remember their previous phone numbers which is needed for record identification. For children and adolescents the identification of the actual user of a mobile phone may be particularly erroneous as subscriptions may be held in the name of the parents and phones may be redistributed within families. Inquires of all these factors may also be subject to recall bias. These kinds of problems are not relevant for time trend analyses since the increase in use and ownership of mobile phones among children and adolescent is quite clearly defined on the population level. Thus, in the special situation of mobile phone use and brain tumors with (i) a steep increase in exposure prevalence, (ii) the availability of high quality registry data and (iii) the limited number of other environmental risk factors, results of ecological study are considered more informative than in most other situations. Moreover, ecological studies are not prone to participation bias as has been discussed in the context of adult brain tumor studies (INTERPHONE 2010). In addition to the epidemiological research, neither animal nor cellular data found consistent evidence for an association or could identify a biological mechanism for carcinogenesis of RF-EMF despite considerable research efforts (Repacholi et al. 2012).

In conclusion, available data on the childhood brain tumor risk in relation to mobile phone use is still relatively scarce. Overall these data do not suggest an increased brain tumor risk from using mobile phones. However, some uncertainties remain with respect to heavy mobile phone use, with early life exposure, or with long latency or with rare histological subtypes. Given the high potential public health consequences of even a small individual risk, further monitoring of childhood brain tumor incidence rate time trends is warranted.

References

Aydin D, Feychting M, Schüz J, Andersen TV, Poulsen AH, Prochazka M, Klaeboe L, Kuehni CE, Tynes T, Röösli M (2011a) Impact of random and systematic recall errors and selection bias in case-control studies on mobile phone use and brain tumors in adolescents (CEFALO study). Bioelectromagnetics 32:396–407

Aydin D, Feychting M, Schüz J, Andersen TV, Poulsen AH, Prochazka M, Klaeboe L, Kuehni CE, Tynes T, Röösli M (2011b) Predictors and overestimation of recalled mobile phone use among children and adolescents. Prog Biophys Mol Biol 107:356–361

Aydin D, Feychting M, Schüz J, Tynes T, Andersen TV, Schmidt LS, Poulsen AH, Johansen C, Prochazka M, Lannering B, Klaeboe L, Eggen T, Jenni D, Grotzer M, Von der Weid N, Kuehni CE, Röösli M (2011c) Mobile phone use and brain tumors in children and adolescents: a multicenter case-control study. J Natl Cancer Inst 103:1264–1276

Baan R, Grosse Y, Lauby-Secretan B, El Ghissassi F, Bouvard V, Benbrahim-Tallaa L, Guha N, Islami F, Galichet L, Straif K, WHO International Agency for Research on Cancer Monograph Working Group (2011) Carcinogenicity of radiofrequency electromagnetic fields. Lancet Oncol 12:624–626

Baldwin RT, Preston-Martin S (2004) Epidemiology of brain tumors in childhood – a review. Toxicol Appl Pharmacol 199:118–131

Boice JD Jr, Tarone RE (2011) Cell phones, cancer, and children. J Natl Cancer Inst 103:1211–1213

Challis LJ (2005) Mechanisms for interaction between RF fields and biological tissue. Bioelectromagnetics Suppl 7: S98–S106

Connelly JM, Malkin MG (2007) Environmental risk factors for brain tumors. Curr Neurol Neurosci Rep 7:208–214

de Vocht F, Burstyn I, Cherrie JW (2011) Time trends (1998–2007) in brain cancer incidence rates in relation to mobile phone use in England. Bioelectromagnetics 32:334–339

Deltour I, Auvinen A, Feychting M, Johansen C, Klaeboe L, Sankila R, Schuz J (2012) Mobile phone use and incidence of glioma in the Nordic countries 1979–2008: consistency check. Epidemiology 23:301–307

Dobes M, Khurana VG, Shadbolt B, Jain S, Smith SF, Smee R, Dexter M, Cook R (2011) Increasing incidence of glioblastoma multiforme and meningioma and decreasing incidence of Schwannoma (2000–2008): findings of a multicenter Australian study. Surg Neurol Int 176:783–790

Frei P, Mohler E, Neubauer G, Theis G, Burgi A, Frohlich J, Braun-Fahrlander C, Bolte J, Egger M, Roosli M (2009) Temporal and spatial variability of personal exposure to radio frequency electromagnetic fields. Environ Res 109:779–785

Gati A, Hadjem A, Wong M-F, Wiart J (2009) Exposure induced by WCDMA mobiles phones in operating networks. IEEE Trans Wirel Commun 8:5723–5727

Hardell L, Carlberg M (2009) Mobile phones, cordless phones and the risk for brain tumours. Int J Oncol 35:5–17

Hug K, Grize L, Seidler A, Kaatsch P, Schuz J (2010) Parental occupational exposure to extremely low frequency magnetic fields and childhood cancer: a German case-control study. Am J Epidemiol 171:27–35

Huncharek M, Kupelnick B, Klassen H (2002) Maternal smoking during pregnancy and the risk of childhood brain tumors: a meta-analysis of 6566 subjects from twelve epidemiological studies. J Neurooncol 57:51–57

ICNIRP and (International Commission on Non-Ionizing Radiation Protection) (1998) Guidelines for limiting exposure to time-varying electric, magnetic, and electromagnetic fields (up to 300 GHz). Health Phys 74:494–522

Inskip PD, Hoover RN, Devesa SS (2010) Brain cancer incidence trends in relation to cellular telephone use in the United States. Neuro Oncol 12:1147–1151

INTERPHONE, s. g (2010) Brain tumour risk in relation to mobile telephone use: results of the INTERPHONE international case-control study. Int J Epidemiol 39:675–694

Kheifets L, Repacholi M, Saunders R, van Deventer E (2005) The sensitivity of children to electromagnetic fields. Pediatrics 116:e303–e313

Kheifets L, Ahlbom A, Crespi CM, Feychting M, Johansen C, Monroe J, Murphy MF, Oksuzyan S, Preston-Martin S, Roman E, Saito T, Savitz D, Schüz J, Simpson J, Swanson J, Tynes T, Verkasalo P, Mezei G (2010) A pooled analysis of extremely low-frequency magnetic fields and childhood brain tumors. Am J Epidemiol 172:752–761

Lahkola A, Auvinen A, Raitanen J, Schoemaker MJ, Christensen HC, Feychting M, Johansen C, Klaeboe L, Lonn S, Swerdlow AJ, Tynes T, Salminen T (2007) Mobile phone use and risk of glioma in 5 North European countries. Int J Cancer 120:1769–1775

Lampert T, Sygusch R, Schlack R (2007) Use of electronic media in adolescence. Results of the German health interview and examination survey for children and adolescents (KiGGS). Bundesgesundheitsblatt Gesundheitsforschung Gesundheitsschutz 50:643–652

Little MP, Rajaraman P, Curtis RE, Devesa SS, Inskip PD, Check DP, Linet MS (2012) Mobile phone use and glioma risk: comparison of epidemiological study results with incidence trends in the United States. BMJ 344:e1147. doi:10.1136/bmj.e1147

McKinney PA (2005) Central nervous system tumours in children: epidemiology and risk factors. Bioelectromagnetics S7:60–68

Pollack IF, Jakacki RI (2011) Childhood brain tumors: epidemiology, current management and future directions. Nat Rev Neurol 7:495–506

Repacholi MH, Lerchl A, Röösli M, Sienkiewicz Z, Auvinen A, Breckenkamp J, d'Inzeo G, Elliott P, Frei P, Heinrich S, Lagroye I, Lahkola A, McCormick DL, Thomas S, Vecchia P (2012) Systematic review of wireless phone use and brain cancer and other head tumors. Bioelectromagnetics 33:187–206

Schmidt LS, Schmiegelow K, Lahteenmaki P, Trager C, Stokland T, Grell K, Gustafson G, Sehested A, Raashou-Nielsen O, Johansen C, Schuz J (2011) Incidence of childhood central nervous system tumors in the Nordic countries. Pediatr Blood Cancer 56:65–69

Söderqvist F, Carlberg M, Hansson Mild K, Hardell L (2011) Childhood brain tumour risk and its association with wireless phones: a commentary. Environ Health A Glob Access Sci Source 10:106

Swerdlow AJ, Feychting M, Green AC, Leeka Kheifets LK, Savitz DA (2011) Mobile phones, brain tumors, and the interphone study: where are we now? Environ Health Perspect 119:1534–1538

Wiart J, Hadjem A, Wong MF, Bloch I (2008) Analysis of RF exposure in the head tissues of children and adults. Phys Med Biol 53:3681–3695

Index

M. Hayat (ed.), *Tumors of the Central Nervous System, Volume 13*,
DOI 10.1007/978-94-007-7602-9, © Springer Science+Business Media Dordrecht 2014

C

D

E

MIX
Papier aus verantwortungsvollen Quellen
Paper from responsible sources
FSC® C105338

If you have any concerns about our products,
you can contact us on
ProductSafety@springernature.com

In case Publisher is established outside the EU,
the EU authorized representative is:
Springer Nature Customer Service Center GmbH
Europaplatz 3, 69115 Heidelberg, Germany

Printed by Libri Plureos GmbH
in Hamburg, Germany